PRAISE FOR
WESTMORELAND

"Scalding . . . Sorley, a West Point graduate and retired Army lieutenant colonel, is unsparing in his analysis of Westmoreland."
— *Los Angeles Times*

"The subtitle says it all: 'The General Who Lost Vietnam' . . . Sorley has stripped away Westmoreland's after-the-fact mythologizing, leaving us with a deeply unflattering portrait of an army careerist who unintentionally did much damage to an institution — and a country — that he loved dearly. Westmoreland is a valuable addition to the growing 'revisionist' literature that shows the Vietnam War was winnable if we had fought differently."
— Max Boot, *Wall Street Journal*

"Sweeping . . . [Sorley] pillories the hapless general for what are now seen as horrendous gaffes of counterinsurgency."
— *Time*

"A first-rate biography of a second-rate soldier."
— *Washington Times*

"No American general has ever been more vilified than William C. Westmoreland, our senior military commander in Vietnam from 1964 to 1968 and the only American general to lose a war . . . Lewis Sorley, a distinguished military historian and Vietnam veteran himself, offers a stinging assessment . . . Napoleon supposedly said, 'Don't give me good generals, give me lucky ones.' This well-researched, engrossing, and hard-hitting biography demonstrates that its subject was neither."
— *Cleveland Plain Dealer*

"An important contribution to the literature of the Vietnam War . . . The research is meticulous and the writing fascinating."
— *Proceedings*

"A military historian's harsh take on the career of the general most associated with America's most controversial war . . . The general's defenders will have their hands full answering Sorley's blistering indictment."
— *Kirkus Reviews*

"A biography as unflinching and unsparing as it is balanced and insightful. Lewis Sorley has added a fine, important volume to our national literature on the Vietnam era."
— Rick Atkinson, author of *The Long Gray Line*

WESTMORELAND

THE GENERAL WHO LOST VIETNAM

★

LEWIS SORLEY

MARINER BOOKS

HOUGHTON MIFFLIN HARCOURT

BOSTON   NEW YORK

First Mariner Books edition 2012
Copyright © 2011 by Lewis Sorley

For information about permission to reproduce selections from this book,
write to Permissions, Houghton Mifflin Harcourt Publishing Company,
215 Park Avenue South, New York, New York 10003.

www.hmhco.com

Library of Congress Cataloging-in-Publication Data
Sorley, Lewis, (1934–).
Westmoreland : the general who lost Vietnam / Lewis Sorley.
p. cm.
ISBN 978-0-547-51826-8   ISBN 978-0-547-84492-3 (pbk.)
1. Westmoreland, William C. (William Childs), 1914–2005. 2. Generals—
United States—Biography. 3. Vietnam War, 1961–1975—Biography.
4. United States. Army—Biography. I. Title.
E840.5.W4S67 2011
959.70434092—dc22
[B]
2011016067

Book design by Victoria Hartman

Printed in the United States of America
DOH 10 9 8 7 6 5 4 3 2
4500525514

For Ginny,
again and always

Truth is like a threshing-machine; tender sensibilities must keep out of the way.

— HERMAN MELVILLE, *The Confidence-Man*

CONTENTS

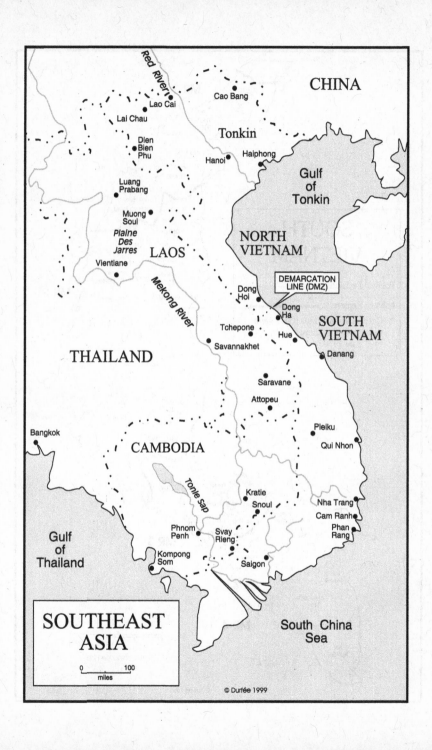

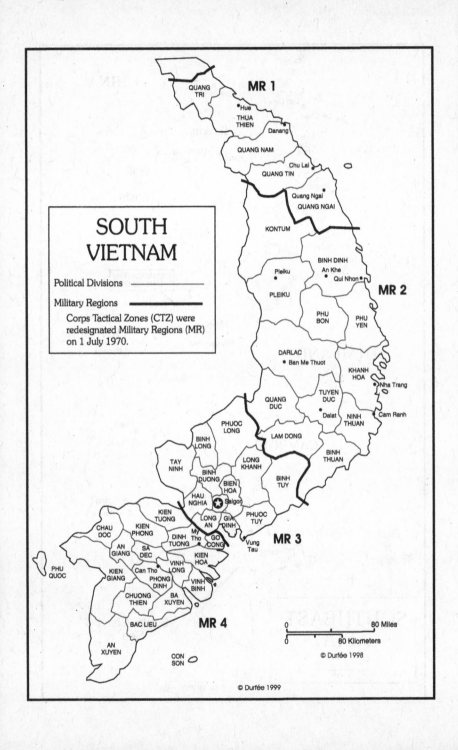

SOUTH VIETNAM

Political Divisions ————

Military Regions ▬▬▬▬

Corps Tactical Zones (CTZ) were redesignated Military Regions (MR) on 1 July 1970.

MR 1

QUANG TRI

Hue
THUA THIEN

Danang

QUANG NAM

Chu Lai
QUANG TIN

Quang Ngai
QUANG NGAI

KONTUM

BINH DINH
An Khe
Qui Nhon

Pleiku

PLEIKU

MR 2

PHU BON

PHU YEN

DARLAC
Ban Me Thuot

KHANH HOA

Nha Trang

QUANG DUC

TUYEN DUC

Dalat

NINH THUAN

Cam Ranh

PHUOC LONG

LAM DONG

BINH THUAN

BINH LONG

TAY NINH

LONG KHANH

BINH DUONG

BIEN HOA

BINH TUY

HAU NGHIA

Saigon

LONG AN

GIA DINH

KIEN TUONG

PHUOC TUY

CHAU DOC

KIEN PHONG

DINH TUONG

My Tho

GO CONG

Vung Tau

MR 3

AN GIANG

SA DEC

KIEN HOA

PHU QUOC

KIEN GIANG

Can Tho

VINH LONG

PHONG DINH

VINH BINH

CHUONG THIEN

BA XUYEN

BAC LIEU

MR 4

AN XUYEN

CON SON

0 _____ 80 Miles
0 _____ 80 Kilometers

© Durfée 1998

© Durfée 1999

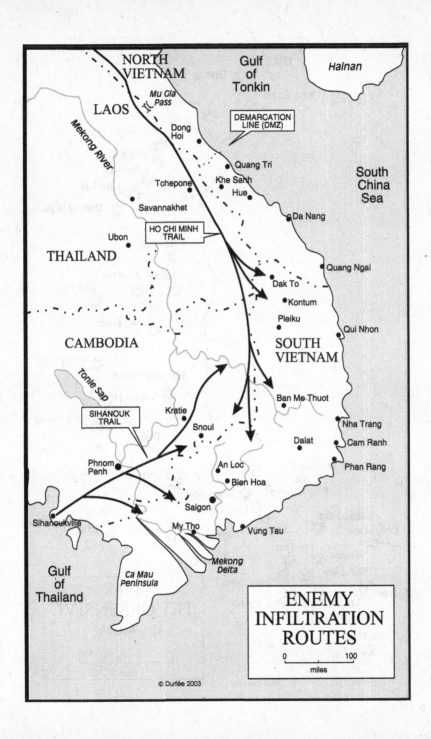

ENEMY
INFILTRATION
ROUTES

0   100
miles

© Durfée 2003

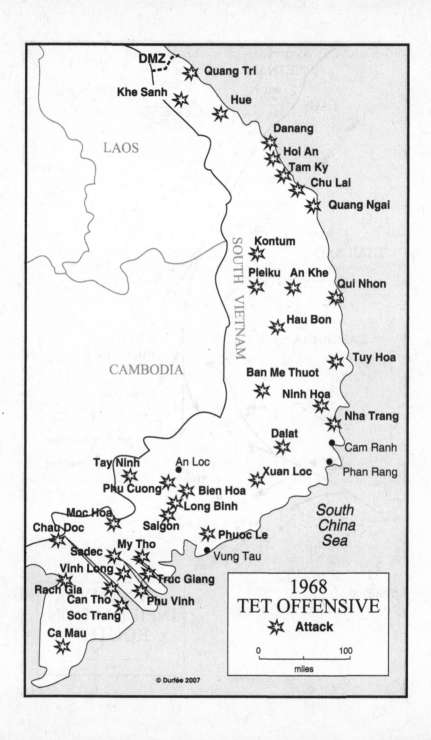

WESTMORELAND

PROLOGUE

The premise of this study is that, unless and until we understand William Childs Westmoreland, we will never understand fully what happened to us in Vietnam, or why.

Westmoreland's involvement in the Vietnam War was the defining aspect of his life. He himself perceived that, and was driven for the rest of his days to characterize, explain, rationalize, and defend that role. His memoirs reflect the fixation. In a long career totaling thirty-six years as an officer, and a string of postings to increasingly important assignments, the four years he commanded American forces in Vietnam, and the aftermath, constitute virtually the entirety of his account, all the rest a meager tenth.

Understanding Westmoreland, a surprisingly complex man, is not easy. Fueled by ambition, driving himself relentlessly, of impressive military mien, energetic and effective at self-promotion, and skillful in cultivating influential sponsors, from his earliest days of service he led his contemporaries, was admired and advanced by his seniors, and progressed rapidly upward.

But Westmoreland also had an extraordinary capacity for polarizing the views of those who knew him—or at least those who encountered him, for not many would claim they really knew this distant and difficult man. Few remained indifferent. Among his admirers, an officer who worked directly for Westmoreland when he was Army Chief of Staff described him as "the most gracious and gentlemanly person with whom I ever served." An officer who was his executive officer in Vietnam regarded Westmoreland as the only man he ever met to whom the term "great" could be applied.

There were others, though, many others, who had a darker view. Among the most prominent was General Harold K. Johnson, a man of surpassing decency and good will. "I don't happen to be a fan of General Westmoreland's," said Johnson. "I don't think I ever was, and I certainly didn't become one as a result of the Vietnam War or later during his tenure as Chief of Staff of the Army." Another officer, one who worked closely with Westmoreland in Vietnam, described him as "awed by his own magnificence."

Westmoreland's own frequent self-characterizations are revealing. "I have been a person who has sought responsibility," he told an interviewer. "I diligently tried to do a good job, not because I was bucking for anything higher, but because I was trying to do a job for the sake of doing a good job. That was my orientation. As a matter of fact, it was throughout my career. It was to do a job for the sake of doing a good job."[1]

Westmoreland took himself seriously, very seriously. There are few photographs of him smiling. Typically he is, instead, and very obviously, posing. While his description in *The Howitzer*, the West Point yearbook, credits him with a good sense of humor, he apparently lost or repressed it as he advanced in age and seniority. Jerry Warner, a teenager when he first met Westmoreland, for whom his father worked, suggested an explanation. Westmoreland, he observed, "had a very keen humorous and affectionate side which he held in reserve and in confidence for his family and those he felt, by extension, were a part of it."

There were other changes over the years. "He was an excellent commander at lower levels," Sergeant Major of the Army Leon Van Autreve said of him. "And his people loved him. But I tell you, after that it was about a hundred and eighty. It's a peculiar thing that you can gain or lose so rapidly the affection of your people." Van Autreve recalled an occasion when Westmoreland as Army Chief of Staff had come to address a gathering of senior noncommissioned officers at

Fort McNair in Washington. "He was getting ready to go outside," said Van Autreve, "and there is a cameraman out there. Now we're all ground pounders and dirt slingers. And this guy [Westmoreland] stands there, oblivious of all of us, and the aide takes his cape and drapes it over his arm and all this sort of thing. Then the aide looks at him and says, 'You're ready,' opens the door, and the flashbulbs start popping. That was the Westmoreland of later years."[2]

Fortunately the historical record of Westmoreland's life is extensive and rich, in part because from his early days he himself made extraordinary efforts to create and preserve it. What it reveals is a man devoted to his profession, and to his own rise in that profession, single-minded in his determination to accomplish the mission as he understands it, skillful in cultivating those who could be helpful to him, faithful in his marriage and loyal to his family, often perceptive in his choice of key associates, limited in his understanding of complex situations, entirely dependent on conventional solutions, and willing to shade or misremember or deny the record when his perceived interests were at risk.

Westmoreland's strengths eventually propelled him to a level beyond his understanding and abilities. The results were tragic, not just for him but for the Army and the nation he served, and most of all of course for the South Vietnamese, who sacrificed all and lost all.

1

★

ORIGINS

WILLIAM CHILDS WESTMORELAND was born on 26 March
1914 in the village of Saxon in Spartanburg County, South Carolina, the son of the textile mill manager James Ripley Westmoreland
and his wife, Eugenia Childs Westmoreland. In later years Westmoreland recalled that he "was born in the South during which time Robert
E. Lee was on the same level as Jesus Christ." His ancestral roots extended back four generations within South Carolina.

His father was well connected politically in the state, his friends including James Byrnes and Strom Thurmond. In the family the son was
called by his middle name, Childs, his mother's maiden name. His
only sibling was a younger sister, Margaret. Westmoreland described
his mother as "extremely religious" and his father as "quite conservative," a man who "didn't take too many chances." But his father, he recalled, "influenced my life more than any other individual."[1]

Later Westmoreland's son-in-law would say of Westmoreland that
"he was raised very, very stiff. His father would turn his head away and
let his son peck him on the cheek." Even so, he was clearly the favorite.
"Your Dad never thought women ever amounted to much in the social
scale," a spinster friend of the family wrote to Westmoreland some

years later. "I think he's mellowed some with age, but he gave Margaret a hard time when she was growing up." For her part, Margaret affirmed that observation: "My brother was my father's life. He was the perfect one. He could do no wrong."

His father, said Westmoreland, "taught me the fundamentals of boxing and I learned to lead with a left, keep the opponent at a distance and take advantage of my right when there was an opening." Later, in Vietnam, Westmoreland would use a boxing analogy to describe his tactical approach to conduct of the war. Apparently he talked a better game than he took into the ring, however, for he never mentioned the summer at Camp Pinnacle in North Carolina's Blue Ridge Mountains when he was matched up against another camper named Harold Cohen. "I landed a blow to his Adam's apple and knocked him cold," remembered Cohen. "I thought I'd killed him."

Westmoreland recalled that during his school days his favorite book was "The Boy Scout Manual" (*Handbook for Boys*). At fifteen he became an Eagle Scout in Troop 1, Spartanburg, the second member of that troop to attain the highest rank in Scouting. That same year he took part in the 1929 World Scout Jamboree at Arrowe Park in Birkenhead, England. There, fifty thousand Scouts from many different nations assembled for the Jamboree. Scouting was still a pretty new thing, having been established in England just twenty-one years earlier, and this was just the third World Jamboree. Scouts paraded before Sir Robert Baden-Powell, Scouting's founder, and the Prince of Wales and, as recorded by Westmoreland in his journal, heard a sermon by "the Arch Bishop of Canelberry." Westmoreland recalled the fun they had exchanging items of uniform, buttons, and badges, and that he himself acquired a Scottish kilt, wearing it into Edinburgh after the Jamboree. They climbed the nearest mountain where, said Westmoreland, "There was a swell wind up there and very cool with kilts on."

Among his fellow Scouts back in Spartanburg Westmoreland made some lifelong friends, particularly Conrad Cleveland, who years later would be best man at his wedding and then, long after that, an important figure in Westmoreland's South Carolina gubernatorial campaign.

Troop 1 was sponsored by the Episcopal Church of the Advent in Spartanburg, where the Westmorelands also attended worship services. They would drive up on Sunday mornings in the family's old

Franklin automobile, then take their customary seats on the right-hand aisle. According to a friend who also attended these services, Dr. W.H.K. Pendleton, the rector, "always said that if the Westmorelands could be there on time, rain or shine, all the way from Pacolet, the rest of us could be too!"

Westmoreland graduated in 1931 from Spartanburg High School, where he was senior class president. Remembered one of his schoolmates, "In spite of the Great Depression, we had a rather isolated, comfortable and secure world in which to grow up."

Years after his retirement, Westmoreland was a speaker at the hundredth anniversary celebration of the small village where he had grown up. He recalled the challenging times the region had seen. The last mill had now closed, leaving something of a ghost town, and Westmoreland confided that he himself had also experienced hard times. "My years away have been fraught with challenges, frustrations, and sadness," he said. "As is frequently the case in life, some jobs have been thankless and some tasks without solution."

AFTER HIGH SCHOOL Westmoreland spent a year at The Citadel, the Military College of South Carolina, where his father had graduated in 1900 and was later for many years Chairman of the Board of Visitors. During this year there began a robust correspondence between Westmoreland and his father, who coached his son on many subjects, often including spelling. In one note he went out of his way to denigrate Margaret. She "is doing very poorly at Ashley Hall and it is a source of great disappointment to us," he wrote. "I am worried as to whether she can do and will not or simply can't do. It is a great source of satisfaction that we have no worry about you."

While he was at The Citadel, an appointment to the United States Military Academy became available, and in 1932 he entered that institution, appointed by Senator James F. Byrnes (who had also been his Sunday School teacher).[2] When word came of Westmoreland's acceptance, his father wrote to him: "You are all set now to make your entire life for yourself and it is up to you." He signed the letter "Wawa," his children's pet name for their father.

Westmoreland had wanted to go to the Naval Academy, but Byrnes counseled that the better choice for him would be West Point, which had "a much broader, less technically oriented curriculum." Westmoreland reported his decision to leave The Citadel and go north to West Point to his Confederate great-uncle, who had fought in the Civil War: "Uncle White, I'm going to the same damn school that Sherman and Grant attended." His uncle was reassuring: "Never you mind, son, Robert E. Lee and Stonewall Jackson went there, too."[3]

Westmoreland continued to receive encouraging letters from his father, including one dated 2 August 1932, perhaps a month after the son had entered the Military Academy: "You do not know [how] happy it makes us all to know that you are making good. Even the small boys and the negroes are interested and proud of it."

Taken altogether, these letters constitute a remarkable record of a father's affection and concern for his only son. "When you need any-thing write me and I will send it to you. There is nothing too good for you." It was also clear that he was deeply missed. "You may feel that you are a long way from home," wrote his father in one of the first let-ters sent to West Point. "We feel that you are but we talk about you about half our time."

AT WEST POINT Westmoreland underwent a second stringent plebe year, made even more difficult by a different kind of communication from home, endless letters of admonition from his father stressing that he simply had to pass the academic courses, since by now, in the depths of the depression and with his sister also nearing college age, the family could not afford to send him to school elsewhere. "Be sure and send me your marks every week because I have a little book in which I keep them. I am also keeping your class standing in the same book," wrote his father.[4]

So incessant was this barrage that in one month alone, December of 1932, Westmoreland received twenty-six missives from his father, twice including two written on the same day. In one of these letters his father unhelpfully observed: "English seems to be your weakest spot. No doubt this comes about from your lack of reading as you grew up." Clearly Westmoreland did not deserve this relentless pressure, as he consistently maintained a respectable academic record, standing 71 of 328 at the end of his first year. His overall class ranking dropped in each successive year, but at graduation he still stood 112 of 276 in gen-eral order of merit, helped by his top military rank but dragged down by economics and government, in which he stood near the bottom of the class.

LATER WESTMORELAND TOLD an interviewer that as a cadet he never walked the area (a punishment for violations of regulations). He received only one major penalty, six demerits and twenty confinements during his yearling year for possession of an unauthorized radio. As for dating, he said, "I was playing the field, so to speak." Little evidence of that remains to posterity. Wrote his first biographer, Ernest Furgurson: "Westy was so busy with his duties that he did not bother to invite a date to the graduation hop."

He tried several varsity sports (not lettering in any), and was Superintendent of the Cadet Sunday School Teachers and vice president of his class.

In his final year at the Military Academy Westmoreland was named First Captain, the senior cadet in military rank—a high honor. His military bearing had undoubtedly been a factor in his selection. He stood five feet eleven inches tall, trim and ramrod straight, with a strong nose, prominent eyebrows, and jutting chin, his handsome countenance enhanced rather than marred by a long scar on the left side of his face, the result of going through the windshield of his father's car in a head-on collision when he was a youngster.

The First Captain at West Point is a genuine cadet celebrity, and properly so. Everyone in the Corps knows who he is, how he looks, what he does. Westmoreland made a fine impression on his classmates, even as a plebe being given the nickname "Chief" by a fellow plebe who viewed him (correctly, as it turned out) as a future Chief of Staff of the Army. It was Ted Clifton, himself later a major general and White House aide, who tagged Westmoreland with the nickname although, he later admitted, "To be honest about it, I hardly knew what the Chief of Staff was in those days." What he did know, or believed, was that Westmoreland was destined for big things in the years ahead.

Apparently Clifton wasn't the only one who held that view. Brigadier General Sam Goodwin later wrote to Westmoreland, after both of them had retired from active duty, to say: "Certainly you know that one of the legends you left at West Point in 1936, a legend repeated to the plebes who were admitted three weeks after you graduated [Goodwin's Class of 1940]: You as First Captain had announced that you would be the Chief of Staff of the Army."

Westmoreland provided "Advice from the First Captain" for publication in *Bugle Notes*, the handbook of advice and information issued to all plebes. "By keeping duty foremost in your mind at all times and on all occasions," he wrote, "you can not fail to develop the most that is in yourself, and to serve West Point and your country in a manner of which you, the Corps, and your country will be justly proud."

Surprisingly for a First Captain, in that senior year Westmoreland amassed a large number of demerits, forty-eight in all (more than he had accrued as a plebe), for such infractions as "giving the command of execution on the wrong beat" while marching and "causing cadet officers and guidon bearers to march out of step by reason of poorly timed command of execution at parade." As a result, in his senior year he ranked eighty-second in the class in conduct.

If his mother is to be believed, Westmoreland somehow escaped punishment for yet another parade-related transgression. She was visiting West Point once, she recalled, when her son forgot his sword for a dress parade. But, she told an interviewer, "her cadet son was able to maneuver so expertly while marching that no one noticed." This seemed highly unlikely, but hearsay evidence from Major General Clay Buckingham, Class of 1949, suggested its accuracy. "One of the stories circulating about him [Westmoreland] back then," said Buckingham, "was about the time when he, as First Captain, led the graduation parade for his class. Apparently in his rush to get ready, he had forgotten to put his sword in its sheath. Recognizing this only after he had gotten out on the Plain in front of thousands, he went through the entire ceremony making all the correct motions without his sword. No one in the audience noticed, but some of his classmates did notice and wouldn't let him forget it." Classmate Major General Gordon H. Austin later confirmed the accuracy of that account. He did not see the episode himself, he said, since he had been marching in one of the rear ranks, "but later everyone talked about it."

Despite such episodes, Westmoreland proved himself a fitting First Captain: dutiful, dedicated, capable if not brilliant, ambitious if not especially social, a father's pride, a model cadet leader. In West Point's yearbook Westmoreland's write-up was admiring: "A fine soldier and true friend is Westy. Modest, generous, tolerant, and possessing a good sense of humor, Westy has made many friends. His executive ability, conscientiousness, high ideals, good judgment and common sense, and his fearless determination — just glance at that chin! — have well fitted him for the position he has held as leader of our class, and as First Captain of the Corps."

IN WESTMORELAND'S CLASS was Benjamin O. Davis Jr., a black who would later become a high-ranking general officer. As a cadet he had been ostracized because of his race. A fellow officer later wrote to Westmoreland about this, saying, "We and those senior to us in the military have much to be ashamed about in those early years of our service." Westmoreland responded that Davis "was a victim of the times." This infuriated his correspondent, who shot back: "That he was. But he was also the victim of individual acts of thoughtlessness, of cruelty, and of cowardice for which individuals ought to answer in this life, and for which, according to our Christian faith, they will assuredly have to answer later."[5]

Another graduate, bringing the matter much closer to home, wrote to Westmoreland about how Davis "was inhumanely discriminated against throughout his cadet days with the silencing and its many peripheral consequences. In Davis's final year at West Point," he reminded Westmoreland, "that discrimination was sustained under your leadership as first captain. . . . I therefore urge you most strongly to issue a public apology to General Davis on behalf of our Army and the United States Military Academy." Westmoreland called that officer and told him that his "hands were tied," that the silence had been imposed on Davis and there was nothing he could do about it.[6]

THERE HAVE BEEN, over the years, a few West Point classes that stand out from the rest. Preeminent was 1915, the class of Eisenhower, Bradley, and other luminaries, with nearly 35 percent of the class becoming general officers. The Class of 1933 (which led the Corps during Westmoreland's plebe year) achieved over 24 percent generals. And Westmoreland's Class of 1936 also turned out to be one of West Point's most distinguished. It produced six four-star generals, seven three-stars, and sixty generals in all, representing nearly 22 percent of the class. Over 92 percent of the class served to retirement (or died while on active duty), a remarkable example of duty performed. For such records to be achieved, of course, the times had to be right—a big war coming up soon after graduation, with fast early promotions, combat experience, professional reputations established early on—and then, also very important, the ability to adjust to peacetime service until the next war came along.[7]

On Friday, 12 June, the Class of 1936—276 strong—graduated and was commissioned. General of the Armies John J. Pershing presented the graduation address. Westmoreland, who on behalf of his class had petitioned the Superintendent to invite Pershing as the speaker, was somewhat disappointed in the outcome, later describing Pershing as "an elderly man" whose address "was not delivered with any fire or enthusiasm." But at least Pershing had said, gratifyingly, that the Corps marched well.

2

★

EARLY SERVICE

COMMISSIONED IN THE Field Artillery branch, Westmoreland was sent to Fort Sill, Oklahoma, where he joined the 18th Field Artillery and gained the usual junior officer experience. He was motivated to pursue a military career, he would tell a later correspondent, "by a desire to serve my country" and to "break out of a parochial environment." In this first assignment he found that the officers were "some good, most mediocre" and the soldiers "could barely read or write." On the positive side were "good horses and mules and one motor vehicle," although the "weapons were antiquated." Those weapons were French 75mm guns, Model 1897, horse-drawn. For communications they still had carrier pigeons and the telegraph using Morse code. Meanwhile Hitler's planes and tanks were just a few years away from invading Germany's neighbors.

Classmate Bruce Palmer remembered some positive things about those early days: "We had a taste of the Old Army, a small, tightly knit band that had survived despite public neglect and non-recognition, and whose older officers and NCOs taught us a lot about what being a professional soldier was all about."[1] Westmoreland's recollections were more negative, describing that Army as "backward" and lacking in "money, weapons, and public support."

Westmoreland's first battery commander was a senior captain, nicknamed "The Stud Duck," who was described by Westmoreland as "virtually incompetent." Allegedly he had failed the basic course at the Artillery School, although how he could have survived that and later advanced to captain is unclear. Things were held together in the battery by 1st Sergeant Bull McCullugh, "a boxing champ who could lick anyone in the outfit."

After a few months on the job Westmoreland wrote to his father dutifully, "I'm working quite hard at present." He cited some of the duties he had been assigned, including stable officer (with thirty-eight horses to supervise). Like most young officers, he was also learning polo, "lots of fun, but plenty of hard work." Nonetheless, he wrote, "Each afternoon, I am free to do as I like."

AT FORT SILL Westmoreland met a pretty and accomplished young horsewoman, Katherine Van Deusen, known as Kitsy, the daughter of Lieutenant Colonel Edwin Van Deusen, then Executive Officer of the Field Artillery School. "I took one look at him, and I've followed him ever since," she later said.[2] Kitsy was then all of nine years old, but she asked Westmoreland (almost thirteen years older) to wait for her until she grew up.[3] He did (after, during his service in Hawaii, a woman he wanted to wed married another officer instead).[4]

Westmoreland also became Scoutmaster of Fort Sill's Boy Scout Troop 37, and evidently a good one. In later years he often heard from youngsters, many of whom had become officers themselves, who remembered him fondly from those earlier days, and some even served under his command in various outfits. Among the Scouts in his troop was Edwin Van Deusen, Kitsy's older brother.

IN THOSE DAYS mounted officers (including horse artillery, not just the cavalry) could acquire private mounts in addition to their issue horses, but only with permission. Westmoreland accordingly asked to be allowed to purchase a mount from the Army Depot at Fort Reno, Oklahoma. That request was approved, and he acquired a three-quarter thoroughbred three-year-old mare named Polly Ann, a bay with black legs below the knees.

Westmoreland recalled that in those days "our daily lives were built around the horse — fox hunts, polo, horse shows." This led to his first brush with Army aristocracy, as Lieutenant Colonel George S. Patton, accompanied by Mrs. Patton, came down from Fort Riley to serve as judge at a horse show. Westmoreland was assigned as his recorder. "As he judged," recalled Westmoreland, "I kept track of his comments and gave him a summary of his evaluation." Westmoreland was suitably impressed, concluding that Patton "knew his horses and how they should be rode."

A slightly more senior officer in the regiment with Westmoreland was Camden McConnell, USMA Class of 1931, who remembered a horse adventure that did not turn out so well for Westmoreland. One duty was to take the school horses to summer pasture. Since horses are well known for their skittishness, the technique was to take a large number of soldiers and only a few horses to establish the camp. Then, when the first contingent of horses had settled in, a few more horses would be added, then a few more, and so on. Eventually the entire herd would be in residence and pretty calm about the whole change of venue. On one occasion when McConnell and Westmoreland were accomplishing this somewhat tricky maneuver, however, McConnell had to leave temporarily to take care of some business elsewhere. That left Westmoreland in charge, with the result that the herd soon "stampeded across the main post with the loss of many horses and destroying the golf course." Fortunately the resulting investigation found —"naturally," said McConnell—no fault on Westmoreland's part because "that's just how horses are."[5]

When Westmoreland was a cadet his father had written to him suggesting that after he graduated, if the Democrats won, "You should have some pull in your advancement in the Army through Jim Byrnes and [John] McSwain," a senator and house member respectively. Almost right away Westmoreland sought to take advantage of some of that political capital, asking to be designated a diplomatic courier and sending that request not through military channels but via Congressman G. Heyward Mahon Jr. When the matter reached the Adjutant General he was not impressed, responding in January 1937 that "the assignment desired by Lieutenant Westmoreland is one requiring considerable maturity and experience," whereas "Lieutenant Westmoreland, having been graduated from the United States Military Academy last June, has less than one year's commissioned service." Not approved.[6]

After three years at Fort Sill Westmoreland was ordered to Hawaii. He accordingly requested authority to ship his private mount, and in due course Polly Ann was assigned passage aboard the U.S. Army Transport *Meigs*. A private named Jerry Reel was put on orders as her attendant for the voyage, while Westmoreland traveled separately aboard another Army Transport, the *Grant*.

IN HAWAII WESTMORELAND found the troops, and duties, much like those at Fort Sill. The artillery was armed with British three-inch guns, and "the division was more interested in unit athletics than training. We went through the motions of training and had athletics and fatigue (police) in the afternoon." Westmoreland gained battery command experience, leading Battery F of the 8th Field Artillery. Then, though a mere first lieutenant assigned as a battalion operations officer, Westmoreland told his father that he "ran a battalion for all intents and purpose[s], since a lieutenant colonel was in command who had little to offer."[7]

There was still plenty of old Army. Westmoreland was invited to act as Honorary Whipper-In on the staff of the Artillery Hunt. Besides serving at the Hunts, he was told, he would assist at hound exercise. Westmoreland also reported in a letter to his mother "a terrible thing"—his horse died very suddenly, apparently the result of an abdominal obstruction. "Very unfortunate for me," Westmoreland observed somewhat coldly, "especially since I had no insurance on her."

In Hawaii Westmoreland received what turned out to be his only formal Army instruction (except for post–World War II attendance at parachute school), acquiring a certificate of proficiency testifying to his successful completion of a special one-month course in mess management conducted by the Quartermaster Corps School for Bakers and Cooks.[8] Surprisingly, Westmoreland later seemed rather proud that he had missed the usual professional development opportunities at the Command & General Staff College at Fort Leavenworth and the Army War College.

EARLY IN 1941, after two years in the islands, Westmoreland left Hawaii for a new assignment with the 9th Infantry Division at Fort Bragg, North Carolina. There, he recalled, "Amazing things began to happen to our Army." One of the most significant was that a military draft brought into the ranks what Westmoreland described as "a cross-section of the young men from our society—good men," quite a contrast to his view of the troops he had been working with.

But Fort Bragg was just a prelude, as by then the Wehrmacht had begun its rampages across Europe and everyone in the Army expected that the United States would eventually join the war—even if the American public opposed that idea right up until the 7 December 1941 Japanese attacks at Pearl Harbor.

3

★

WORLD WAR II

THE ARMY MOBILIZED immediately upon news of the Japanese attack, and within days Germany had also declared war. It would take one more year for Westmoreland to enter the fight. At the age of twenty-eight he found himself in command of the 34th Field Artillery Battalion of the 9th Infantry Division. In North Africa Westmoreland's 155mm towed howitzers caught up with earlier deploying elements when they arrived in Casablanca on Christmas Eve of 1942.

Westmoreland took an unsuspected chance in evading orders upon arrival. He had been designated commander of troops aboard ship during the voyage. When they reached port, he was summoned by General Patton's chief of staff and told to get the troops ashore immediately. Westmoreland "urged a delay in disembarking so the New Year's dinner could be enjoyed, but to no avail," he later recalled. So he went back and gave the order to leave the ship, but designated his own battalion to stay aboard and police up the ship "after enjoying the prepared feast." Only later did he learn that the reason for the order to disembark immediately was that "German bombers were reportedly on the way to strike the docking facilities at Casablanca."[1]

The 34th Field Artillery remained in Casablanca until 17 January 1943 before departing for Port Lyautey and then, with various intermediary stops, finally entered battle on 22 February. Recalled Westmoreland, they were in "almost continuous operation against the enemy" until 10 May 1943. He would later speak of World War II, where "there was some *very* severe fighting, on a sustained basis, contrary to Korea, and contrary to Vietnam, where fighting was sporadic and not sustained."

General Patton had inspected Westmoreland's battalion while they were at Casablanca. As he was going through one of the battery messes he noted how they were using immersion heaters (designed for heating dishwater) to heat C-Rations in the can. "General Patton had never seen this before," said Westmoreland, "and was profuse in congratulating the mess sergeant on his initiative. The mess sergeant's chest was so puffed up that he was rather unbearable to his colleagues for weeks thereafter."

Another Patton visit made quite an impression on Westmoreland and, very obviously, on some of the 9th Infantry Division's staff. "One day," Westmoreland said, "General Patton arrived at the command post with his pearl-handled [actually ivory] revolvers, riding boots, pink breeches, and shining helmet, looking like a military fashion plate." The division commander, Major General Manton Eddy, greeted Patton, with the staff following behind. Patton's first words were, "Manton, I'm getting goddamn sick and tired of the lack of progress this division is making. I want you to get off your ass and start moving." Then, pointing to the staff, Patton continued, "And I want you to get these bastards out to the front lines and get them killed." Well, said Westmoreland, "those words had a sobering effect on the division staff. That evening they hauled the G-1, the personnel staff officer, out on a stretcher as a psycho case and he never returned."[2]

Westmoreland and his troops were also exposed to a version of the notorious Patton speech, the one made famous by George C. Scott in the movie *Patton*, describing what he saw ahead and what he expected in training and attitudes. "Men," said Patton, "all this stuff you've heard about America not wanting to fight, wanting to stay out of the war, is a lot of horse dung. Americans, traditionally, love to fight." Westmoreland said the film well represented the original except that Patton delivered it in a high, squeaky voice. It was nevertheless effective. Patton was "erect and always immaculately dressed," said Westmoreland. "His language was abrupt, profane, and gory. He got and kept the attention of his troops, and he and his speech became matters of constant conversation and discussion for weeks thereafter."

ALTHOUGH HE EVENTUALLY served in three wars, Westmoreland was never decorated for valor. He had a fine hour, though, in leading the 34th Field Artillery Battalion during fighting in Tunisia, where they earned great distinction.

In February 1943 the unit encountered and dealt brilliantly with a challenging combat situation. Over a period of four days the entire battalion made a forced march of 735 miles from Tlemcen, Algeria, over the Atlas Mountains to Thala, Tunisia. The move began in a snowstorm, proceeding "in bitter weather over tortuous and almost impassable mountain roads," and ended a hundred hours later when the troops went immediately into battle against elements of the German Afrika Corps, helping to stop the enemy's breakthrough at Kasserine Pass. That feat resulted in the Presidential Unit Citation for the battalion, a high honor in recognition of valorous service. The citation noted that, "although enemy forces were entrenched only 2,500 yards distant and there were only three platoons of friendly infantry in front of the artillery, the unit maintained constant and steady fire with such deadly effect that enemy tank units were dispersed and driven back."

Westmoreland himself received a highly commendatory letter from Brigadier General S. Leroy "Red" Irwin, at that time commanding general of the 9th Division artillery. "Your handling your battalion during the march," he wrote, "and during the action at Thala, was a splendid example of leadership, and was characterized by personal initiative, courage and coolness under very adverse conditions. You are commended for a superior performance of duty."

WHEN THE FIGHTING moved to Sicily, Westmoreland made an opportunity for himself and his unit that was later to pay personal dividends. The 82nd Airborne Division was then assembling in southern Sicily, and Westmoreland (apparently entirely on his own initiative) went to the division command post to see what was going on. A meeting was under way at which Major General Matthew Ridgway, then the division commander, and Brigadier General Maxwell Taylor, the division artillery commander, were discussing future operations. As Westmoreland later explained: "Upon learning that the division was short of transportation, I spoke up, identified myself as an outsider, and stated that I had an excellent artillery battalion that could give medium artillery support to the division as it moved and at the same time provide a large number of trucks to assist the lightly equipped parachute division."

According to his account, a call was made to the corps commander, who approved a request that Westmoreland's battalion be attached to the 82nd Airborne Division. Nothing is said concerning how the losing division commander felt about this freelancing on Westmoreland's part, but it was the beginning of a long and close relationship between Westmoreland and Taylor, who rose to far greater prominence and became Westmoreland's principal mentor and patron.[3]

Westmoreland remembered the next few weeks as a fast-moving situation. The motto of the 34th Field Artillery was "We Support," and they were living up to it. The battalion's trucks were being used around the clock. Often at the front of the column, Westmoreland recalled, would be General Ridgway, General Taylor, and Lieutenant Colonel Westmoreland. When resistance was encountered, Westmoreland contacted his executive officer and had him move the guns forward as fast as possible. Westmoreland would station guides on the road to lead the guns into position, then go to the nearest high ground to establish an observation post, determine the location of the friendly infantry elements, and proceed to adjust the fire of the battalion as soon as the first gun could drop trails.

"On one occasion after I had done this," said Westmoreland, "General Taylor was surprised to find me on the top of a hill overlooking the enemy's position, and he asked me what I was doing there." Westmoreland responded that his battalion was in the process of moving into position, that he had just finished adjusting its fire, and that they were ready to attack any target. He proudly remembered Taylor's comment as he departed the position: "Westmoreland, that was a workman-like job."[4]

In Sicily Westmoreland's jeep hit a mine. The vehicle was destroyed, but only one of the four occupants was injured, and that man only slightly. Westmoreland said that sandbagging the vehicle had saved its occupants from more serious effects. He himself was blown free, "shaken up but not wounded." According to subsequent accounts by Westmoreland, that was but one of a number of near misses. Earlier, in Tunisia, "a shell hit my vehicle but without harm to me," he re-called. Later, "on the Roer River in Germany, just as I got out of my jeep and entered a company command post, a mortar shell struck my vehicle. In the Remagen bridgehead on the Rhine a shell demolished a latrine moments after I had departed."

As the fighting progressed on Sicily, with the 34th Field having re-joined its parent unit, the 9th Division was "pinched out" by the 3rd Division on the north and the British on the south. The 34th Field had by then fired 4,393 rounds, including 901 on a single day during forty-three fire missions at Cerami on 4 August. While there the unit was taking very heavy incoming fire from artillery and Nebelwerfers (mortars) firing high explosive shells. The only solution was to dig in, and keep digging in, for five days. Moving through his B Battery posi-tion one day, Westmoreland heard a chief of section say to one of his cannoneers: "Jones, if you dig that slit trench two inches deeper, I'll try you for desertion!"

Then came a well-deserved respite. Rhapsodized the division his-torian: "These were days of vino, marsala, and vermouth; of grapes and melons and almonds; of gaily-painted donkey carts and swims in the blue Tyrrhenian Sea; of visits to Palermo and Monreale and the dark catacombs."[5]

DURING MUCH OF his battalion command time Westmoreland's ex-ecutive officer was a major named Otto Kerner. They got along well and stayed in touch in later years. Kerner remembered that in the bat-talion Westmoreland's officers and men referred to him as "Super-man" and that it was deserved, "a title that he earned by his deeds and capacity for deeds." Kerner was from Illinois and would later become governor of that state and then a federal judge.

Westmoreland frequently exhibited a measure of disdain for book learning, an outlook he recalled in an oral history interview after his retirement. "I had never been to a school and I didn't know the doc-trine except what I read in the field manual, and I wasn't too impressed with it," he maintained. "I figured I could do better and worked up some SOPs for maneuvering the battalion, controlling fire and some firing techniques." His interviewers, eager to learn more about those fundamental doctrinal differences, asked Westmoreland to describe them. "Well, I'd have to reflect back," he responded. "I think I might come forth with them, but it's pretty tough to do off the top of my head."[6]

Years later Westmoreland would say that, if he could cite one thing that had contributed "to any success that I may have had as a commander, it was that as a young officer I developed a keen appreciation of communications. As a battalion commander, I used to talk to my battalion as a group about once a month. I used to talk to the batteries and the gun squads individually, and brief them—even in combat—on the situation. I was constantly talking to my troops."

Westmoreland took away some enduring lessons from his command experience, including a conviction that the important determinants of troop morale were food, mail, and medical care. "These were the things that I derived from my World War II experience to which I gave consideration," he recalled. "Actually, I had them written on a slip of paper which I kept in my wallet."[7]

IN LATER YEARS Westmoreland heard from many of those who had served in his battalion, especially the citizen-soldiers who left the Army after the war. These letters were uniformly complimentary of the way Westmoreland had treated them and of how he had commanded the outfit. "You were kind to me and I appreciated it then and appreciate it even more now," said a 1962 letter from Dr. W. A. Wilkes, who had begun by saying "I have intended writing to you every year since leaving the Army in 1945." Wilkes gave a short summary of his postwar professional accomplishments, saying he did so "only to let you know that I put every ounce of energy that I have into my work. This I know you can appreciate and sanction wholeheartedly. I know that is the way you go about being a soldier."

Another former officer, in civilian life an editor and publisher, wrote to recall a division review at war's end and that there were tears in the eyes of some of the officers: "It was the end of something terrible and magnificent."

IN NOVEMBER 1943 elements of the 9th Division headed for England, there to refit and train to take part in the upcoming Normandy landings. Westmoreland gave frequent talks to the green troops just joining the division. He advocated periodic showdown inspections to get rid of the excess gear the troops accumulated, maintaining that in doing so he had found one man with a Beautyrest mattress and another with a four-poster bed. "As surprising as it may seem," he continued, "the American soldier will loot. They will molest native women. When they have not seen a steak for several months, they will kill cattle without authority. The hospitality of the alcoholically generous liberated people is difficult for soldiers to refuse."

Also while in England, Westmoreland rendered one of his earliest geostrategic judgments, expressed in a letter to his father dated 10 May 1944 and written while he was hospitalized with malaria contracted in Sicily: "I'm convinced that S. A. [South America] is the continent in which our future lies. Certainly we should get out of this mess in Europe and Asia as soon as we have stabilized the situation on these two vast and unsettle[d] areas."

By this time Westmoreland had relinquished command of his battalion and become executive officer of the division artillery. That put him working for Brigadier General Reese "Hooks" Howell. From the first it looked like trouble, as Westmoreland told his father one of his most important duties was going to be "to prevent the general from leaping before he looks (which is a habit which he regrettably possesses) and to attempt to place his decisions on as sound a plane as possible. I pray I am worthy to the job."

Westmoreland said later that Howell was one of the most difficult personalities he had ever served under—rude, abrupt, and arrogant in dealing with his subordinates and very jealous of what he considered his prerogatives. Eventually Westmoreland found the relationship intolerable and asked to be transferred to another job. When a slot opened up for division chief of staff, he recalled, "General Howell happily released me."

It was while still executive officer of the division artillery, however, that Westmoreland landed in Normandy on D+4, four days after the initial landings. The division moved to cut off the Cotentin Peninsula, driving to Cherbourg, then taking part in the St. Lo breakout and closure of the Falaise Gap. By 28 August 1944 it was across the Marne and driving east. Westmoreland characterized the fighting as "*very* severe" on "a sustained basis," saying they "lived in foxholes for months and months and months at a time."

While in the executive officer's job, near the end of July 1944,

Westmoreland was promoted to full colonel.[8] It was an early—and, as it turned out, temporary—wartime promotion. "The eagles feel very heavy at this point," he told his father. "Hope I can carry the load."

In a prematurely optimistic forecast of early August 1944, Westmoreland wrote to his father that "this campaign is going very well. . . . All indications are that it shouldn't be long now—Jerry appears to be disorganized and demoralized." Six weeks later reality had set in. "Germany is a cold place at this writing," Westmoreland told his mother. "All had hoped it would be over before winter arrived, but we don't feel so hopeful at this time. It appears that the war is far from over and that the last fight will be the toughest."

In mid-October 1944 Westmoreland became the division's chief of staff, just in time for the really terrible fighting—and losses—in the Huertgen Forest battles. He was extremely critical of the generalship at higher levels that inflicted such an ordeal on allied troops based on their "questionable decision to clear the enemy from that vast forest instead of bypassing it."[9] It was a "costly blunder" resulting in thousands of casualties, he stated, and "my division was chewed up twice." In his view, "The forest could have been bypassed, and should have been. Too few generals saw, at first hand, the situation."[10]

Beginning in mid-December 1944 the division held defensive positions until, at the end of January 1945, it jumped off again in a drive across the Roer and to the Rhine, where it crossed the unexpected windfall of a bridge seized intact at Remagen by the 9th Armored Division. Next it helped seal and clear the Ruhr Pocket before moving farther east to Nordhausen for an attack in the Harz Mountains. There, Westmoreland recalled, they entered the concentration camp only recently liberated by the 104th Infantry and 3rd Armored Divisions. By that point the end was near, and the division occupied positions along the Mulde River and held that line until V-E Day.

As chief of staff of an infantry division, Westmoreland was gaining valuable experience outside his own field artillery branch, and also observing a model leader in the person of Major General Louis A. Craig, the division commander. He and Westmoreland got along splendidly, and the two maintained contact for many years after the war. When General Craig moved up to command XX Corps, he sent Westmoreland a very warm handwritten note recalling "all the work, pressure, and anxiety that were nearly continuous" during their shared service and thanking him for his friendship. For his part, Westmoreland later stated that he had served no officer he admired more than General Craig, and under no officer had he learned more.[11]

The 9th Infantry Division, nicknamed the "Old Reliables," had a tough and very successful war. Major General J. Lawton Collins, Commanding General of VII Corps, documented some of it in a highly complimentary letter. "After crossing the Seine, the Marne, and the Aisne rivers in rapid succession," he observed, "the Ninth again came to grips with the retreating enemy in the edge of the Ardennes Forest east of Hirson and drove him across the Meuse. The division's successful crossing of the Meuse in the vicinity of Dinant, in the face of strong opposition, was one of the most difficult tasks of this war." And, continued Collins, "During these extensive operations, the Ninth Division advanced almost 600 miles against enemy opposition, captured over 28,000 prisoners and participated in three major campaigns with not more than five days out of action in a period of over four months. This outstanding record is one of the finest in the European Theater."[12]

The famed war correspondent Ernie Pyle was equally admiring, saying the 9th Infantry Division performed like "a beautiful machine." As he remembered it, the division moved so fast it got to be funny. He was based at the division command post, which at one point displaced forward six times in seven days, prompting a soldier whose job was taking down and putting up the tents to volunteer that he'd "rather be with Ringling Brothers."[13]

NEAR THE END of the war elements of the 9th Infantry Division met the Russians, as Westmoreland reported to his father: "Twice we were entertained by the Russians. Both parties were rip-snorters. Vodka flowed like water; it was difficult to imbibe moderately and walk out under your own power. I succeeded. At the last party we were served by female soldiers wearing boots and pistols. After the meal they became our dancing partners and were very graceful dancers."

V-E Day did not mean the end of European service for Westmoreland, as the 9th Infantry Division was kept overseas for occupation duty. Westmoreland told his father about this, adding, "I'm certainly going to try to get out of this deadly existence."

Fortunately for him, as it turned out, he stayed on, receiving another excellent opportunity for command, this time of an infantry unit. Even though his branch was field artillery, he became regimental commander of the division's 60th Infantry Regiment.[14] Writing to his father, Westmoreland reported that the regiment, with a normal strength of 3,200 men, was now over 5,000 and that life was good. "I have a very nice house with all the conveniences. My executive officer, supply officer, and operations officer live with me. I have a Chinese orderly, my own mess with very good cooks, two horses and a good stable. Shortly I hope to acquire a sedan. At the moment I ride a jeep." General Patton had recently come to inspect the division, and Westmoreland and the other regimental commanders lunched with him. Concluded Westmoreland: "He is quite a bird."

A key mission of the occupation forces was caring for large numbers of refugees, including those housed in sixty displaced persons camps and a civilian detention facility. Some of these people were from Estonia, Latvia, and Lithuania, recalled Westmoreland, "all fine people." The regiment's area of responsibility covered about a thousand square miles. With headquarters in Ingolstadt, Germany, Westmoreland was able to do some very effective, imaginative, and compassionate things for these unfortunate victims during the next seven months. Reopening the schools, winterizing the primitive facilities in which they were housed, and improving sanitation were key missions, along with guarding captured stocks of arms and munitions, conducting security patrols, and training.

"Occupation duties are becoming complicated," Westmoreland wrote to his father in early October 1945, "and in many instances confused. The deployment of troops [back to the United States] is progressing rapidly due to pressure from home." That was making things quite difficult for the occupying force, "which in effect is no longer an effective army but merely a group of more or less random troops."

THE PRINCIPAL SOURCE of troop morale during the occupation period, according to one of the unit's officers, was the Red Cross Donut Girls. "Shortly after Westy's arrival," said Dick Vestecka, "the troops of the regiment were getting more donuts and coffee than any similar unit in the entire European theater." Instead of staying the customary one or two days with a unit, these girls were staying for a week and even two weeks with the 60th Infantry, and there were fifteen such teams taking turns. "The lure," concluded Vestecka, "was Colonel Westmoreland—single, tall, dark and handsome."

Westmoreland was very critical of the rapid demobilization of the American army in Europe, writing to his mother in January 1946 that "what was a wonderful army over here has been literally torn to pieces." And, he added, "it goes without saying that the U.S. has lost considerable prestige as a consequence of the stupid and selfish attitude on the part of the American people. It is difficult to blame the people, however, when one realizes that they have been fed a lot of distorted reasoning & propaganda by our newspapers and Congressmen."

WHEN IT WAS TIME for Westmoreland to rotate back to the United States his regiment gave him a raucous farewell party. The program provides the flavor of the event: "Sixtieth Infantry Dancing Officers Club in Honor of Colonel Wm. C. Westmoreland. Presenting the Famed 'Hungarian Orchestra' and a Floor Show 1930. Snacks at 2300. Drinks. Eggnog Served from 1930 to 2030. At 2030 Your Choice Coke Highball, Cognac Special, Boiler Maker, Red and White Wines. Farewell Westie."

Westmoreland's copy of the program was signed by many people, perhaps most notably "Arthur R. Woolley The Best God damned Bn Cmdr you ever had and damn it I can still lick you." It appears Woolley may have sampled the eggnog before signing.

At the end of January 1946 Colonel Westmoreland was appointed, on paper, Commanding General of the 71st Infantry Division, with the sole mission of getting the remnants of that outfit back to the United States for inactivation. Westmoreland went on ahead, traveling home by air.

4

★

AIRBORNE DUTY

TWICE EARLIER WESTMORELAND had tried to transfer to airborne duty, but each time his request was turned down. Now he finally had that chance. After attending parachute and glider school at Fort Benning, in July 1946 he took command of the 504th Parachute Infantry Regiment, part of the 82nd Airborne Division. That put him right back at Fort Bragg, the place where he'd launched for World War II. The division commander now was the famous General James Gavin, who had earned a brilliant reputation during the war, participating in four combat jumps. The command climate in the division proved very much to Westmoreland's liking. "I found Gavin an excellent division commander to work for," he reflected. "He gave general guidance as to what he wanted and left his commanders alone to exercise their own initiative."

Westmoreland later recorded a puzzling comment: "I was given command of two infantry regiments but as a matter of principle did not transfer from the F.A. [field artillery]." He did not explain what principle he had in mind, but it may have had something to do with overcoming early resentment by infantry officers of his having been given the regiment, as Westmoreland later commented: "I don't say that an artilleryman is not necessarily a good division commander. But at a lower level, I believe, if you are not an infantryman but an artilleryman, you start with a certain disadvantage in dealing with echelons below."[1] Westmoreland overcame any such disadvantage, becoming widely respected as an effective commander of an airborne infantry regiment. There he continued his emphasis on talking to his men frequently, at least once a month "in personal communication, eyeball-to-eyeball, with every man in that unit."

Melvin Zais, later to reach four-star level himself, remembered the "splendid job" Westmoreland did in commanding the regiment. "He was ambitious without clawing and scratching," said Zais. "He wanted to do well, but he didn't step on anybody to do it." Other traits that would perhaps become more pronounced in later years were also apparent, however. "He carried himself like a general. But Westy did not have a very good sense of humor, nor did he have a light touch. Westy took himself quite seriously. He knew he was going to the top and worked hard at it. He put every ounce of energy into it."

AFTER A YEAR of regimental command, Westmoreland was moved into the position of division chief of staff, still working for Gavin. While that was a repeat assignment, since Westmoreland had served as 9th Infantry Division chief of staff during much of the war in Europe, the different type of unit and fast pace of operations made it a worthwhile professional experience.

Westmoreland gained particular approval on the occasion of a parachuting accident. A planeload of jumpers en route to the drop zone were killed when their C-82 aircraft lost power in both engines and crashed. Westmoreland was then on the drop zone, having just completed a jump. Anticipating the possibility of mass panic and very low morale among the troopers who had not yet jumped, he went immediately to the loading area. He found that the troopers there were eager to jump, but had been held up because the tower had grounded all aircraft. "Colonel Westmoreland immediately took command of the situation and ordered the next lift to prepare to load," reported a regimental commander. Then, using his authority as division chief of staff, Westmoreland ordered the tower to clear the aircraft for takeoff. He then organized the next lift, boarded the aircraft with the troopers, and led one stick in his second jump of the day.[2] Major General Byers, commanding the division, wrote to Westmoreland to commend and thank him for his actions, saying that they had "unquestionably served to minimize the harmful effects of the crash on morale of the personnel who were scheduled for jumping that day."

Private First Class Tom McKenna had an interesting encounter with Westmoreland soon after arriving in the division. He was just back from parachute and glider school at Fort Benning when the division chief of staff came down to inspect the new troops. Stopping in front of McKenna, Westmoreland asked him, "How many jumps do you have, son?" That surprised McKenna, since everyone made exactly five jumps in parachute school, but he answered, "Five, sir." Westmoreland said, "Great sport, isn't it?"

About two weeks later McKenna was detailed to spend Sunday morning riding in a C-82 to help the crew chief pull in the static lines after all the jumpers had exited the aircraft. The object was to get the maximum number through their required pay jump, so the rides were really short. On one of those flights Westmoreland sat across from McKenna and, remembered McKenna, "He was obviously sweating out the jump." McKenna had a small camera with him and took Westmoreland's picture. When the flash went off, Westmoreland looked over at him, and McKenna asked brightly, "Great sport, isn't it, sir?" Westmoreland only nodded.[3]

HAVING HELD THE post of division chief of staff for an unusually long tenure of three years, Westmoreland had also served three division commanders, all very impressive and successful officers. After Gavin came Major General Clovis E. Byers, who won a Distinguished Service Cross and two Silver Stars in World War II and wound up a lieutenant general,[4] and then the famous curmudgeon Major General Williston B. Palmer, who followed command of the airborne division with command of an armored division, and then of a corps in Korea, eventually achieving four stars as the Army's Vice Chief of Staff.[5]

Westmoreland adjusted to these very different personalities and operating styles smoothly and effectively, making himself indispensable to each in turn. An officer who served with Westmoreland in the division later sent him a glowing tribute: "While in the 82d you demonstrated superior staff ability and, without trying to kid anyone, I feel that you exerted more influence and were more instrumental in maintaining the high state of training, discipline and morale which existed than any commander who was there during my tour. In your tactful, diplomatic way you actually commanded the division. And I must say did a fair job."

WESTMORELAND DEVOTED A single paragraph of his 425-page memoir to his courtship of and marriage to Kitsy Van Deusen. She seems to deserve much more. It is widely recognized that she was essential to his career success, yet he could not or would not acknowledge it. "I attribute a lot of his Army success to Kitsy," said one close friend of both.[6] Wrote David Halberstam: "She was, in any real sense, his ambassador to the rest of the world. Because he is so formal, it always had been her job to serve as the intermediary between him and others, to humanize him."[7] Many others shared that view, perhaps even Kitsy. "No one in the family gave our marriage a chance of succeeding," she later told a reporter. "Not because of the age difference—my father was 12 years older than my mother—but because Wes was so serious and they knew I wasn't."[8]

Kitsy was attending college in North Carolina when she heard from her parents that Colonel Westmoreland was stationed at nearby Fort Bragg. As she later told the story many times, she called him, and Westmoreland asked if she was grown up yet. "Come see for yourself," she responded. He did, and she was.[9] Six months later the colonel and the college girl were married in St. John's Church, Fayetteville, North Carolina. He was then thirty-three, she twenty.[10] Almost two years later, while they were still at Fort Bragg, a daughter was born. They named her Katherine Stevens and called her "Steven," later softened to "Stevie."

An officer at Fort Bragg remembered a visit to the post by members of the West Point Class of 1948. They went through a receiving line that included General Byers, the division commander, Westmoreland, the division chief of staff, and their wives. The receiving line dialogue went like this: "Good evening, General Byers. [Pause.] Good evening, Colonel Westmoreland. [Pause.] Kitsy!!" She had been at

Cornell and had arranged dates for the West Pointers passing through there, and they all remembered her with great fondness.

AFTER HIS FOUR YEARS with the 82nd Airborne Division, West-moreland was assigned to Fort Leavenworth, Kansas, first, briefly, as an instructor at the Army's Command & General Staff College and then in the postwar Army War College, temporarily located there. This was noteworthy, since Westmoreland was not a graduate of ei-ther school, although he had been awarded constructive credit for the staff college course. He then moved with the War College when it relocated to a new permanent home at Carlisle Barracks, Pennsylva-nia.

Said Westmoreland of his instructor duties, "I had cognizance over airborne operations, airmobile operations, and the use of helicopters. In addition I had cognizance over psychological warfare and irregular warfare. I was also involved in strategic planning."

Early on in the assignment he gave his father a status report: "You can see they have me on the hop and I am finding it difficult getting oriented in my new field of activity and adjusted to the point where I can start producing." While he did not say so, his new assignment as an instructor had to have felt like a major comedown after being chief of staff of an airborne division. His sponsor in the instructional assign-ment was Colonel Walter "Dutch" Kerwin, with whom he would have much to do in later years. "He was a good instructor," said Kerwin, "but mainly what he did was talk about World War II."

Westmoreland had been one of those successful young officers who received an early wartime promotion to colonel followed by a postwar reduction to lieutenant colonel. That came for him in July 1947. Once again promoted to colonel in 1951, he wrote to his mother about it: "After four years I'm back where I was before." His ambition was never far from the surface.

General Craig wrote to congratulate Westmoreland on getting his eagles back. "I am far more interested in the next step," he said. "I don't know anybody around your time who has a higher equity in the right to be promoted. Don't forget to keep your head." He added a bit of wisdom that would prove prescient: "An impeccable reputation can be a very dangerous quality in the possessor and, as rank increases, nothing is more disturbing to mental balance than a group of people [apparently referring to subordinates] who always say yes."[11]

IN JUNE 1950 the Korean War erupted. Two years into it, Westmoreland joined the fight as Commanding Officer of the 187th Airborne Regimental Combat Team. He wrote to the officer he was to succeed to express his delight at escaping from "such a stagnant assignment" as the War College and another "dull year ahead."

5

★

JAPAN AND KOREA

B Y THE TIME Westmoreland became involved in it, the Korean War had already dragged on for more than two years and degenerated into a static exchange of shelling and patrols while fruitless armistice talks continued interminably.

Westmoreland's new command, the 187th Airborne Regimental Combat Team (nicknamed "Rakkasans"), was something of a fire brigade for the allied forces. Stationed in Japan as theater reserve, when Westmoreland joined it had already made two combat deployments to Korea, each time subsequently returning to camps in Japan to refit. They were in Korea for the third time when Westmoreland took command on 29 July 1952.[1] His initial observation was that "the men had long been in reserve in Japan and needed refresher training."[2]

Ten days later Westmoreland wrote to Kitsy, "I have seen all elements of the combat command—some several times—and am favorably impressed with the outfit. Trap [his predecessor, Brigadier General Thomas Trapnell], as would be expected, did a wonderful job and surrounded himself with a group of good people that I am fortunate to inherit." To his classmate and close friend Colonel Robert Fergusson, Westmoreland wrote that he had "found since arrival that the unit is in first class shape."

Even so, Westmoreland soon relieved several officers for cause, including the regimental surgeon who, he said, "became worthless, apparently was taking dope, and upon relief shot himself in the leg." On the positive side he found his living conditions quite comfortable, writing to his parents that he had a van "containing a bed, desk, wardrob[e], wash basin, shower and—believe it or not—Frididare."

Within the first few weeks Westmoreland issued a revealing order. Stating that "a fundamental of effective command is inspection of the unit and troops . . . frequently, systematically and methodically," he noted that the chain of command was often overlooking this duty. He thus prescribed some minimums. Each battalion commander or his executive officer was to inspect daily a major element of his subordinate commands and attachments. Each company commander would inspect daily all organic and attached elements. And platoon leaders were to inspect three times daily their fighting positions, living quarters, and available individuals of the platoon, and do that within specified time periods: once before 0900, once after 1300 hours, and once between the hours of 2200 and 0400. That pretty well spelled it out.[3]

WHEN THE 187TH Airborne was introduced into the Seoul area, Westmoreland found it "fairly quiet and stabilized," with the action "basically trench warfare with a lot of scouting and patrolling." Scarcely two weeks into Westmoreland's command, Corporal Lester Hammond Jr., a radio operator in A Company of the 187th, earned the Medal of Honor for actions while part of a six-man reconnaissance patrol that penetrated nearly two miles behind enemy lines.

Westmoreland himself had a very close call, potentially even more serious than his various World War II near misses, although it was in training rather than a combat situation. He and some members of his staff were on a hilltop near Taegu, observing an exercise during which mortars were to deliver preplanned supporting fire. Through some miscalculation, the incoming rounds landed not in the designated target area but where Westmoreland and the other observers were located. A lieutenant standing right next to Westmoreland was severely wounded, and several others were hurt, but once again he came through unharmed. Investigating, he found that an administrative warrant officer, not qualified in the use of mortars, was commanding the support platoon. He was relieved posthaste.

Westmoreland had surprisingly limited combat experience during the Korean War. He commanded the 187th for a total of about fifteen months, of which only six were spent in Korea and the remaining nine in Japan.

At the beginning of September Westmoreland wrote his parents that "it is an unpleasant war. Everyday we have our losses and the attrition adds up as time goes by." On 9 September he told Kitsy, "we have been in combat one month today." By early October he was writing to her that they had been "in the line as a part of the 7th Division for almost two months. Our casualties were not heavy, however we had our portion. It would not surprise me if we had more men hospitalized from our training jumps ahead than we suffered in combat." Now they were "out of the line and at Taegu preparing to engage in some airborne training preliminary to returning to Japan."

Westmoreland was eager to build up the number of his parachute jumps, and now he had his chance. "He once made thirteen jumps in one day," in a training situation near the Han River, recalled Brigadier General Weldon Honeycutt, who as a lieutenant in Korea was Westmoreland's pathfinder platoon leader. "He'd go up with Sergeant Wolfe and Sergeant Bowser and Sergeant Card. As soon as they made a jump they'd get back to the airstrip and get on the next plane going up, then do that over and over again. We were competing to see who could get the most jumps in one day." One result of that marathon was that, only weeks after taking command of the unit, Westmoreland was awarded the Master Parachutist Badge, recognizing his completion of sixty-five jumps. He was now, as the airborne informally styled the accomplishment, a "Master Blaster."

In late October 1952 the 187th returned to its stations in Japan, known as Camp Chickamauga, outside Beppu, and Camp Wood, near Kumamoto, both on the island of Kyushu.

IN THE 187TH Westmoreland was promoted to brigadier general, but not before he had qualified (as a colonel, the highest rank entitled to the designation) for the Combat Infantryman's Badge. When the promotion came through, he of course received many congratulatory messages. Surely one of the most significant came from Lieutenant General Maxwell D. Taylor, then a Deputy Chief of Staff of the Army. "Sincere congratulations and best wishes on the newly acquired star," he wrote. "It is partial payment for that fine artillery support which you rendered the 82nd Airborne in Sicily."

The promotion allowed Westmoreland to bring his family to Japan. He wrote to Kitsy, advising that she could ship a washing machine, dryer, and freezer if she so desired, adding that "they may come in handy and could be sold over here for a profit." Another potential money-making scheme involved bringing a car. "Strangely enough the Japanese market is for 4 door *black* sedans. Such type brings the highest price," he advised. Since their car was not black, Westmoreland suggested this: "As I remember your Dad has a 4 door 1950 Chevrolet. If he would have it painted black, perhaps we could swap him even-steven and you would have what we want."

Kitsy made the move before Christmas, after which she wrote a warm letter to her mother-in-law, known as Mimi, to thank her for a special Christmas gift: "I did wear your lovely slip, and how I needed it. Childs gave me an ice-cream freezer. Could have cried when it wasn't a nylon slip. So when I opened your lovely present I could at least look the freezer in the eye. Many, many thanks."

Kitsy also asked for some help from her husband's father: "I wish Wawa would drop him [Westmoreland] a line, especially now he is buying stocks. He is going to some broker in town, but Childs still thinks he can make a killing on the market!"

Shortly before his promotion to brigadier, Westmoreland had received an important letter from Major General Louis Craig with some detailed advice. "You are probably hitting the real crisis of your career at this time," he observed. "If you pass through it with the same credits that you've earned heretofore, you are made—unbreakably so." And: "I'm sure, in your case, that you will never let ambition get ahead of character—which sometimes breaks a reputation on the rock of principle. These youngsters under your command need that sort of reassurance."

AFTER EIGHT MONTHS in Japan the 187th went back to Korea on 21 June 1953, only weeks before the armistice, when the Chinese communists launched a major attack with the 68th Army. The troopers flew to landing fields near the front rather than parachuting in. Westmoreland wrote to his father that "for a while we were engaged in a pretty good fight on the fringes of the Chinese offensive against the

South Korean Army." He later recalled seeing "the famous Capital Division [a Republic of Korea unit] break and stream through my command [while retreating in panic]. We held the shoulder of the breakthrough, which was a very big one. We made a relief of the 7th Regiment of the 9th Division in the middle of the night." Then "we had a very hairy several days."[4]

The Korean War armistice took effect on 27 July 1953. Leading up to it, there were small but vicious fights as both sides sought to grab more territory before the end of hostilities. At that time Westmoreland had his regiment on line above Kumwha, with the 1st Battalion occupying positions on what was known as Hill 604. The forces were disposed along a ridgeline to protect against just such Chinese efforts to seize more terrain. Late one afternoon he got a call telling him to withdraw the 1st Battalion and move it to a new position some three or four miles to the rear.

Westmoreland was extremely reluctant to carry out this order, since his 3rd Battalion was out in front of Hill 604, and would thus be left more exposed and endangered by the withdrawal, besides which the 1st Battalion was holding critical terrain at the head of the Chorwon Valley.[5] He protested the order to Major General William L. Barriger, commander of the 2nd Infantry Division, to which his outfit was then assigned. Barriger, said Westmoreland, "was known as the 'Bull' and he lived up to that reputation. He had a bark for everybody which was a cover-up for his inadequacy as a division commander." Barriger gave Westmoreland an ultimatum—carry out the order or be relieved of his command. Westmoreland complied, but stated that he was carrying out the order under protest. The move was made in total darkness during a driving rainstorm and, maintained Westmoreland, "jeopardized [my] headquarters and lines of communication." "It was a hectic night which I will never forget—confusion reigned. But the next morning we were able to regain our tactical integrity."[6]

Apparently no adverse consequences resulted from this controversy, on or off the battlefield. General Barriger, for his part, held no grudges. Instead he sent Westmoreland a cordial letter, recalling that "from 14 July 1953 until the signing of the Armistice, the 187th Regimental Combat Team . . . was given the mission of defending the approaches to Kumwha, some of the most important terrain in the division sector." That mission was accomplished extraordinarily well, he said. "During this period of fourteen days not one enemy soldier crossed your lines except as a PW [prisoner of war]." An interesting aspect of the matter was later revealed in the 187th's command report for July 1953: General Barriger had ordered Westmoreland to withdraw the battalion "at the direction of the corps commander."

Major Frederick Kroesen, later a four-star general, was in command of the 1st Battalion during that controversial withdrawal and got to know Westmoreland well. He came away with mixed views. "Westy was a great commander of airborne troops," he recalled. "They responded to him very well." But "we questioned his judgment on some occasions. We talked about his ambition, which clouded some of the things he wanted to do. I remember, in one of those discussions, someone saying he'd court-martial his wife if he thought it would get him another star."[7]

Barriger was not the only senior commander Westmoreland had problems with. On an earlier deployment to Korea the 187th had been attached to the 7th Infantry Division, then commanded by Major General Wayne Smith, described by Westmoreland as "a short roly-poly man with a loud mouth who the troops called 'Shaped Charge Six' because of his protruding belly. General Smith could be described as a dishonest showman," said Westmoreland, "and it was disgusting to me to see some of the demonstrations that he would stage for visiting delegations from the United States."[8]

THE TERMS OF the armistice included a seventy-two-hour period during which supplies and equipment could be recovered from the positions then occupied by the opposing forces in what was to become the Demilitarized Zone. Westmoreland kept detailed records of what his troops accomplished, noting that they policed up and stockpiled 562 miles of commo wire, 103,813 sandbags, 4,148 pickets, 7,688 pieces of lumber, and 26,020 logs. "This salvage operation tops anything that I have seen," he proudly told the unit.

The 187th redeployed to Japan aboard a Navy transport, the USS *Pope*. Westmoreland recalled that as his troops boarded the vessel he stood at the gangplank and inspected every man. "The officers and NCOs were so effective I did not have to correct one man," he said. "I was proud." Later the ship's captain told Westmoreland these were the finest troops he'd ever seen. "Who have you been carrying?" Westmoreland asked. "Marines!" the skipper replied.[9]

WHILE THE OUTFIT was in Korea, Westmoreland had promoted as its motto "Every Man a Tiger." "When the fighting stopped and we were ordered back to Japan, I had tigers on my hands," he realized, "and Beppu was no place for tigers." He launched a "detigerization" program—with what he admitted was mixed success.

One project involved selecting verses to be sung to the tune of the "Rakkasan March." A board of judges had chosen various candidate verses on which the troops were asked to vote. Another diversion was a "gigantic" intramural sports program. Westmoreland also devised a card to be issued to each member of the outfit. It included these definitions, developed by him personally: Trooper: Trained, Reliable, Observant, Proud, Efficient, Rakkasan. Drooper: Disgraceful, Rowdy, Overbearing, Obnoxious, Poor Example, Rakkasan. His message: "Be a Trooper, Not a Drooper."

"THE TROOPS LOVED Westmoreland," said Charlie Montgomery, later Westmoreland's military stenographer. "They really did. He knew their names, his 187th compatriots." Many of those troopers wrote to Westmoreland in later years, always in very warm terms, one saying he did so "to thank you for the soldierly example you had set for me during my early years as a very impressionable and slightly scared junior officer." Sergeant James Costa wrote a memoir of his service in which he said of Westmoreland, "He is a great American, patriot, statesman, and combat general with a lot of heart for his soldiers."[10]

Corporal Arno Land was one of those who served in the 187th under Westmoreland. He was having a problem with an allotment from his pay that was supposed to be going to his mother. One day, after a skirmish with a North Korean unit, Corporal Land was walking down a road en route to the regimental command post to see if he could get some help with his problem. He had just started out when he encountered the regimental commander coming forward in his jeep. Westmoreland stopped and asked what the young soldier was doing. Land explained the problem. Westmoreland got on the radio, talked to someone in his headquarters, then got out of his jeep and told Land to get in. The regimental commander proceeded to walk to the forward positions while Corporal Land was driven to the rear. At the command post, Land was met by a senior NCO who handled his situation. Some years later Land, by then a sergeant first class, related this story in a discussion about leaders with a young officer recently assigned to the 101st Airborne Division. "Corporal Land never again had a problem with the allotment to his mother," recalled Lieutenant Colonel Robert Frank, who had heard this story as a lieutenant, "and that's why he held his commander in such high regard for the rest of his life."[11]

DURING HIS MANY months in Japan Westmoreland worked hard at building good relations with the Japanese communities near where his troops were stationed, among other things by helping orphanages. He proposed establishment of a Joint Governor's Council for Oita Prefecture, serving on it himself and hosting its meetings at the Camp Chickamauga Officers Club. This council served, said Westmoreland, as a forum in which to "discuss mutual problems, determine corrective action, and make wise recommendations with the long range objective of promoting and maintaining a healthy Japanese-American friendship." He likewise set up a Beppu City Committee and an Oita City Committee and served as a member of both.

In Beppu Oita, Sizuko Ogo ran Orphanage Eiko-en, a facility his mother and father had established and which Westmoreland had helped when he was stationed in Japan. "If you had not helped us," Ogo wrote to him many years later, "many children would have died at that time." Now, he reported, he was carrying on the work of his parents, and "your photograph is hanging on the wall of my office and it always reminds me of the good old days."

SURPRISINGLY, GIVEN HIS reputation as a moderate drinker, when Westmoreland headed back to the United States for his next assignment he decided to ship home several cases of liquor using a tax-avoidance scheme. "In Virginia there is a rather high tax," he wrote to a fellow officer, whereas in the District of Columbia one could obtain a permit without cost. "In my case," he explained, "since I was a member of the Fort McNair Club I had mine shipped to me in care of the club," which was in the District and where, he said, "I had no difficulty in getting a permit from the authorities."[12] Of course he did not reside at the club or anywhere else within that jurisdiction.

En route home from Japan, Kitsy had an emergency operation in Kyoto which delayed their travel for several weeks. During that time Westmoreland was placed on temporary duty with the nearby U.S. 3rd Marine Division. "It was interesting to serve with the Marines and I found them to be a fine outfit," he wrote to his friend Bob Ashworth. "However, they are decidedly second best to the airborne troops in the Army. They can in no way hold a candle to the 187 in the matter of appearance, discipline, esprit de corps or training."[13] Westmoreland wrote a similar account to another friend, observing that "although it was interesting duty, frankly they did not show me much."[14]

TOWARD THE END of his command tour and thereafter, when he was back in the United States, Westmoreland worked energetically—and successfully—to promote award of a Republic of Korea Presidential Unit Citation to the 187th Airborne Regimental Combat Team. He had an in with Colonel Carl Schmidt at the U.S. Military Assistance Advisory Group to the ROK (Republic of Korea). That officer was willing to push such a recommendation. At the end of December 1953 he wrote to Westmoreland, "I have seen nothing of a proposed ROK citation for the 187th RCT. Perhaps you might remind the present commander that it would be very easy to have one approved once it is submitted." Westmoreland then wrote to Major Nicholas Psaki on the staff of the 187th, noting a letter he had sent to Colonel Russ (the Deputy Commander) recommending expedited action on such an application. "If we can get this in before Colonel Schmidt is reassigned," he stressed, "it would pay the outfit great dividends." Somehow it all worked out, and the award was authorized.

IN LATER YEARS Westmoreland again served with a number of people who had been with him in the 187th. Among the most prominent was Edward "Fly" Flanagan, who in Korea led the RCT's artillery battalion and became a lifelong admirer of his commander. Flanagan recalled that Westmoreland's "demeanor, personality, and physique were Hollywood's answer to a modern brigadier, but he was more than a model. He had the intelligence, moral quality, and judgment to fit the physical appearance." He remembered gratefully that "this commander checked his units without smothering the commanders, cared for the troops and their needs, fought with higher commanders when he thought that his men were being grossly ill-employed in combat, understood his mission, carried it out with skill, and backed up his battalion commanders when necessary."[15]

Later Westmoreland would describe command of the 187th as "the most satisfying experience of my military service." And in a television biography he reaffirmed his pleasure in that experience and others that had come before it, saying, "The highlight of my career was commanding paratroopers." Now another major comedown lay before him, the life of a Pentagon staff officer. Apparently he was not much looking forward to that new posting, writing to a friend that he hoped to see him on his next assignment, but "don't wish you the Pentagoose."

6

★

PENTAGON

WESTMORELAND'S FIRST HIGH-LEVEL staff assignment took him to the Army Staff in the Pentagon, where he was designated Deputy Assistant Chief of Staff G-1 for Manpower Control. This put him, entirely unschooled, into the complicated and crucial business of Army personnel policy. Early in the new assignment he wrote a rather plaintive letter to a staff officer in his old outfit: "It is considerably frustrating attempting to fit into this huge and complicated operation without benefit of G-1 [personnel] background or Indian-level training in the Pentagon."

A friend's letter offered an explanation for the assignment. "Saw Willie Palmer [General Williston B. Palmer, now Vice Chief of Staff of the Army] about three weeks ago. Asked him why he hadn't got you in his office. Told me that he felt the boys who are to run the Army should get a G-1 education and that he had wanted you prepared."

Kitsy went to North Carolina to be with her parents while she continued recovering from surgery. Westmoreland leased an apartment pending her arrival at his new station. Soon it developed that she was also pregnant, and some complications after she came to Washington caused her to spend another period of time with her parents. In early June she finally made it back to Washington. "I have the road well reconnoitered to Walter Reed," Westmoreland wrote to his friend Bob Fergusson. In July 1954 a son was born. They named him James Ripley Westmoreland and called him variously "Rip," "the Ripper," and "the Ripster." A year and a half later there came a second daughter, whom they named Margaret Childs Westmoreland, thus completing the family.

EARLY IN THE new assignment Westmoreland became involved in intense work on the Fiscal Year 1955 budget. "Since reporting 10 days ago," Westmoreland wrote to a fellow officer, "I have been in a complete fog and still have only a vague idea of what people are talking about. If I could master this new language that has been created in this Pentagon atmosphere, and particularly in G-1, it would be a big help."

Three weeks into the new job Westmoreland wrote to his classmate "Bev" Powell, commanding the 11th Airborne Division artillery, that he could "now appreciate why G-1 is the target of such widespread criticism in the Army. A great percentage of their problems are of such a complicated nature that it is a question to select the better of many bad solutions. Consequently many people are naturally unhappy with any decision made." It seemed that, even this early on, General Palmer's plan was working.

ANOTHER ISSUE ENGAGING Westmoreland's attention during this period was reliance on a drafted army. Speaking at a military convention in mid-1954, he told the attendees that "each volunteer soldier we can add will reduce the draft calls and, by reducing the number of men serving two years of duty, will reduce wastefully rapid turnover and costly training." Thus, he concluded, "we will get more defense per man per dollar if we can get long-term volunteers and eliminate short-term draftees." He favored the all-volunteer approach on philosophical grounds as well, asking, "What then is the American way? Is not the volunteer theory a part of the very fabric and structure of our society?" That outlook would be worth recalling when, more than a decade later, an all-volunteer Army became a central issue when he was Army Chief of Staff.

Westmoreland developed a close and friendly relationship with Congressman Gerald Ford, who invited him to be a speaker in Grand Rapids, where he stayed with Ford's parents. When Ford was named to West Point's Board of Visitors, Westmoreland and Kitsy accompanied him and Mrs. Ford to the Military Academy. Later Westmoreland accompanied Ford on a trip to Europe, a ten-day jaunt that included stops at Heidelberg, Stuttgart, Berlin, Amsterdam, and Paris, where, Ford claimed, they stayed out all night together.

"I had not served there earlier, and I didn't want to serve there," Westmoreland said of the Pentagon in an oral history interview. But apparently the experience of doing so proved valuable, since a year or so into his G-1 duties he expressed to a friend a much different outlook. "In my opinion one has not received a full measure of education in the Army and the inner workings of our great democratic society until one serves a tour in Washington." In the meantime he had testified before Congressional committees, worked on budget development, and responded to innumerable outside inquiries and petitions, all undeniably valuable experiences for those destined for greater responsibilities. But the frustrations were also very great. "Problems here in the Pentagon seem to repeat themselves on an annual cycle and never seem to get completely solved," he lamented.

The source of much of his frustration was Secretary of Defense Charles E. Wilson, who was determined to reduce the Army's strength as a cost-cutting measure. Wilson ordered a succession of hundred-thousand-man cuts, increasingly draconian in their impact. "To back up his decisions," wrote Ernest Furgurson, "Wilson sent down a memorandum asserting that, given proper leadership and training, the Army could maintain its nineteen divisions with the number of men it had been allowed. General Matthew B. Ridgway, then chief of staff, felt personally insulted." Westmoreland came up with a scheme that, while admittedly "phony as hell," was intended to placate Wilson. It reorganized commands in Alaska and Panama under division designations, keeping the remaining seventeen divisions at or near their authorized strengths. "Wilson, who was most interested in having the nineteen divisions on paper as evidence that he could cut manpower without hurting capability, accepted it."[1]

DURING HIS ASSIGNMENT to Army G-1 Westmoreland was sent on temporary duty to attend Harvard University's three-month Advanced Management Program. "Although I feel extremely ill-prepared to participate in such activity," he told a fellow officer, "I am nevertheless keenly looking forward to the opportunity. Perhaps after being exposed to such a business atmosphere I will be able to better understand what they are talking about here in the Pentagon." It turned out to be an important experience for him. He got along well with his classmates, who were mostly up-and-coming civilian business executives, and maintained contact with a number of them in later years. In correspondence Westmoreland told one that he "attached much greater importance to the associations and friendships that I made than I did the brief academic exposure." But many years later he maintained the opposite, that "those 'new leaders' that I met had no influence on me although I liked them. That experience had no influence on my outlook." By that point more successful, Westmoreland was also less willing to acknowledge that he had ever needed help.

In a very generous gesture, another of the handful of military participants in the Harvard program, the Navy's Rear Admiral Robert Morris, wrote to Army Chief of Staff General Matthew Ridgway. He was doing so, he said, to let Ridgway know "what an excellent impression the Army's 'student' made on all hands. Brigadier General Westmoreland's contributions to the discussions were thoughtful, organized and well presented. West Point, the Army, and military affairs generally were well represented by this splendid officer."

AFTER A YEAR and a half in Army G-1, Westmoreland received a plum new assignment. General Maxwell Taylor, who in July 1955 became Army Chief of Staff, selected him as his Secretary of the General Staff. That "gatekeeper" function put Westmoreland in continuous contact with Taylor as one of the Chief's principal advisors and manager of the flow of paper, scheduling of briefings and conferences, and virtually all other aspects of his complex official life. Westmoreland said the SGS "attempts to shield [the Chief of Staff] from minutiae and to present actions in such a way as to simplify the making of decisions."[2]

A significant Taylor initiative was creation of what was called the Pentomic division. Westmoreland described how it had come about: "On one occasion [President Eisenhower] was conferring with General Taylor and he said, 'You have to do something to sex up the Army'—I am not sure he used that word—to give the Army more public charisma. And that brought about the Pentomic division."

Westmoreland watched Taylor personally work it out, remembering how he would come in with diagrams he had drafted at home and give them to Westmoreland to send to the staff. "I was never enthusiastic about the Pentomic division concept," he recalled, "but General Taylor felt that he had to do something, something new, to give the Army a modern look."[3] That look involved reorganizing divisions into five battle groups (replacing regiments and battalions), each commanded by a colonel, which could (it was theorized) be widely dispersed on a nuclear battlefield for force protection, then rapidly concentrated for offensive action.

"General Taylor and I had many theoretical discussions" on this concept, said Westmoreland. "I didn't agree with the Pentomic organization and later played a role in changing it. It was not sound in my opinion. It was not sufficiently flexible, nor was it a sufficiently cohesive system that would stand the stresses and strains of battlefield pressure. It was also short of infantry and short of artillery." (As things turned out, Taylor subsequently sent Westmoreland to command the 101st Airborne Division at just the time it was designated to reorganize as the Army's first Pentomic division.)

The situation Taylor inherited as Chief of Staff was a difficult one. "The Army was feeling sorry for itself," said General William DePuy, at that time a colonel assigned to an element of Westmoreland's office. "Because Ike thought he knew all about the Army, it was getting short shrift."[4] One major element of this was downgrading the importance of ground forces generally and increasing reliance on a nuclear deterrent to discourage Soviet aggression. Along with the Pentomic division concept, the Army now began to talk about and plan for the use of tactical nuclear weapons. While he was SGS, Westmoreland went on the lecture circuit to make the case. "We must not work under any illusion that atomic weapons will not be used should it be to our advantage to do so," he argued. "Atomic weapons can be used with discrimination without resulting in a mutual thermonuclear exchange."[5]

DESPITE HIS MISGIVINGS about the Pentomic concept, Westmoreland was eager to implement Taylor's plans and programs, sometimes perhaps too eager. When Taylor remarked on one occasion that he didn't think it was a good idea for people assigned to the Army Staff to send Christmas cards to others in the local area, Westmoreland issued a directive on his own initiative forbidding such exchanges. That created something of a stir, not to mention much hilarity at Westmoreland's expense. Nor was the matter forgotten when, years later, Westmoreland himself became Chief of Staff and General Taylor wrote advising him to "be sure to get a SGS with a well-developed sense of public relations." On the file copy of Taylor's letter, sent from the White House, someone has written in red pen: "Referring to Xmas card letter."[6]

WESTMORELAND WAS NOTHING if not conscientious and exceptionally hard-working as Taylor's SGS. He would routinely get to the office at 7:00 A.M., where his cable man (who got there at 5:00) would have on his desk all the overnight cable traffic. Westmoreland screened the mass of material, selecting those cables he thought General Taylor should see, then went through them and underlined the key passages. Thus, he recalled, "in 15 minutes he could go through them and really grasp what went on. And then I would have breakfast at the Pentagon. I did that for almost three years in that job."

After Westmoreland had been SGS for a year and a half, General Taylor promoted him to major general. Westmoreland was forty-two, and said to be the youngest major general in the Army. He received the customary flood of congratulatory letters, noteworthy in his case for the large number coming from very senior officers, evidence of the extent to which he was now known throughout the Army and how well he had established himself with the top echelons. One of the most interesting came from Charley Askins, then with the MAAG in Indochina: "It is people like yourself who keep our Army preeminent, and despite the Air Force, the Republicans, the Congress and the mamas, will see it remains so."

Westmoreland annotated these letters in red pencil to show the salutation he wanted on the replies and what letter to send—Type A, Type B, Type C, in descending order of warmth and gratitude. His classmate Phil Gage, who reported class news to the West Point alumni magazine *Assembly* and was a near worshiper of Westmoreland, received only a Type C response.

One of Taylor's speechwriters at that time was Colonel Winant Sidle. He would later, as a major general, become Westmoreland's public affairs officer. Now he observed that Westmoreland's relationship with Taylor was "a little fearful, because [Westy] was not a great grammarian, and Taylor demanded perfect papers." On the personal side, observed Sidle, the relationship "was not close. Taylor was cool. Westmoreland had no chitchat. He was only comfortable when he had a topic to address."[7]

One positive approach Westmoreland displayed here, as he would in later assignments, was paying attention to people who had been largely neglected by his predecessors. Cary Shaw Jr., retiring from his post as a civilian personnel administrator, wrote to thank Westmoreland for his attentions. He had served for sixteen years in the Department of the Army, he said, and known scores of general officers. "But there are only a few, yourself included, who have impressed me in an extraordinary manner by demonstrating that rare combination of managerial skills and human understanding which is so essential to effective administration." Shaw closed by reminding Westmoreland of their first conference, when Westmoreland had extended the discussion for an hour and three-quarters. "That was more time than I had spent with all of your predecessors over a period of ten years," said Shaw.

THROUGHOUT HIS CAREER Westmoreland stayed in touch with his Boy Scout roots, serving while a young officer as a Scoutmaster at Fort Sill and then Schofield Barracks, then later in other positions. He was impressive enough in the aggregate that he was awarded the Silver Beaver and later the Silver Buffalo, a very high honor, by the Boy Scouts of America. Clearly the ideals of Scouting continued to mean a great deal to him. As SGS he wrote to Conrad Cleveland Jr., one of his closest friends in his youth and ever after and a fellow Scout in Troop 1 in Spartanburg: "In this troubled world in which we are living, I believe that the Boy Scout movement has taken on a higher degree of importance than has ever been the case in the past. The fact that the Scouts teach citizenship, courtesy, consideration of their associates, and leadership make them, as an organization, one of the great assets of our country."

7

★

DIVISION COMMAND

IN APRIL 1958, after nearly four and a half years in the Pentagon—"laborious years," he called them—Westmoreland received a
splendid new assignment as Commanding General of the 101st Airborne Division, the outfit Maxwell Taylor had commanded during
World War II and one of the Army's elite units. "At General Westmoreland's assumption of command," remembered Lieutenant General Charles Bagnal, then a young officer in the division, "we had to
pass in review in reverse order so that the 101st Airborne Division
patches [worn on the left shoulder of the uniform] would show. We
had to learn 'Eyes Left.'"

There had been considerable apprehension within the division,
and particularly its staff, when they learned that Westmoreland was
coming to take command. Major General Thomas L. Sherburne Jr.,
who had led the 101st since its reactivation in May 1956, was highly
esteemed. The contrast with Westmoreland, thought an aide to the division commander, was going to be traumatic. In place of the broad-
gauged and imaginative Sherburne they were going to get "a man who
fulfilled every stereotype."[1] That outlook presaged a commander expected to be doctrinaire, rigid, and pedestrian in his approach.

Westmoreland inherited a top-notch organization, "the finest out-fit I ever served in," according to a young officer in the division during Sherburne's tenure and who then served for a short time as an aide to Westmoreland. "The soldiers were all volunteers, and most were Regular Army. They were committed. It was a fantastic division, really crackerjack."

JUST THREE WEEKS after Westmoreland took command, there occurred a tragic and controversial event: a mass parachute jump during Exercise Eagle Wing in which high winds resulted in seven men being dragged to their death on the drop zone. As many as 137 others were injured.

Brigadier General Weldon F. Honeycutt, then a captain and Westmoreland's senior aide, remembers it this way: "We're all chuted up and ready to go. The lead elements jumped, and the wind gusted up about 8–10 miles per hour more, just when the first lift was about half-way down. They told Westmoreland not to jump, but we went anyway. He got dragged all across the place. I collapsed his chute."[2]

Major General John Singlaub, then a lieutenant colonel, was the Division G-3 (Operations Officer). "Westmoreland was a part of the jump," he recalls. "He was on the second wave. The question was whether we should jump or not. I said we should go in. It would be a disaster if, after the accident, the division commander didn't jump."[3]

Westmoreland addressed this accident in his memoirs, saying that he jumped after the first unit on that same day, then devoting much of a very brief account to his own difficulties: "As I came to earth, an unanticipated wind approaching twenty miles per hour dragged me joltingly across the ground. Even though I managed on occasion to get to my feet, gusts threw me down again and again. The wind dragged me several hundred yards before others who had landed ahead of me were able to collapse my parachute." Then: "Only later did I learn that the wind had dragged seven men to their deaths."[4]

There is no word of concern or compassion for those unfortunates or their families. What follows immediately is this: "Since training even under adverse conditions is essential to a unit's preparations for battle, I wanted to continue the maneuver the next day; but with wind conditions for a jump still doubtful, I decided to make the first jump myself and on my experience base a decision whether to proceed. After a hard landing, the wind again seized my parachute and dragged me across a rough field until some of my men succeeded in collapsing the chute." So, says Westmoreland, he called off the additional jump.[5]

Aides of Westmoreland's at the time agree that he made such a jump but that—as one described it—"he sort of dressed up his role later as he told it (and the calls were coming in)."[6] Those calls, a virtual flood of them, were from senior government officials and members of Congress, questioning Westmoreland's judgment in allowing the jump to take place under the prevailing conditions. That subsequent jump, and Westmoreland's characterization of it, concluded his aide, "was in fact a large part of his assurance later to the Pentagon and politicians that the decision had been 'within limits' as far as good judgment was concerned (together with the need to prepare for tough contingencies in time of war)."[7]

At this time Westmoreland's classmate General Bruce Palmer Jr. was serving in Washington as Assistant Secretary of the General Staff, in the same office where Westmoreland had been assigned before going to the 101st. "When they had that terrible jump accident at Fort Campbell," recalled Palmer, "Secretary of the Army Wilber Brucker wanted to relieve General Westmoreland for poor judgment." Also, said Palmer, "the Under Secretary of the Army, a man named Milton, was all for relieving Westmoreland." Hugh M. Milton II was a retired reserve major general and quite influential with President Eisenhower, so his views carried unusual weight for a sub-cabinet officer. Palmer said he tried to dissuade the two officials and that Colonel Fred Weyand, Brucker's executive officer, also argued against relief. "I never told Westy that I had helped keep him from losing his job," said Palmer.[8]

For several days it was unclear what Westmoreland's fate would be. Martin Hoffmann, an aide, sat in the commanding general's office and made notes of numerous meetings and telephone calls, many with members of Congress. The young officer noted Westmoreland's "technique of pushing. He never gave the slightest indication of any fault on his part. He had a way of creating a truth in his own behalf that, while improbable, was not totally incredible. He had a way of filling that gap in ways large and small." Of course eventually Westmoreland got off the hook for the disastrous jump, an outcome Hoffmann saw as significantly determined by "opportunistic interventions by influential members of Congress."[9]

Westmoreland's final word on the accident was rendered in an oral history interview conducted in 1978. "It turned out to be a pretty good maneuver despite the initial tragedy," he said. "So my indoctrination and introduction to the 101st was very traumatic and very sad. The event became national news."[10]

ESTABLISHING WARM AND CLOSE relationships with influential citizens in nearby communities was a high priority for Westmoreland, and he did very well with that, organizing and hosting various events at which the activities of the division could be showcased. A county commissioner from Tennessee who at Westmoreland's invitation attended a Fort Campbell event, one featuring squirrel stew, was so impressed that he wrote to Secretary of the Army Brucker to say of Westmoreland: "It is phenomenal the knowledge and background that this general has of the personnel in his command." Another local citizen was so taken with Westmoreland's performance on a special project that he too went all the way to the top, writing to the Army Chief of Staff to report that "the recent Girl Scout cookie sale [at] Fort Campbell set an almost unbelievable record," an outcome he attributed "chiefly to the example set by General Westmoreland."

David Halberstam, then a young reporter in Nashville, saw Westmoreland for the first time when he was making a speech at some civic affair. Westmoreland made it a practice on such occasions to have an enlisted man accompany him with the division mascot, an eagle (of course). "With almost anyone else it might have seemed hokey," recalled Halberstam, "but with Westmoreland—his bearing so proud, his posture so ramrod straight—it seemed quite natural. He already was marked as a rising star in the Army."[11]

Concerned to improve the competence of junior leaders in the division, Westmoreland organized what he called Recondo School (a name derived from a combination of "reconnaissance" and "commando"). He put Major Lewis Millett, a Medal of Honor holder, in charge and directed a program that stressed patrolling, map reading, survival, and other fundamentals of fieldcraft.[12] Rifle squad and fire team leaders were among the first to be enrolled.

Westmoreland still liked to talk to the troops, and to keep them informed about what was going on in the division and why they were doing it. Observed one young officer, "Some soldiers might not know their battle group commander, but they all knew their division commander."

Colonel Thomas McKenna had a dramatic recollection of one of these occasions after an act of terrible violence. "General Westmoreland liked to stand on the hood of a jeep and talk to the troops assembled around him," said McKenna. "When I was in the 101st at Fort Campbell and he was division commander, some dumb troopers raped a local woman and there was bad town-post feeling. Westmoreland came to our battle group on his rounds, stood on a jeep hood, and gave a speech that included, 'When the sun goes down the airborne trooper's brain shrinks to the size of a pea and his penis grows to three feet long!'" Added McKenna, "That sort of colorful speech has disappeared now that there are so many women in the Army."[13]

A CERTAIN FASCINATION with and belief in the usefulness of statistical indicators surfaced during Westmoreland's command of the division. He later described how he had worked up a system where "we had some type of crude measure for everything, and every month I would have a council with all my commanders and I put on the chart and I had them all compared." Those sessions must have been an ordeal for some of the officers involved, as Westmoreland acknowledged. "They didn't like that. They didn't like to be exposed when they fell short. But I had data compiled and called it the command data system."[14]

Ever since his short course at the Harvard Business School, Westmoreland had made it a point to seek efficiencies in management and budget wherever he was assigned. At Fort Campbell he christened such a program "Operation Overdrive." Shortly before finishing his command tour in the 101st Westmoreland made the rounds of several Army headquarters and schools, briefing his findings based on Overdrive. These presentations included stops at the Army Management School, at Headquarters of the Continental Army Command, and even at the Bureau of the Budget. Ticking off items on a list of twelve conclusions he had reached, Westmoreland asserted that work measurement was practical and productive in technical service post activities and troop units, that personnel administration was complicated and needed major revision, and that production suffered from "too much emphasis on police and eyewash."[15]

ONE OF WESTMORELAND'S young aides had a sister at Vanderbilt University in Nashville. Kitsy invited her to come and bring some friends on the weekend and to stay at the Westmorelands' quarters. She would then arrange dates for them with young officers on the post. One Sunday morning when they were guests Westmoreland came down to breakfast wearing a coat and tie, the better to make a good impression on the young ladies. Kitsy teased him about that for months thereafter.

WESTMORELAND'S CONCERN FOR the well-being of his soldiers was genuine and almost without limit. During his tenure new branch exchanges and clubs were opened in the various barracks areas, along with a new craft shop, and branch libraries and an additional chapel were not far behind. A free post bus line for troop use was established, and a new theater was under construction. All this was designed to keep the troops not only busy and happy, but close to home. In his annual "State of the Division and Post Message" at the end of 1958 Westmoreland addressed a very serious and troubling problem. "It is a shocking and saddening fact," he wrote, "that in 1958 we lost the equivalent of a rifle company strength through deaths or injury in private automobile accidents."

HAVING ARGUED WITH Maxwell Taylor about the Pentomic division concept, an argument he lost, Westmoreland then found himself obliged at Fort Campbell to implement the plan, the first Army division to do so. A later external critique described the multiple difficulties involved: "Predictably, Pentomic inherited many of the problems inherent in airborne units, most notably a dependence on other organizations for logistics and firepower and insufficient manpower, weaponry, and logistics for sustained combat operations. These problems were compounded by inadequate technology, particularly in communications equipment, fire control systems, artillery, and air transportable multipurpose vehicles, which collectively and perhaps individually made it all but impossible to wage the mobile, firepower-intensive warfare that justified Pentomic's organization."[16]

Discussion of the Pentomic concept and its implementation led to a small lesson for Westmoreland in the workings of the chain of command. A meeting at Fort Bragg involved Lieutenant General Robert Sink, Commanding General of XVIII Airborne Corps, and the commanders of his two subordinate divisions, Major General Hamilton Howze of the 82nd Airborne and Westmoreland at the 101st Airborne. When the briefer had finished and was gathering up his papers, General Sink turned to Westmoreland and Howze and asked, "Aircraft here? Meet you in my headquarters in an hour and fifteen." Howze said, "Yes, sir." But Westmoreland responded, "I'm on my way to Washington. I have to meet Congressman So-and-so." The briefer watched what happened next: "Sink *slowly* picked up his hat, put it on his head with the three stars, and said, 'I'll see you in my office in an hour and fifteen minutes.'"[17]

During this period Westmoreland spoke at an annual meeting of the Association of the United States Army, a prominent forum, where he adhered to the official position: "We find the Pentomic divisional organization well adapted to our mission. It is strategically mobile, flexible, and equipped with weapons and vehicles capable of supporting a variety of missions."[18] Privately, he continued to denigrate it. "The day I left [the division]," he said in an oral history interview, "I signed a letter saying I had worked with the concept [of the Pentomic division] extensively, believed it to be unsound, and recommended that it be scrapped. That was done."[19]

SERVING WITH PARATROOPERS was highly satisfying to Westmoreland. "These are the men with whom I have identified myself for many years," he said in a graduation address at the Army Chaplain School, "and I have the deepest respect for them. They are men who know what they are doing, and know the importance of it and take pride in their ability to accomplish the mission. There is no room in the paratroops for those who equivocate. Paratroopers must be forceful, affirmative, and enthusiastic."

Westmoreland had brought seemingly limitless energy to commanding the division, setting a soldierly example, exhorting others to give it their best, and interesting himself in every aspect of division and post activities. "'When you work for Westmoreland,' said one of his longtime aides, 'you are always tired. But you're always satisfied that you've done your job the best it can be done.'"[20] For his part, Westmoreland later recalled, "I never worked so hard in my life as I did during my days at Fort Campbell." But he was content with the results, stating later his conviction that "we had the number one trained and ready unit in the United States Army."[21]

In a later self-evaluation he emphasized the qualities he had brought to the division command assignment. "I'm a very conscientious man and have always taken responsibilities very seriously," he said. "Every time I was given a job I put myself into it totally. I lived it. When I went to sleep at night, I thought about it. I woke up thinking about it. My life was totally enmeshed in the job I was given. And I worked long hours and had very little time for extracurricular activities." He was correct, of course, but those were not the only qualities later assignments would require.

In the judgment of a former aide-de-camp to Westmoreland who later knew him well at several stops along the way, "Westmoreland was not equipped to be Chief of Staff. Division commander was his best role."[22] Now, in the summer of 1960, Westmoreland was to leave the division for a much different, and differently demanding, post, that of Superintendent of the United States Military Academy at West Point.

8

★

SUPERINTENDENT

IN EARLY MAY 1960 Westmoreland received a very important letter, handwritten on Army Chief of Staff stationery with a "Dear Westy" salutation, from General Lyman Lemnitzer. "The purpose of this note," it began, "is to let you know on an *eyes only* basis that I have recommended, and Secretary Brucker has approved, that you be assigned as the next Supt., U.S.M.A. [Superintendent of the Military Academy at West Point], effective upon completion of Gar Davidson's tour of duty about 1 July 1960. This is in recognition of your outstanding ability, your fine record and the splendid job you have been doing."

General Lemnitzer apparently felt it necessary to add a comment anticipating Westmoreland's disappointment that his next assignment would not include a promotion: "While this position will not carry three-star rank (all Academies are now 2-star) it is one of our most important assignments and we feel that you are the best qualified to handle it."

Westmoreland seemed a surprising and unlikely choice for the position. He had no graduate degree, had not even attended any of the Army's own advanced educational institutions, and had no faculty experience at West Point (although he once received, but turned down, an invitation to come back as an instructor in the Department of Mathematics). He was realistic in writing to his longtime mentor and friend Major General Louis Craig, acknowledging that the new assignment "will indeed be a challenge and I am frank to state that I do not feel particularly qualified for the job at this juncture."[1]

General Williston Palmer, now retired, who understood Westmoreland's intense ambition, wrote: "I cant see that it can hurt you at all to be the Supe for a couple of years; although I would never, if I had been the boss, have sent you there. There are bigger jobs for which you are ready. Still, as I say, it cant hurt you and perhaps it will be advantageous to sit in dignity and detachment for a while reappraising the world scene."

On the way to his new assignment Westmoreland called on President Eisenhower at the White House. Ann Whitman, the President's personal secretary, was away at the time, but later reported being told "by some of the younger (and more impressionable) girls around here that he made a terrific impression and I am only sorry that I did not get an opportunity to meet him!"

When Westmoreland arrived at West Point, he was not known as a scholar. Wisely he decided to concentrate on other aspects of the job, in particular post administration and efficiency. But he also had to deal with the Academic Board, so he began going to classes and labs. He would get the homework, borrow the textbooks, and sit in on a complete session each time he went. Over time he thus gained substantial knowledge of the various departments and their course offerings, as well as of their teaching faculty, giving him an advantage in board sessions where every professor knew all about his own department, but relatively little about what was going on in the others.[2]

Westmoreland's work during his first year received strong endorsement from an important source, retired Brigadier General Chauncey Fenton, President of the West Point Alumni Foundation. "Dear Westy," he wrote, "since 1925 I have heard 19 'State of the Union' messages to the officers at West point by 8 different Superintendents in the fall of each year. This morning I was privileged to listen to the most interesting, the most comprehensive—yet most concise, and the most effectively presented such discourse of them all."

The next summer Westmoreland emphasized a Spartan life when he spoke to members of the new Class of 1965 during their Plebe Hike. "I think that it is quite common that there be rain on plebe hikes," he observed, "and I think there should be rain on plebe hikes." Reflecting on the entirety of their first summer as cadets, he added: "It has been said that the purpose of new cadet barracks is to make Mama's boy go home." And: "Second only to honor, I believe the mark of a West Pointer is that he never quits."[3]

WITHIN WEEKS OF taking command he sent a contingent of West Point staff officers to Fort Campbell to be briefed on Operation Overdrive. Westmoreland also drew on some of those who had been his classmates at the Harvard Business School back in 1954, asking them to visit West Point and advise him on such operations as the Cadet Mess, the printing plant, and possible renovation of an old ordnance compound for use as a First Class Club (for the seniors).

After his first year as Superintendent Westmoreland arranged the publication of a pamphlet, "West Point Points the Way in Post Efficiency," shotgunning it to a wide range of influential people. Citing his "command emphasis on management improvement," it stressed concentration on industrial or commercial-type activities, "which consume the bulk of our manpower and dollar resources," as areas that could produce the greatest benefits. Among the approaches described were many of those Westmoreland had introduced at Fort Campbell, including work measurement, engineered time standards, monitoring of employee performance, and work simplification.

Matters of efficiency and economy were tracked at quarterly business reviews. As Westmoreland approached the end of his second year in the job, he heard from D. H. Rohrer of the General Electric Company after this corporate executive attended one such review. "The progress that has been made at West Point in the last two years in the industrial field is unbelievable," he attested. "Your officers have control of their businesses, and know where they are going and how they are going to get there."

The workers at the various industrial facilities appreciated and responded to Westmoreland's interest and support, in particular at the printing plant. A staff officer later wrote: "Your picture still hangs in a place of honor. Since you were the first Superintendent to ever really visit the Printing Plant and take an interest in it, they refuse to accept the fact that you are no longer here."

Westmoreland was also very proud of an efficiency of a different kind, telling the assembled staff and faculty on one occasion that he had caused the Military Academy Band in the aggregate to lose an impressive number of pounds.

He himself was always very conscious of maintaining his own physical fitness, with racket sports being a favorite activity. Major General Neal Creighton, then a captain on the faculty at West Point, was on the roster of those occasionally scheduled to play tennis with Westmoreland. After several such sessions he received a call from Westmoreland's aide, Captain "Sam" Wetzel, with a surprising suggestion. "I understand that some of your matches with the boss have been pretty one-sided," said Wetzel. "Don't you think it would be better if the general won a few more games?" He then booked Creighton for another game with Westmoreland a few days later, suggesting that he "make it come out 6–3, 6–3 in your favor, and both the General and I will be a lot happier." Creighton did as instructed, later noting that he "found it very difficult to accomplish the score rigging that Sam desired, but I came close."[4]

SUPERINTENDENTS AT WEST POINT have considerable power and authority, especially over cadets. But there are limits, as Westmoreland found out in the course of one ill-considered initiative. At one point "he announced from the poopdeck [the balcony where the Officer in Charge and the Brigade Staff sit during meals] in the Cadet Mess that *Playboy* was licentious and lascivious, and he would not permit cadets to receive it," recalled Richard Chilcoat, First Captain of the Class of 1964. "Two weeks later he had to back down—he could not interfere with the U.S. Mail." Apparently the Tactical Department had been withdrawing copies of the magazine before they were delivered to cadets, a practice that when exposed was hastily discontinued.

In another initiative that at first was seen as controversial, though it eventually worked out well, Westmoreland sought to establish a source of funding for programs and facilities needed at West Point that were not provided for in appropriations. His concept became the Superintendent's Fund, to which friends and alumni could contribute money for special projects. Among the enhancements thus funded were an amphitheater at Trophy Point, historical panels at the Cadet Library entrance, an upgraded ski slope, and an electric carillon for the Cadet Chapel.

DURING HIS THREE YEARS as Superintendent Westmoreland worked hard at cultivating General Douglas MacArthur, then living in Manhattan. Soon after arriving at West Point, Westmoreland wrote to MacArthur asking if he might come to call and enclosing a similar letter from Kitsy making the same request of Mrs. MacArthur. "Since you were one of the Academy's most distinguished Superintendents I would like to ask you if you would permit me to call upon you at your convenience to discuss with you, in a general way, West Point and your perspective of the job that I have inherited," he wrote. "I am fully aware of my inexperience in an academic environment and it would be helpful for me to have a discussion with you." In due course the Westmorelands were invited to lunch at the Waldorf Towers apartment where the MacArthurs were spending their retirement.

In responding to Westmoreland's request, MacArthur had reminisced a bit. "I know just how you feel at assuming the Superintendency," he wrote. "I shall never forget my first meeting with the Academic Board. Every member had been an instructor or professor when I was a cadet. But you need have no fears. You have a host of friends backing you such as Babe Bryan and Earl Blaik. Such men do not pick the wrong man."

Westmoreland had the good fortune to be the host when MacArthur gave his famous "Duty, Honor, Country" speech to the Corps of Cadets, and marveled that he had done so "speaking without notes." Apparently Westmoreland had not yet realized, as Mrs. MacArthur later revealed, that her husband had spent weeks memorizing and rehearsing the speech with her as the audience. Westmoreland sent printed copies of the speech, some autographed by MacArthur, to hundreds of people.

MacArthur was passionate about Army football, as were many other "old grads," and Westmoreland—like every Superintendent —received much "good advice" and intense scrutiny of his management of that sport. In fact, he reported, when he called on President Eisenhower (Class of 1915) at the White House on his way to take over at West Point, Ike urged him to "buck up the football team." During his superintendency the football program was moderately successful, compiling winning records of 6-3-1, 6-4-0, and 6-4-0. But where it really counted, in Army-Navy games, the team's record was 0-3, including a humiliating 20-point thrashing in his final season.

Major General Carl McNair was once, as a more junior officer, escorting a couple of congressmen to West Point. They attended the football game on Saturday afternoon and were seated in the Superintendent's Loge. Of course the Westmorelands were there, with Kitsy "dripping minks and looking very stylish," remembered McNair. Every time Army made a good play Kitsy would leap to her feet, clapping and cheering. Her husband, ever serious and self-contained, never moved from his seat, just reached up and pulled her down by a mink tail.

Kitsy was widely liked and admired by the West Point community, where she involved herself in many aspects of post life. Cadet Mark Sheridan met her while he was hospitalized for repair of torn ligaments. "An incredibly beautiful and sweet woman appeared at my bedside," he recalled. She asked how he was doing and how he felt about West Point. "All my life," he said, "this is where I wanted to go and this is the school I wanted to graduate from." Then she asked, "What do you think of the Supe?" "With all due respect, ma'am, I think he's a prick." Kitsy: "That's my husband." Pause. "Pardon me, ma'am, that's my morphine talking."[5]

Secretary of the Army Stephen Ailes saw clearly Kitsy's importance to her husband. "Westmoreland had no sense of humor," he noted. "But he was blessed with a really wonderful wife." On one visit to West Point, said Ailes, everything went wrong. The Westmorelands' dog went to the bathroom in the middle of the guest room, and his wife had to clean it up. Then, when the visit was over and they'd been taken to the airport, the aircraft had a problem and they couldn't leave as planned. So they trooped back to West Point, just when the Westmorelands thought they were rid of their guests. "We had drinks, beans and hot dogs in the kitchen," recalled Ailes, "and it was the highlight of the visit." Kitsy handled it all with grace and aplomb.[6]

In June 1962 President John Kennedy came to West Point to deliver the graduation address. His stay included a visit to Quarters 100.

Westmoreland wanted his children to dress up for the occasion. Rip was in his baseball uniform, up a tree, and vetoed the change of clothes. The President graciously went to the tree to speak with him.

The speech Kennedy gave was widely reported at the time and has been quoted ever since, particularly the lines anticipating "another kind of war, new in its intensity, ancient in its origin—war by guerrillas, subversives, insurgents, assassins, war by ambush instead of by combat, by infiltration instead of by aggression, seeking victory by eroding and exhausting the enemy instead of engaging him." That graduating class, and many others, experienced exactly what the President foretold in Vietnam and on other battlefields.

A CRITICAL OBJECTIVE pursued by Westmoreland as Superintendent was expansion of the size of the Corps of Cadets, which was then fixed at just under 2,500. By contrast, the Naval Academy and the Air Force Academy were authorized about 4,400. It was not lost on Westmoreland that this disparity had an adverse impact when it came to fielding winning football and other sports teams. At the 1962 Army-Navy football game in Philadelphia, President Kennedy sat with Navy during the first half, then moved to the Army side at halftime. Navy was dominating the game. Sitting beside Westmoreland, the President asked him: "Why can't you compete?" Westmoreland cited the disparity in authorized strengths. According to one account, the President's response was: "You get a paper on my desk tomorrow morning." Said Lieutenant General Jack Norton, "I talked to Westy. He jumped on it. He was proud of it."[7]

Westmoreland immediately began planning for "an expansion of the Corps of Cadets involving virtually a doubling of the size of the Corps" and building a constituency in support of the new initiative. "With the basic new plan in hand," he said, "the first thing I did was go to see General MacArthur. He was enthusiastic about expanding the Corps, since he believed it was in the national interest to do so. I then flew out to Palm Springs in California to see President Eisenhower, then retired, and had lunch with him. He was also very pleased with the concept and endorsed it. I then cleared it with General Omar Bradley before we made it known to the alumni."[8]

The expansion Westmoreland recommended and lobbied so hard for did not actually take place until after he had moved on, but it is clear that he deserved much of the credit—and much of the later blame—for laying the groundwork and generating the political support required to bring it about. Wrote Theodore Crackel in his history of the Military Academy, Westmoreland's commitment to expansion "became the central focus of his superintendency."[9]

Lieutenant General Charles Simmons, who was a young faculty officer at the time, observed that "the people who precipitated that change never gave any thought to the issue that they might turn out a different product. They just assumed they would be able to turn out more of the same product. In that they proved to be very mistaken."[10] A decade or so after expansion, the rapidly increased size of the Corps was determined to have been a contributory factor in a major honor scandal involving widespread cheating on a graded project by members of one class.[11]

THE MILITARY ACADEMY during Westmoreland's superintendency was still an all-male institution, as it would remain until 1976. Soon after that Westmoreland revealed his opposition to coeducation in an oral history interview. "I was aware that there would be political pressures," he acknowledged, "but I thought they would be resisted. I think the academies, the Department of the Army, and the Department of Defense got caught napping on this one and, in my opinion, if they had mounted a countercurrent against it, it was a matter that could have been defeated."[12] He discussed it again during a 1980 appearance on a Larry King radio program. "Are you happy about that?" asked King, referring to the presence of women at West Point. "Actually, very frankly, I am not," Westmoreland told him. "I was against this as a matter of principle, but they're there now, and I must say I admire the performance of those that have been able to endure the curriculum and the rigors of the system."

These views were consistent with Westmoreland's outlook while Superintendent. West Point's ice hockey coach Jack Riley learned about that one time when his daughter Mary Beth said to Westmoreland, "When I grow up I'm going to come to West Point." Westmoreland's reply was blunt: "Over my dead body!"

THE COMMENCEMENT SPEAKER for the Class of 1963, whose graduation concluded Westmoreland's tour as Superintendent, was—not surprisingly—Westmoreland's longtime patron and mentor, General Maxwell Taylor. At the end of the ceremony Westmoreland administered the oath of office to the new graduates. His copy of the oath, now in his papers, was annotated, after the final words ("so help me God"), with "*Hand down!*"[13]

An unfortunate flap marred Westmoreland's last full day as Superintendent. Apparently he had been returning from the tennis courts and two First Classmen who were sunning themselves on the bleachers did not see him, so they did not leap up and salute. Cadet Richard Chilcoat, soon to be appointed First Captain for the coming year, was then "King of the Beasts," the senior cadet on the detail training new plebes. That night at 11:00 he got a call from Westmoreland, who said he wanted to address the detail the next morning. Chilcoat had everybody in ranks at 6:00 A.M., expecting a stirring departure speech. Instead Westmoreland, standing on the stoops of barracks, harangued them for fifteen minutes on their lack of discipline, lack of military courtesy, and so on. Then, remembered Chilcoat, "He stomped back to his quarters and, an hour later, drove away. This had a permanent, and *negative*, effect on the Class of 1964." Kitsy's comment to her husband (as she later told a friend): "You never give up, do you?"

Westmoreland left West Point still a two-star officer, but a short stint as Commanding General of XVIII Airborne Corps at Fort Bragg advanced him to three stars, a rank he would hold only briefly before again being promoted. His tenure at West Point had been notably successful in the views of most, as attested to by a friend in Nashville who informed him that, "from the reports I get, you stand almost as high at the Academy as Kitsy."[14]

Beginning in late autumn at his new post, Westmoreland followed his customary practice by distributing copies of his final annual report as Superintendent to a large number of prominent people, both in and out of the military establishment. These included not only the Chairman of the Joint Chiefs of Staff and the Army Chief and Vice Chief of Staff, but such personages as Dwight Eisenhower, Bernard Baruch, John J. McCloy, DeWitt Wallace, and dozens of retired generals.

9
★

VIETNAM

IN THE SPRING of 1962 a Westmoreland admirer had written to General Maxwell Taylor, advocating longer tenure for Westmoreland at West Point. "I certainly agree with your evaluation of Westmoreland, who is a superior Superintendent," Taylor replied. "Unfortunately, the fact that he is constantly in demand elsewhere makes it a continuing battle to keep him at the Academy." One year later, in the summer of 1963, Westmoreland was given command of the XVIII Airborne Corps, headquartered at Fort Bragg, North Carolina.

There, on 31 July 1963, he was promoted to lieutenant general and became, as a caption in his photo album states, "For one day—the youngest lieutenant general in the Army." A day later Westmoreland's younger West Point classmate Creighton Abrams was also promoted to three stars.

When he took command of XVIII Airborne Corps, Westmoreland was very much on home ground. He was a veteran of extended service in both of the corps' two constituent divisions, the 82nd Airborne and the 101st Airborne. But the assignment at Fort Bragg proved to be little more than a way station. In just over six months he would again be reassigned. At least one significant task was accomplished, though, during his short tenure. Secretary of the Army Stephen Ailes recalled that the Army hierarchy had decided to do something about the way black soldiers were treated off base at various posts. Business and civic leaders were to be told that if the Army was not backed on this initiative, the base in their area might have to be closed down. Then the Army leadership called in three commanders, including Westmoreland, and explained the plan. Ailes knew that Westmoreland was from South Carolina, but remembered his response: "I understand that 100 percent." Then, said Ailes, Westmoreland had a plan in his hands by close of business the next day, and in two weeks such things as the availability of housing for minority soldiers had changed for the better around Fort Bragg.[1]

Soon Westmoreland received a call from General Earle Wheeler, the Army Chief of Staff, asking that he come to Washington the next morning. When Westmoreland walked into his office, Wheeler got right to the point. "I don't know whether you've heard it or not, but you are going to Vietnam." Westmoreland later said: "My assignment, as described by General Wheeler and announced to the press, was that I was to be deputy to General Harkins. However, without definitely saying so, General Wheeler relayed to me the impression, which was quite clear, that I was being sent eventually to take over the command."[2]

Captain Stephen "Dick" Woods, Westmoreland's senior aide, had accompanied him to Washington. "When he got back on the airplane," Woods recalled, General Westmoreland "confided that he was telling only me, Mrs. Westmoreland, and his corps chief of staff that he would soon be going to Vietnam as the deputy COMUSMACV and, when General Harkins departed, his successor."

In 1964 President Lyndon Johnson had been offered four candidates to succeed Paul Harkins as Commander, U.S. Military Assistance Command, Vietnam. Besides Westmoreland, the others were Harold K. Johnson (assigned instead to be Army Chief of Staff), Creighton Abrams (who was made Vice Chief of Staff), and Bruce Palmer Jr. (posted as Deputy Chief of Staff for Military Operations).[3] Probably Defense Secretary Robert McNamara made the actual choice, as LBJ had been President for only a short time and would not have known the candidates, but it was undoubtedly the influence of General Maxwell Taylor that swung the assignment for him.[4]

Brigadier General Amos "Joe" Jordan made a valiant attempt to head off this appointment. He had heard that Westmoreland was being considered for the position. "I was so concerned about this that I went to the Secretary of the Army, Cy[rus] Vance," he said. Jordan knew Vance well and felt that he could approach him on such a matter. He also knew Westmoreland well, having served at West Point as a permanent professor while Westmoreland was Superintendent. "I had extensive contact with him during those years," Jordan told Secretary Vance, "and can tell you it would be a grave mistake to appoint him. He is spit and polish, two up and one back. This is a counterinsurgency war, and he would have no idea of how to deal with it." Vance heard Jordan out, then replied, "Joe, you're too late. We've already made the decision."[5]

Predictably, Westmoreland got plenty of advice about the new assignment. "There is need for a man of your drive in Viet Nam at the present time," wrote retired Brigadier General Willard Holbrook Jr., "but please don't allow yourself to be made a scapegoat for a situation for which there may be no solution—we need you too much for high positions in the Army in the years to come." Tom Lambert of the *New York Herald Tribune* was apparently familiar with the situation on the ground in Vietnam, for he wrote, "Don't let any momentary dismay on arrival there get you down."

By far the most significant communication Westmoreland received at this juncture, however, came from Major General William Yarborough, then Commanding General of the U.S. Army Special Warfare Center at Fort Bragg. Yarborough sent Westmoreland an eight-page letter, "one classmate to another," telling him that he "was both surprised and disappointed at your rapid departure from Fort Bragg. Had I received any inkling that you were going to leave, I would have made every effort to have passed to you all of the materials, studies and impressions concerning the situation in the Republic of Viet-Nam that we have been gathering for some time."[6]

Not having had that opportunity, Yarborough provided a number of important observations in his letter. "I cannot emphasize too greatly that the entire conflict in Southeast Asia is 80 percent in the realm of ideas and only 20 percent in the field of physical conflict," he stressed. "Under no circumstances that I can foresee should US strategy ever be twisted into a 'requirement' for placing US combat divisions into the Vietnamese conflict as long as it retains its present format." Also:

"I can almost guarantee you that US divisions . . . could find no targets of a size or configuration which would warrant division-sized attack in a military sense." Instead: "The key to the beginning of the solution to Viet-Nam's travail now lies in a rising scale of population and resources control." And finally: "Nothing is more futile than a large-scale military sweep through Viet Cong country, since always there must be left behind a tangible symbol of governmental power and authority."[7]

Those views proved uncongenial. Westmoreland had already formed a different outlook on the task ahead, one reflected as early as his service for Maxwell Taylor when Taylor was Army Chief of Staff. Westmoreland recalled "the obsession that President Kennedy and General Taylor had with our ability to fight small wars and to counter Khrushchev's strategy involving 'wars of national liberation'" and that this "had a major impact on the attitude of the Army and its preparation for commitment in Southeast Asia."[8] As he made clear many times in the years to come, Westmoreland had no intention of being captured or driven by such an outlook in his conduct of the war in Vietnam.

En route to Vietnam Westmoreland arranged a visit to West Point, where he addressed the Corps of Cadets. From there he made his way to New York to call on General MacArthur, who favored him with an hour-and-a-half-long monologue and some really terrible advice: Treat the South Vietnamese officers you will be advising "as you did your cadets." And then some even worse: "Do not overlook the possibility that in order to defeat the guerrilla you may have to resort to a scorched earth policy." Recalling the visit in his memoirs, Westmoreland gave no indication that he had any problem with either suggestion.

Naturally he got a number of preparatory briefings in the Pentagon and elsewhere before departing. He then wrote to Congressman Gerald Ford, saying, "the only positive conclusion that I drew from these multiple briefings and consultations is that I will encounter a most complex situation."

On 17 January 1964 Westmoreland also met for twenty-five minutes with the Joint Chiefs of Staff, where General Earle Wheeler, then Army Chief of Staff, told him that the "most important" thing was that the "war can be lost in Washington if Congress loses faith."[9] Commenting on this in his analysis of decision making during the early years of the war, H. R. McMaster concluded that Wheeler was "in essence telling Westmoreland to be really careful of what he says and to portray the war in the most favorable light. . . ."[10]

On the way back to North Carolina the small Army aircraft carrying Westmoreland and Kitsy, along with Westmoreland's aide, Captain Woods, had problems with a stuck nose wheel, which collapsed when they touched down for a skidding landing on a foam-covered runway. No injuries were incurred, just a little warm-up for the combat zone.

Surprisingly, or perhaps not, Westmoreland did not meet with Lyndon Johnson before heading out to Vietnam. Immediately after becoming President the previous November, according to Bill Moyers, LBJ had stated his outlook on the war and its prosecution pretty clearly: "I want 'em to get off their butts and get out in those jungles and whip hell out of some Communists. And then I want 'em to leave me alone, because I've got some bigger things to do right here at home."[11]

It is hard, if not impossible, to square such pronouncements with significantly more grandiose aspirations set forth by the administration in almost the same season. Then–Lieutenant General Harold K. Johnson took notes during a 2 March 1964 meeting of the Joint Chiefs of Staff, recording a comment by Chairman Maxwell Taylor as follows: "Was intent of White House that RVN [the Republic of Vietnam] should be used as laboratory, not only for this war but for any insurgency."[12]

WESTMORELAND'S AIDE-DE-CAMP, Captain Woods, was due for company command, but he accompanied Westmoreland as far as Hawaii, where he turned him over to the new aide, Captain Dave Palmer (Woods's West Point classmate) for the remainder of the trip to Saigon. "When I first arrived in Vietnam," said Westmoreland much later, "I frankly was astonished at the state of affairs."[13] For the first couple of weeks or so he was accompanied solely by his new aide, with no other staff or assistants. They were billeted in rooms on the top floor of the Rex Hotel in downtown Saigon. When they got back to the hotel after their first day of work, Palmer started toward the elevators. "Let's take the stairs," Westmoreland suggested, so they did, double time. Several floors up, remembered Palmer, "I began thinking to myself, 'I'm *armor* — what am I doing here?'"

Soon Kitsy arrived in Saigon — such was the state of the war at the time — along with the two younger children, Rip and Margaret. Stevie remained in boarding school in Washington for the time being. Westmoreland had decreed that another family member, Hannah the dog, would not make the trip, but Kitsy had other ideas. Hannah got a little sidetracked en route and spent an additional day in Bangkok, but eventually she too made it to Saigon and was taken back into the family circle.

WITHIN DAYS OF Westmoreland's arrival a coup led by Nguyen Khanh ousted Duong Van Minh and his cohorts, who had themselves overthrown President Ngo Dinh Diem (and killed him and his brother) only the past November. Dealing with a shifting cast of leaders became a continuing challenge for Westmoreland. It would take years for a reasonably effective South Vietnamese government to be established, years during which the war effort would suffer accordingly. After Khanh, according to General Bruce Palmer's tabulation, "nine more changes in power occurred during the ensuing ten months, August 1964 to June 1965." At that time General Nguyen Van Thieu became Chief of State and Air Vice Marshal Nguyen Cao Ky the Premier. Subsequently, when a constitution had been adopted and elections held, Thieu was chosen President, a position he held until the very last days of the war.

"Aware that I was to assume Paul Harkins' position," said Westmoreland, "I spent as much time as possible during my early months traveling in South Vietnam in order to get to know the country, the people, the military forces, and the nature of the fighting."[14] Then in late March there arose another challenge to Westmoreland's planned succession to the top job. H. R. McMaster describes the episode in *Dereliction of Duty*. The Joint Chiefs of Staff Chairman was then General Maxwell Taylor, and several members of the JCS — Admiral McDonald, General Greene of the Marine Corps, and Air Force General LeMay — had come to believe that Taylor was misrepresenting their views to the President. These officers confided their misgivings to Brigadier General Chester Clifton, an Army officer then serving as President Johnson's military aide, after which Clifton prepared a talking paper and briefed the President. "Clifton told Johnson," noted McMaster, "that the Chiefs felt that Westmoreland was a malleable personality who would fare no better than Harkins."[15]

For his part, Westmoreland was not favorably impressed by either the staff at MACV or their working habits. "The quality, I thought, was no better than average and in some cases below average," he judged, adding that "that applied to the entire headquarters." He found the outlook to be "business as usual," with people uniformly taking siestas after lunch and enjoying a regular social life. "So it was not a headquarters geared to an emergency situation," he concluded. "The American community had fallen into the sleepy habits of the Vietnamese."[16]

In March 1964 Westmoreland revealed in a letter to his father that he was receiving inquiries about whether he had any interest in becoming president of The Citadel. Surprisingly, given his new assignment and impending move to the top position in MACV, he told his father, "I have been thinking this over carefully and do not find it an easy decision to make." Hard decision or not, he turned it down. It is hard to imagine how he could have done otherwise. Ten days later Kitsy wrote to family members, "I still hate to think of passing by the Citadel job. What a good life that would have been for all of us." A few days after that she wrote to Westmoreland's parents, again referring to her husband's decision to turn down the Citadel invitation. Westmoreland's hand had been tipped by a public announcement that he was about to succeed Harkins as COMUSMACV. "But," Kitsy added, "I will be frank to admit it is with a great heartache I say this."

WESTMORELAND AND CAPTAIN PALMER were making periodic visits to different parts of the country, including one in late April 1964, when they stayed overnight in Hue. The next day, traveling in an Army Caribou aircraft, they visited a small camp farther north. As Kitsy reported in a family letter, when they were preparing to depart and taxied to the end of the runway, they came under fire. Both pilots were slightly wounded, she said, and two accompanying Vietnamese were wounded quite badly. "Evidently bullets were everywhere, and we can thank God that they missed West. The pilots must have been great, as they kept right on and got out of there in a hurry even though they were wounded."

Another visit was to Malaysia, a trip suggested by the British counterinsurgency expert Robert Thompson, who accompanied the group. Westmoreland and a few others spent several days being briefed on the techniques the British had used to defeat communist insurgents there.[17]

En route from duty at West Point to the XVIII Airborne Corps, Westmoreland had taken into his entourage Charlie Montgomery, then an E-6 Army enlisted man and eventually a warrant officer. General MacArthur's advice to Westmoreland had included the suggestion that he keep close records in Vietnam, and he did. "I took 160 words per minute shorthand, all verbatim, from him," recalled Montgomery. "I went with him to three presidential conferences: Manila, Midway, and Honolulu. He dictated a memo for the President and told me he had to have it in 30 minutes. I did it in 20 with no errors, then went out and got drunk." Montgomery stayed with Westmoreland for the rest of his active service and beyond, an essential member of the team and processor of the voluminous and valuable chronological material and commentary later known as the Westmoreland "History." Said Montgomery of that extended narrative, periodically dictated in pieces: "Everything in there is true. He did it as we went along—no revisionism. He had character beyond reproach."[18]

That first Honolulu conference convened on 1 June 1964. Although Harkins was still the MACV commander, Westmoreland attended in his stead and played a significant role in the discussions. "I made several presentations, sat in on policy conferences, and gave them some of my ideas on how we should proceed," he said. Charles MacDonald, later to serve as ghostwriter for the Westmoreland memoirs, noted that in Honolulu "General Harkins' deputy and heir-designate as commander of MACV, Lt. Gen. William C. Westmoreland, was considerably less pessimistic than the advisers from Washington."[19]

ON 20 JUNE 1964 General Harkins departed, headed back to the United States and retirement, and Westmoreland moved up to succeed him. Right away he scheduled a talk to officers on the MACV staff. Among the items in his handwritten notes for the occasion: "Expect all hand[s] to put in a minimum of 60 hrs per week—even in Hq.

All will be happier—and more efficient." Elsewhere he recorded an additional consideration: "Keeps them out of trouble."

The conventional view of the war, even now, is that it was micromanaged from Washington. There are many stories of how, at Lyndon Johnson's White House "Tuesday Lunches," he and other top (mostly civilian) officials would even select and approve individual bombing targets in North Vietnam and make other such detailed determinations about aspects of the war. But those decisions had to do with actions taken *outside* South Vietnam. Within South Vietnam, the U.S. commander had very wide latitude in deciding how to fight the war. This was true for Westmoreland, and equally true for his eventual successor.

General Bruce Palmer Jr. described how it was not only the Joint Chiefs of Staff, but the political leadership as well, who gave Westmoreland his head. "Washington never made any basic decisions on the strategic concept," he said. "And that left Westmoreland in Vietnam to invent his own strategic concept, which he did. In effect, what he was doing was a war of attrition." This was not, Palmer concluded, a good thing. "There were many weaknesses in this strategy which in numerous interrelated ways played into the hands of the enemy." For one, "chasing around the countryside was futile."[20]

ON 1 AUGUST 1964, after the usual round of musical chairs in the Army hierarchy, Westmoreland was advanced to the four-star rank of full general. Maxwell Taylor, now Ambassador Taylor, pinned on Westmoreland's new insignia of rank.

Almost immediately Westmoreland had difficulty, although he was apparently not fully aware of it, with Army–Air Force relations. "This problem continues to plague me," General Earle Wheeler said in a 17 September 1964 letter to Westmoreland. He cited allegations in Stennis subcommittee hearings "of a lack of cooperation and coordination between Army and Air Force officers in Vietnam," charges that he said were "neither vitiated by the passage of time nor did they emanate from the Pentagon." General Curtis LeMay and other senior Air Force officers, said Wheeler, were not convinced that air power was being used properly and to maximum effect in Vietnam. In addition, a range of senior Navy and Marine officers, to include the Chief of Naval Operations, the Commandant of the Marine Corps, and the CINCPAC, "believe that the Army has undertaken a deliberate program to undercut the Air Force in Vietnam in the interest of proving the concepts embodied in the air assault division," a new type of Army division then under development.[21]

Wheeler now put it to Westmoreland very bluntly. "I draw from your messages to me the impression that you consider that the corrective measures you have taken in Vietnam have put out the doctrinal fire there and that the blaze is flaring only in the Pentagon. I assure you that such is not the case." And, "to be exceedingly frank with you," Wheeler told Westmoreland, he feared two things—a Congressional investigation that would get into roles and missions issues, and a Congressional investigation of the conduct of the war in Vietnam. Emphasizing the latter possibility, Wheeler said that "things are not going too well in Vietnam. . . . Should things go worse and for any reason we lose—or let's say we don't win—I foresee" such an inquiry.[22]

IN EARLY DECEMBER Westmoreland sent a fateful message to all senior advisors, *directing* optimistic outlooks on their part. "As advisors we must accentuate the positive and bring best thought to bear to work out solutions to problems in a dynamic way," he instructed. "Frustration and stagnation are occupational hazards to which larger staffs are subject and which must be prevented by appropriate command attention."[23] Negativism would not be tolerated, a dictate certain to have an impact on the reliability of reporting.

WESTMORELAND'S FATHER, James Ripley Westmoreland, died in South Carolina on 14 December 1964. Westmoreland went home alone to attend the funeral, held four days later, Kitsy remaining in Saigon with the younger children. Back in Vietnam, Westmoreland wrote to his mother of how much he had admired his father, "a man of courage, honesty, intelligence and integrity!"[24] The press reported that the elder Westmoreland had left an estate exceeding $1.2 million in value and that half the income from the investments would go to his widow, the other half to be divided between his children.

After his father's funeral Westmoreland visited Washington briefly, but surprisingly again had no contact with Lyndon Johnson or Secretary of Defense Robert McNamara, even though absolutely crucial decisions on the possible deployment of U.S. ground forces to Vietnam were then in the offing. At the time, said Westmoreland, "I assumed that that was attributable to the President's desire to maintain a low profile on the war. When I learned later that at the time of my visit major new steps for escalating the war were under consideration, I deemed it odd that neither the President nor the Secretary had sought my views."[25]

ON CHRISTMAS EVE 1964 the Viet Cong carried out a successful terrorist attack on the Brink Hotel, a large structure in Saigon being used by U.S. forces as officers' quarters. Westmoreland had now been in command for half a year. He and Ambassador Taylor proposed some retaliation against the enemy, but Lyndon Johnson would not agree. In a cable to Taylor the President wrote, "I also have real doubts about ordering reprisals in cases in which our own security seems, at first glance, to have been very weak."

Moreover, he continued, "I do not want to be drawn into a large-scale military action against North Vietnam simply because our own people are careless or imprudent. This too may be an unfair way of stating the matter, but I have not yet been told in any convincing way why aircraft cannot be protected from mortar attacks and officer quarters from large bombs." That put the matter squarely in Westmoreland's court. Years later Westmoreland was still incensed, writing in his memoirs about how Washington had cited reasons against reprisal "that may have seemed cogent thousands of miles away but in Saigon were absurd."[26]

At year's end, President Johnson cabled Ambassador Taylor that he had little faith in the efficacy of bombing and that it seemed to him that "what is much more needed and would be more effective is a larger and stronger use of Rangers and Special Forces and Marines or other appropriate military strength on the ground and on the scene." He would soon order just that. Meanwhile in Saigon Westmoreland and his staff conducted a preliminary study of the forces required, reporting what one analyst called "the startling requirement" for 34 maneuver battalions and the necessary support personnel, a total of about 75,000 people. As things were to evolve, of course, that would be only the beginning.

Westmoreland described 1964 as "a hectic year—characterized by constant political turmoil and Viet Cong military successes."[27] During Westmoreland's predeparture visit with General MacArthur, the old soldier had said, "I know you realize that this new assignment carries with it great opportunities, but it also is fraught with hazards."[28] A cycle of escalation, enemy resilience, and arguments with Washington was beginning to suggest some of those hazards.

This picture of Westmoreland, fond of uniforms from his early days, was inscribed "I am ready now" in the family archives. *Margaret Clarkson Collection*

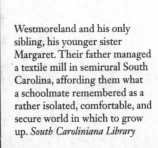

Westmoreland and his only sibling, his younger sister Margaret. Their father managed a textile mill in semirural South Carolina, affording them what a schoolmate remembered as a rather isolated, comfortable, and secure world in which to grow up. *South Caroliniana Library*

LEFT: An Eagle Scout at age fifteen, Westmoreland attended a World Scout Jamboree in England. It was for him a life-changing experience, providing his first look at the wider expanse beyond the confines of his youth. *South Caroliniana Library*

BELOW RIGHT: First Captain of the Corps of Cadets, Vice President of the class, and head of the Cadet Sunday School Teachers. *The 1936 Howitzer*

BELOW: Westmoreland's first assignment was in the 18th Field Artillery at Fort Sill, where he found they were still armed with "antiquated" horse-drawn 75mm guns and that the soldiers "could barely read or write." *South Caroliniana Library*

ABOVE: Equitation was an important part of Army life before World War II. Westmoreland became an enthusiastic horseman, learning to play polo and acquiring a private horse in addition to his issue mount. *South Caroliniana Library*

LEFT: During combat in Tunisia early in World War II, the 34th Field Artillery earned a Presidential Unit Citation under Westmoreland's vigorous command. *South Caroliniana Library*

Serving in the 9th Infantry Division, Westmoreland became executive officer of the division artillery and then, promoted to colonel, chief of staff of the division as it took part in the Normandy invasion and subsequent fighting across Europe. *South Caroliniana Library*

Assigned to Fort Bragg after the war, Westmoreland married Katherine "Kitsy" Van Deusen, an Army brat he first met at Fort Sill when she was nine years old. *South Caroliniana Library*

During the late stages of the Korean War, Westmoreland (here with Lt. Gen. Maxwell Taylor) commanded the 187th Airborne Regimental Combat Team and was promoted to brigadier. *U.S. Army photo by Jack Phillips*

Moving up in Army ranks, Westmoreland was given a plum assignment as Commanding General of the 101st Airborne Division at Fort Campbell. *South Caroliniana Library*

LEFT: During 1960–1963 Westmoreland served as Superintendent of the United States Military Academy at West Point, where he is shown with Kitsy and their children, Stevie, Margaret, and Rip. *South Caroliniana Library*

BELOW: After a short stint in command of XVIII Airborne Corps, Westmoreland was ordered to Vietnam as Deputy Commander of U.S. forces. En route he stopped in South Carolina to visit his father, mother, and sister. *The State*

In June 1964 Westmoreland moved up to be Commander, U.S. Military Assistance Command, Vietnam. Ambassador Maxwell Taylor and Kitsy pinned on his fourth star. *South Caroliniana Library*

Initially families were allowed in Saigon, but as the war expanded they were evacuated. Kitsy and the children (Margaret, Rip, and Stevie) moved first to Hawaii, then to the Philippines, during three and a half years of separation from Westmoreland save what short visits he could arrange. *South Caroliniana Library*

LEFT: Major General (later General) William DePuy was an important associate, the architect of the "search and destroy" tactics employed by Westmoreland and then Commanding General of 1st Infantry Division. He impressed Westmoreland with his belief in "the importance of massive firepower" in the conduct of the war. *U.S. Army Military History Institute*

RIGHT: Beginning in 1965, in response to repeated Westmoreland requests, large contingents of U.S. ground forces were sent to Vietnam. Here, in what he called "a favorite stance," Westmoreland addresses troops from the hood of a jeep. *U.S. Army Center of Military History*

LEFT: Kitsy, like many waiting wives, was concerned that her husband was missing the years in which his children were growing up. "Deep down," she concluded, "I know that's exactly what is happening." Here, she and the children prepare for Christmas, superintended by Hannah, the family dog. *South Caroliniana Library*

10

★

FORCES BUILDUP

COMMENTING ON THE Christmas Eve 1964 Brink Hotel bombing and Lyndon Johnson's subsequent invitation to request U.S. ground forces, Westmoreland maintained that "all of us," referring to Ambassadors Taylor and Alexis Johnson and himself, "were conscious of the momentous nature and probable long-range consequences of introducing American ground troops and were anxious to avoid it if at all possible." Given the unfolding events of the following months, this assertion is difficult to credit, at least in Westmoreland's case. To the contrary, he seemed eager to get such forces, and then get more and more of them, while simultaneously pressing for more and more freedom of action in their employment. While he never said as much, and would no doubt have vigorously denied such a suggestion from others, the conclusion is inescapable that Westmoreland decided early on that he could take over the war effort, get the job done promptly, then hand it back to the South Vietnamese and depart in glory.

Certainly that was the way it looked to Ambassador Bui Diem, South Vietnam's longtime emissary in Washington. Despite Americans often saying that it was the South Vietnamese who would in the final analysis have to win or lose the war, he reflected, "the United States *took over the war* and tried to do everything the American way with almost no consideration as to whether or not such a strategy would meet the complexities or local conditions of the war."[1]

In a later period, under a different President and a different U.S. commander in Vietnam, plans were laid for "Vietnamization" of the war, implicit acknowledgment that earlier the war had been "Americanized." Ambassador Ellsworth Bunker, who served during both periods, agreed. "It seemed to me we started late in training the Vietnamese, and that we had a lot to make up. In the beginning, I think we had misjudged the war and thought that it would be a short-term proposition that we could finish ourselves."[2]

The authors of the *Pentagon Papers* saw it the same way. "Written all over the search and destroy strategy [Westmoreland's approach to conduct of the war, which we will get to very shortly]," they wrote, "was total loss of confidence in the RVNAF [Republic of Vietnam Armed Forces] and a concomitant willingness on the part of the U.S. to take over the war effort."[3] Ambassador Robert Oakley, who served in Vietnam as a more junior diplomatic officer during 1965–1967, confirmed that judgment. "I remember the total disdain of General Westmoreland for the ARVN," he said. Westmoreland's outlook on them was: "You can't fight. We don't want you in the way."[4]

This neglect was costly, as General Richard Stilwell noted in his oral history. "The key problem," he said, "as seen by the American authorities, was to introduce and provide continuing support of the US forces. As a result, there was a reduced effort in expanding, training, and equipping the Vietnamese Army. The development of the ARVN was on the back burner for 3 or 4 years, and we lost some precious time."[5]

Lieutenant General Julian Ewell viewed the thing whole. "The conventional theory was that we, being strong and mobile, were taking care of the main forces, and somebody else would take care of the rest. But that was not what was needed," he observed. "The enemy operated at many levels, and to defeat him you had to beat them up wherever they were found."[6]

Meanwhile Ambassador Taylor was trying vainly to stanch the tide of Americans coming to Vietnam. "We could perhaps improve on our use of them," he cabled the President in early January 1965, commenting on the employment of U.S. personnel already in Vietnam, "but we definitely do not need more."[7] That view was destined to be overwhelmed by urgent warnings from Westmoreland that, absent new deployments, the South Vietnamese would inevitably be defeated, and soon.

Brigadier General James Lawton Collins Jr., the Army's Chief of Military History, wrote (ironically in one of the Vietnam monographs commissioned by Westmoreland when he was Chief of Staff) that "for a while many U.S. leaders felt that American troops could defeat the insurgency alone." Meanwhile, he said, the following problems resulted from that approach: "By 1966 U.S. forces had been given first priority for men, money, and materiel, and the basic mission of strengthening the Vietnamese armed forces became a second priority. This change immediately lowered the quality of advisory personnel and the availability of the more modern equipment for the Vietnamese Army."[8]

Westmoreland's determination to take over the war was also reflected in the division of combat responsibilities he engineered. He assigned his new U.S. ground forces the task of dealing with enemy main force elements while the South Vietnamese were relegated to support of pacification. In reality, of course, theirs was the more challenging role, as rooting out the enemy's covert infrastructure and strengthening the South Vietnamese governmental apparatus were far more difficult than straightforward combat operations.

Westmoreland also favored U.S. forces over the South Vietnamese when it came to close air and artillery support, helicopters and tactical airlift, and naval gunfire. Lieutenant General John Tolson recalled, for example, that "in 1965 there were just not enough airmobility assets to go around. Partly by design and partly by default, most of the airmobile assets ended up in support of U.S. forces, to the detriment of overall Army of the Republic of Vietnam operations." In fact, he added, "during this phase, the Army of the Republic of Vietnam forces actually had less helicopters available to them than in the period before the buildup."[9]

THE DECISION TO send U.S. ground forces to Vietnam evolved over the winter and early spring of 1964–1965. Recalled General William DePuy, "Westmoreland sent a message to Washington that said over the last few months we'd been losing almost a battalion a week, and a district town every month. He gave the government six months to live unless something was done. It was that opinion, and that sense of alarm, that underlay the deployment of U.S. combat troops."[10]

In early February 1965 it was decided that U.S. families should be evacuated from Vietnam on very short notice. Said Westmoreland proudly, "The President told us to get them out in a week to ten days. We did it in six."[11] He then had to cable General John Waters in Hawaii with a plea: "Because of short notice many do not know where they want to go. Can you help?"

Earlier Westmoreland had suggested a subterfuge of a sort. Wives with children could be evacuated first, no new dependents would be authorized to come to Vietnam, and wives without children would depart as their spouses' tours ended. "With this plan," he told a senior Pentagon official, "the disappearance of U.S. dependents from the scene would be so gradual as to pass almost undetected by the Vietnamese."[12] That scheme, unlikely in any event to escape the notice of the fiercely attentive South Vietnamese, was turned down in Washington.

Kitsy and the three children went to Honolulu to live in a hastily acquired house. Not long after, while driving, Kitsy heard a radio report that her husband had been assassinated. Only after a frantic cable was sent by CINCPAC headquarters to Saigon was it was determined that the report was incorrect, the result of a misinterpreted AP bulletin saying merely that Westmoreland was on an enemy assassination list.[13]

Honolulu eventually became untenable for the family. Kitsy was the victim of verbal attacks by strangers, and someone vandalized their home. In August of the following year, the family, less Stevie, moved to Clark Air Base in the Philippines, while Stevie returned to Washington to finish her senior year at the National Cathedral School for Girls. It could then be argued that Westmoreland had the better living conditions. He was reportedly furious when columnist Jack Anderson portrayed him as "leading some type of country-club existence" in Saigon, but in fact that was not so far off the mark. Ensconced in a comfortable villa with an attentive house staff, riding in an air-conditioned limousine with a police escort, eating fine meals prepared by his personal chef, playing tennis at the Cercle Sportif, working in an office outfitted with executive furnishings, Westmoreland was effectively insulated from the war in the jungle, even when he was helicoptered into various base camps and command posts for whirlwind visits.

ON 6 APRIL 1965 the White House issued National Security Action Memorandum 328, documenting the decision to introduce U.S. ground forces into the fighting in South Vietnam. Ominously, noted the *Pentagon Papers*, "missing from NSAM 328 was the elucidation of a unified, coherent strategy." The large buildup of American ground forces in Vietnam began slowly in March (two Marine battalion landing teams) and May 1965 (the Army's 173rd Airborne Brigade), rising to a flood in July, when division-sized forces began to arrive. In June 1965 Westmoreland had cabled Wheeler to say "we need more troops, and we need them quickly."[14] Now they were coming in large numbers and at a rapid pace.

Early in this process Westmoreland joined the deceptions and dissimulations that LBJ and McNamara were resorting to in misleading the public about the extent and even intent of American involvement in the war. Westmoreland sent to Washington a concept paper on proposed offensive use of U.S. forces in combat, a departure from the essentially defensive employments authorized to that point. In this document he described the use of the additional forces as being not only in support of the South Vietnamese, but also for independent "deep patrolling and offensive operations." In that cable, as William Conrad Gibbons noted in his superb collection of documents and commentary on the war, "Westmoreland proposed . . . that the 'public stance' on the use of U.S. forces in combat should be, in part, that U.S. forces were providing combat support to the South Vietnamese rather than conducting their own offensive operations."[15] Washington agreed, and the deception had begun, almost before the first ground forces arrived in the combat theater.

Robert McNamara later recalled what he described as "a constant turmoil over Vietnam between mid-June and mid-July" of 1965. "Every few days," he said, "we received a message from Max [Taylor] or Westy reporting further arguments for more troops. We attended one meeting after another. I spent countless hours with the Joint Chiefs in 'The Tank' [the JCS conference room] debating Westy's shifting plans and requirements."[16]

Westmoreland said that "McNamara frequently made the point to me that the economy of the country [the United States] could afford to support as many troops as I wanted, and for me not to be concerned about it." In mid-June 1965 Westmoreland dictated for his history notes the view that "so far we are not cost accounting the war and I don't believe we will."

On 7 June 1965 Westmoreland dispatched a cable noting "the conflict in Southeast Asia is in the process of moving to a higher level." He asked for a total of 175,000 U.S. troops, with the prospect of "even greater forces if and when required." McNamara, calling this a "bombshell," was cornered. "Of the thousands of cables I received during my seven years in the Defense Department," he recalled, "this one disturbed me the most. We could no longer postpone a choice about which path to take."[17]

While there is much to criticize LBJ for in his conduct of the war, one cannot help having some sympathy for the dilemma posed by the often wildly conflicting advice he was getting from his senior aides and advisors, including those in uniform. General Wheeler, his senior military advisor, was often just flat wrong in what he told the President. When these major U.S. ground force deployments were under consideration in July 1965, for example, LBJ worried that North Vietnam would respond by pouring in more men of its own. He need not be concerned, soothed Wheeler, because the "weight of judgment" was that the enemy "can't match us on a buildup."[18] That turned out to be one of the classic misjudgments of the war, comparable in magnitude and consequences to General MacArthur's assurances to President Truman that Chinese forces would not enter the Korean War.[19]

IN MID-JULY 1965 McNamara spent several days in Saigon, discussing the troop buildup and other requirements. General William Rosson, then serving as MACV Chief of Staff, dated the "transformation of the conflict into a US war" from that fateful visit. "During the months that followed," he observed, "little more was heard of the thesis that this was a war that must be won or lost by the Vietnamese themselves. Accent instead was on what the US required to fight and win. Programs dealing with GVN [Government of Vietnam] military expansion and improvement slid into the background."[20]

Westmoreland's troop requests were the product of intense negotiation between the field commander and the civilian leadership in Washington, represented primarily by McNamara. This resulted in a more or less continuing deception on the part of the administration, one in which Westmoreland took part and later (after Tet 1968) sought to use as cover. A further troop increase would be in the offing. McNamara would travel to Saigon to discuss it. Westmoreland would suggest a number. Jawboning would follow, and a new (sometimes smaller) number would emerge. Westmoreland would then formally ask for, and get, that number, enabling him, the Pentagon, and the White House to all maintain that the field commander was being provided with everything he requested. A related aspect of McNamara's managerial technique was later revealed by his assistant Adam Yarmolinsky, who said that McNamara "regarded his trips as theater, and, in fact, the [trip] report was usually drafted before he left and then revised in light of what assessments they made of what people told them."[21]

Almost right away there developed a major controversy over how to deploy and employ the one truly innovative unit entering the war, the new 1st Air Cavalry Division. "General Westmoreland's first reaction," said a battle history of the war, "was to split the division, sending each of its three brigades to a different part of the country." The division commander, Major General Harry W. O. Kinnard, was dead set against that, pointing out to Westmoreland that "the whole point of airmobility was to keep the closely integrated force together to maximize its impact."[22] Kinnard recalled vividly his initial encounter with Westmoreland after arriving with an advance party. "It was quickly apparent that he had not had the time to track what had been going on in airmobile developments," he said. "He didn't know zilch about it." He remembered that Westmoreland had begun by saying, "Harry, I know exactly how to station your division." He wanted to split it up and parcel out the brigades in scattered locations. Kinnard explained why that wouldn't work, how the division had been organized and equipped to operate as a whole. Finally he talked Westmoreland out of the penny packet scheme, but there were more troubles ahead.[23]

"Westy had a micromanagement approach which was totally contrary to the style I liked," said Kinnard. "He'd say, 'Take your division into Happy Valley and operate for 48 hours.'" And that would be based on something Westmoreland had heard, but not current intelligence. Kinnard said to Lieutenant General "Swede" Larsen, his next higher level commander, "I really don't know how to do what Westy said—'Operate for 48 hours.'"[24] Westmoreland said in his memoirs only that he "tried to be flexible with plans and afford leeway to the local commander."

For his part Westmoreland clearly disdained Kinnard, belittling his ideas in a condescending and almost nasty way in dictated history notes. "General Kinnard of the 1st Cavalry Division arrived in town with his advance party and I had an extensive discussion with him," he began. "General Kinnard's ideas on how his division would be employed are not realistic to the environment in Vietnam. I am sure he will reorient his thinking after he has an opportunity to see at first hand the nature of the conflict." Then, after Kinnard had suggested another possible concept of employment: "I explained to General Kinnard that such a plan was not in the cards in the foreseeable future because of complex political and other considerations." And when Kinnard broached a plan the Army Chief of Staff, General Johnson, had suggested to him: "I pointed out that this is a much-discussed plan but, in my opinion, completely impractical in the foreseeable future."[25]

That about covered matters, leading Westmoreland to dictate this coda: "This discussion served to point out the difficulty that senior officers who have not served in Vietnam experience in attempting to understand the situation and the practical problems faced by our military units in fighting the Viet Cong and countering the well-developed covert infiltration from the north."[26] After the 1st Cavalry Division had been in-country for a time, said Westmoreland in his history notes, he "congratulated General Kinnard on the successes being achieved by elements of his division in Pleiku and made the passing observation that they were apparently getting over their prima donna complex."[27]

The new air cavalry division had experienced a severe shock even before deploying. On 28 July 1965 President Johnson addressed the nation, stating that he was authorizing 50,000 more U.S. troops for Vietnam. The 1st Air Cavalry, at Fort Benning, was among the units to be sent. The division commander and a select few staff members, having been alerted to this impending announcement, listened expectantly for the President to state that he was also calling up reserve forces and extending the terms of those already in service. No such word was uttered. As a consequence this unique unit, with its highly skilled cadre of specialists, experienced the immediate loss, recalled General Kinnard, of "over 500 highly skilled pilots, crew chiefs, mechanics" and so on, those men who had insufficient time remaining in their terms of obligated service to be eligible for assignment overseas.[28] Said Lieutenant General Hal Moore, then a lieutenant colonel battalion commander, "the Commander-in-Chief sent the First Cav Division to war under-strength. I lost over 150 men." That situation was not corrected, said Moore, and as a result he was never up to strength the whole time he was a battalion and brigade commander in Vietnam.[29]

IN HIS APOLOGIA regarding the Vietnam War, Robert McNamara described the series of troop requests submitted by Westmoreland and their effects on people in Washington. In the autumn of 1965, he recalled, the communists had matched our initial increase in forces, also strengthening their air defenses and upping the quantity of men and materiel sent down the Ho Chi Minh Trail. Authorized U.S. forces then numbered 175,000, but in early September Westmoreland asked for 35,000 more, resulting in a 210,000 authorization for the end of that year. Another authorization allowed a total strength of 275,000 by July 1966. Then, in mid-October 1965, Westmoreland sent in a revised estimate of the additional forces needed, boosting the 275,000 figure to 325,000 and raising the possibility of even more later. And all this came, said McNamara, "with no guarantee" that even at that level "the United States would achieve its objectives."[30]

Even after the war Westmoreland apparently remained insensitive to the difficulties his troop requests had caused (even as scaled down by McNamara) for the military establishment in the United States. To the politicians the problems were even more agonizing. "Westy's troop requests troubled us all," said McNamara. "We worried that this was the beginning of an open-ended commitment. The momentum of war and the unpredictability of events were overwhelming the Joint Chiefs' calculations of late July [1965] and Westy's predictions of early September. I sensed things were slipping out of our control."[31] Said Westmoreland: "I wasn't worried about where they got the troops anyway."

In late November 1965 Westmoreland requested 200,000 more troops for 1966, "twice his July 1965 estimate." That would, said McNamara, bring the total of U.S. forces in Vietnam by the end of 1966 to 410,000. "The message came as a shattering blow," McNamara remembered. "It meant a drastic—and arguably open-ended—increase in U.S. forces and carried with it the likelihood of many more U.S. casualties."[32]

McNamara again flew to Saigon, taking Wheeler with him, to confer with Westmoreland and Admiral Sharp, in from Honolulu. The meetings, said McNamara, "confirmed my worst fears." In South Vietnam political instability had increased, pacification was stalled, and South Vietnamese army desertions were up sharply. "The U.S. presence," concluded McNamara, "rested on a bowl of jelly." All this, he remembered, "shook me and altered my attitude perceptibly." Back in Washington, he met with the President, who asked: "Then, no matter what we do in the military field, there is no sure victory?" "That's right," McNamara told him. "We have been too optimistic."[33]

WESTMORELAND'S REPEATED REQUESTS for troops were essentially approved—until the spring of 1967, when he asked for 200,000 more troops and got only a pittance. The mood in Washington had changed, as illustrated by a 4 May 1967 memorandum from McGeorge Bundy to the President: "I think there is no one on earth who could win an argument that an active deployment of some 500,000 men, firmly supported by tactical bombing in both South and North Vietnam, represented an undercommitment at this time. I would not want to be the politician, or the general, who whined about such a limitation."[34]

Yet the massive escalation brought no real change in the war's dynamics. Someone recalled the story of the Texan selling watermelons alongside the road who bought a hundred melons for a hundred dollars, sold them for a dollar apiece, and wondered why he hadn't made any money. His conclusion: "We've got to get a bigger truck." Westmoreland reached that very conclusion, figuratively speaking, over and over again.

WITH LARGE NUMBERS of troops flooding into the combat zone, a basic question involved the policy on tour length. Westmoreland decided it should be one year. He defended this policy on the grounds that it was inevitable that the war would drag on, "which was the basic reason for the one-year tour." When he came back from the war zone and appeared on the television program *Face the Nation*, he argued that he didn't believe "we had any alternative because of considerations of morale, and the necessity of sharing the burden of the war in consideration of the fact that those of us in policy positions during the early days saw this as a long war." Thus, he concluded, "I don't believe it was a mistake, I think it was necessary."[35]

Later Westmoreland reminded his command historian, "In 1965, when we committed US troops, I insisted on a 12-month tour." At a commanders' conference in April 1966 Westmoreland told his subordinates that "I continue to be a proponent of the one-year tour, but the price we pay is in the teamwork, proficiency, and competence of tactical units."

The costs in human terms were great, as established by studies conducted by Thomas Thayer. He found that both the one-year tour and the six-month command tour "apparently had the effect of raising the toll of U.S. combat deaths. Twice as many troops died during the first six months of their tour as in the second half. After the first month, the number of deaths decline as the tour progresses, without exception. Thus, the longer one stayed alive after arriving in Vietnam, the better one's chances for survival, presumably as the result of a learning curve, which then had to be repeated for each new arrival." Russell Glenn reported confirmatory findings, including longer-term effects. "The 12-month rotation policy," he stated, "touted for its benefits provided to the individual soldier, seems in fact to have been a significant element in causing greater numbers of men to lose their lives and may have increased neuropsychiatric casualties."[36]

Westmoreland had apparently received negative views on these tour policies from some of his senior subordinates, as evidenced by a letter he sent to Lieutenant General Jonathan Seaman, then commanding II Field Force, Vietnam, in September 1966. "We have discussed many times the manifold problems that accrue from our one-year tour policy," said Westmoreland. "I have taken the position that the plus factors involving morale more than offset the problems generated by the turnover. I am confident that this is sound reasoning." A decade later, in his memoirs, Westmoreland wrote—in a rare admission of fallibility—"it may be that I erred in Vietnam in insisting on a one-year tour of duty for other than general officers."[37] And Lieutenant General Julian Ewell noted, upon completion of his tour as a division commander in Vietnam, the negative effects of the policy. "One hears that we have in Vietnam the most professional army ever fielded by the U.S.," he wrote (and indeed, that claim was often made by Westmoreland). "The fact is that due to turbulence all units fluctuate between order and chaos and tend to be about average."[38]

Belatedly, Westmoreland seems to have become aware of the multiple implications of the policy. "The turnover of personnel that has evolved from the one-year tour has been our greatest liability," he said in a later *Washington Post* interview. "It has brought about a situation of personnel instability. Our company commanders, first sergeants and squad leaders are rotating their assignments to the extent that they were never able to get a grip on their organizations."

When he was Chief of Staff, Westmoreland—perhaps looking for some vindication of his stubborn insistence on the policy—had the personnel people undertake a study of the matter. That produced only more bad news. "Most personnel problems of the past six years are traceable to the 12-month Vietnam tour," said the staffers who had had to cope with the fallout of that approach. "The greatest effect of the 12-month Vietnam tour was personnel turbulence. It was felt throughout the Army over a continued period of time and had a decided adverse impact upon units, missions and individuals."[39]

THE ARMY, not permitted to call up its reserve forces, scrambled to meet the multiple troop requests.[40] With the 1965 decisions on sending more U.S. ground forces to Vietnam in the offing, on 24 July 1965 the service secretaries and their senior military associates met with McNamara to be told of the impending (28 July) announcement by the President of additional forces, including the new airmobile division, he had approved for dispatch to Vietnam. Then, in a reversal of what they had been anticipating, McNamara said this would be done without calling the reserves. "This came as a *total* and *complete* surprise," recalled Army Chief of Staff General Harold K. Johnson, "and I might say a *shock. Every single contingency plan* that the Army had that called for any kind of an expansion of force had the assumption in it that the reserves would be called," and those plans had been approved by OSD (the Office of the Secretary of Defense). Johnson spoke up: "Mr. McNamara, I haven't any basis for justifying what I'm going to say, but I can assure you of one thing, and that is that without a call-up of the reserves that the quality of the Army is going to erode and we're going to suffer very badly. I don't know at what point this will occur, but it will be relatively soon. I don't know how widespread it will be, but it will be relatively widespread." Johnson recalled bitterly that McNamara just looked at him, made no response whatsoever, and continued with his remarks.[41]

Not only then but afterward LBJ steadfastly refused to call up reserve forces. Eventually Westmoreland became aware of his aversion to doing so and accepted it. Visiting the LBJ Ranch in August 1966, Westmoreland told the President, according to notes made at the time, "We're going to win this war for you without mobilization."[42] When the next day the President presented Westmoreland at a news conference, one reporter asked whether he and the President had discussed the status of the reserves and whether they might have to be called up in the near future. "The source of the units and the manpower is not a matter that I have to be—fortunately—concerned about," Westmoreland responded. "I have no cognizance of the matter of the reserve forces."[43]

Later he said flatly that "those of us in Vietnam did not appreciate the intensity of the debate going on in Washington relative to the callup of reserves."[44] Even if this assertion is accepted, it is difficult to understand. There was a constant flow of high-level visitors to the war zone, so much so that Westmoreland often complained about all the coming and going and how time-consuming it was for him to deal with. Senior defense officials, including McNamara, Wheeler, Sharp, Harold K. Johnson, and Krulak, were back and forth almost nonstop. Added to that were numerous members of Congress and their staffers, plus high-powered media representatives supplementing the many accredited journalists on more or less permanent assignment to Saigon. In the aggregate, these people would seem to have constituted a rich source of information and insight into all manner of military and domestic issues impinging on conduct of the war. But Westmoreland said not.

Thus Westmoreland apparently never realized, even a decade after the war ended, the situation with respect to reserve forces. In 1986, for example, he maintained that "it was General Johnson's decision to meet my relatively modest requirements by cadreing the Army rather than by insisting on a reserve callup," a perspective so widely off the mark that it raises fundamental questions of Westmoreland's awareness of the context in which the war was being fought.

WESTMORELAND ULTIMATELY PROVED unable to accomplish the self-assumed mission of winning the war with American troops, but in the course of trying he managed to get massive American ground forces sent to Southeast Asia, even so complaining later that getting those forces had been like "pulling teeth."[45] "It was impossible to execute the strategy that had been adopted by my government without additional forces," Westmoreland argued. "I would ask for a request that would be cut back. I'd ask for the same request again, and to include what I thought was needed, and that would be cut back. And the thing just dragged on and on and on. In other words I got piecemeal reinforcements rather than receiving what we could have received; necessary forces that could have done a lot of things if the political constraints were going to be changed based on pressures that were being applied by my cables, and presumably by the Joint Chiefs of Staff."[46] Instructing Charles MacDonald on what to say about troops in the memoirs, Westmoreland said: "Note we were always cut short of what we asked for."

Westmoreland's multiple troop requests eventually ratcheted the authorization up to 549,500, a figure that was approached but never actually reached, the deployments peaking at 543,400 at the end of April 1969. That was, of course, after Westmoreland had left Vietnam, but resulted from the arrival of men who were in the pipeline as a result of his earlier requests. There were no troop requests from Vietnam in the post-Westmoreland years.

11

★

SEARCH AND DESTROY

Now our strategy," explained Westmoreland, "was not to *defeat* the North Vietnamese army. It was to put pressure on the enemy which would transmit a message to the leadership in Hanoi — that they could not win, and it would be to their advantage either to tacitly accept a divided Vietnam, or to engage in negotiations."[1] Consistently inconsistent, on other occasions Westmoreland asserted that our aim *was* to defeat the enemy, as he maintained in an interview: "Our purpose was to defeat the enemy and pacify the country, and the country couldn't be pacified until the enemy was defeated."[2]

His approach to achieving that was to wage a war of attrition, using search and destroy tactics, in which the measure of merit was body count. The premise was that, if he could inflict sufficient casualties on the enemy, they would cease their aggression against South Vietnam. In his single-minded pursuit of this objective, Westmoreland essentially ignored two other crucial aspects of the war, improvement of South Vietnam's armed forces and pacification.

Westmoreland describes in his memoirs how and why he came to adopt an attrition strategy. There is no doubt that he himself decided on it even though, as he stresses, it was not for him the strategy of choice, merely the best remaining option — or, as he viewed it, the only other option—after his preferred approach involving operations in Laos, Cambodia, and North Vietnam had been ruled out by Lyndon Johnson's insistence on "no wider war."

Westmoreland confirmed the fact that, other than determining the pace of "gradual escalation" and the use of certain weapons, "the President never tried to tell me how to run the war. The tactics and battlefield strategy of running the war were mine. He did not interfere with this. He deferred to my judgment, and he let me run the war or pursue tactics and battlefield strategy as I saw fit." Also: "I, in effect, had a carte blanche in the devising and pursuing tactics and battlefield strategy of the war."[3]

Implementing the attrition strategy, Westmoreland prescribed search and destroy tactics. What this meant in practice was a series of large unit sweeps, often multibattalion and sometimes even multidivision, frequently conducted in the deep jungle regions next to South Vietnam's western borders with Laos and Cambodia, designed to seek out enemy forces and engage them in decisive battle. That proved possible only with the enemy's cooperation; otherwise, as Andrew Krepinevich tellingly observed, "search and destroy was like Whack-a-Mole."[4] General Alexander Haig contributed another dramatic characterization, calling Westmoreland's tactics "a demented and bloody form of hide-and-seek."[5]

The early conduct of search and destroy operations fell to the 173rd Airborne Brigade, which in May 1965 deployed from Okinawa to Vietnam under the command of Brigadier General Ellis Williamson. The experiences and reactions of that unit to the Westmoreland way of war were instructive, or should have been. From the beginning "Butch" Williamson was not an admirer of that particular tactical approach. "After our thrashing around in the jungle on our first large-scale operation," he said, "I was convinced that large operations of that type were not the way to go except when we knew in advance what the objective was. Having thousands of men fight in the high grass, brier bushes, large overlapping layers of vegetation, etc. is not productive to the effort expended."[6]

In early August 1965 Williamson issued one of his periodic *Commander's Combat Notes*, saying, "I hope that we have conducted our last 'search and destroy' operation. I am thoroughly convinced that running into the jungle with a lot of people without a fixed target is a lot of effort, a lot of physical energy expended. A major portion of our effort evaporates into the air."[7]

As early as March 1965, when the U.S. troop buildup was just beginning, senior Washington officials had begun to perceive the problem with Westmoreland's approach. McGeorge Bundy wrote to the President on 6 March, describing a discussion he, Robert McNamara, and Dean Rusk had just had. "Last night Bob McNamara said for the first time what many others have thought for a long time — that the Pentagon and the military have been going at this thing the wrong way round from the very beginning: they have been concentrating on military results against guerrillas in the field, when they should have been concentrating on intense police control from the individual villager on up."[8] What is baffling is that these senior civilian officials nevertheless allowed Westmoreland to doggedly pursue his flawed approach for year after bloody year. "Westy believes the war begins and ends with killing VC," reported a senior defense official in frustration after a trip to Vietnam.[9] But nothing was done to change course.

McNamara described matters as bleak even at this early juncture. "We had no sooner begun to carry out the plan to increase dramatically U.S. forces in Vietnam than it became clear there was reason to question the strategy on which the plan was based," he observed. "Slowly, the sobering, frustrating, tormenting limitations of military operations in Vietnam became painfully apparent."[10]

Westmoreland's nominal military superior, Admiral U. S. Grant Sharp, the Commander-in-Chief, Pacific, had a clearer view of the nature of the war and its imperatives than did his field commander. Sharp cabled General Wheeler in September 1965 to observe that "this is a counterinsurgency war . . . the primary object is to restore security to the population. . . . If we are to succeed we must do a number of things at the same time and do them differently than we did in past conflicts."[11] Nonetheless Sharp did nothing to redirect Westmoreland.[12]

Even Ambassador Lodge saw the problems inherent in Westmoreland's approach. "Let us visualize meeting the VC on our own terms," he suggested at a July 1965 meeting. "We don't have to spend all our time in the jungles."[13]

General Fred Weyand was of the same mind, explaining: "In South Vietnam, the sole basis for effective, meaningful operations is specific intelligence information; without it, the commander is left groping almost aimlessly."[14] Such groping characterized much of what constituted search and destroy operations.

But perhaps the most devastating judgment on Westmoreland's approach was rendered by Edward Murphy, author of a history of the Marine Corps in the Vietnam War. "One major flaw existed with Westmoreland's plan," he wrote. "Just months earlier he had pleaded for the deployment of U.S. troops, arguing that the ARVN were incapable of protecting only a few air bases. How he now expected them to handle the much more difficult task of providing security for more than twelve thousand hamlets remained unexplained."[15]

Meanwhile Westmoreland was rendering highly optimistic reports, even though just months earlier he had forecast dire consequences unless major U.S. ground forces were dispatched to the war zone. "In summary," he said in his COMUSMACV Monthly Evaluation for March 1965, "current trends are highly encouraging and the GVN may have actually turned the tide at long last."[16]

By late June 1965, with combat operations heating up and many more troops on the way, Westmoreland rendered a contrastingly gloomy report. "The struggle has become a war of attrition," he wrote. "Short of decision to introduce nuclear weapons against sources and channels of enemy power, I see no likelihood of achieving a quick, favorable end to the war." Having planted that arresting idea, he went on to emphasize his need for still more troops.[17] Then, despite having himself so recently raised the matter, Westmoreland dictated in his history notes for 23 November 1965: "Congressman [Hays] came in to see me and I was surprised to hear him ask about the possibility of using atomic weapons in-country. The thought had been [so] remote that it caught me by surprise."

THE FIRST MAJOR battles between U.S. ground forces and the North Vietnamese Army took place in October 1965 in a region known as the Ia Drang Valley. Elements of the 1st Cavalry Division slugged it out with three NVA regiments. This was the battle later made famous in the book by Lieutenant General "Hal" Moore and Joe Galloway, *We Were Soldiers Once . . . and Young*. Westmoreland's reaction to what took place there, or in any event to his interpretation of the battle, confirmed his belief that he could deter further enemy aggression by imposing unacceptable casualties on the NVA and the Viet Cong.

Moore, who as a lieutenant colonel commanded the battalion most centrally involved in the Ia Drang battle, recalled with some bitterness an incident during the fighting. Around midnight one night he was passed an astonishing message: "General William Westmoreland's headquarters wanted me to 'leave X-Ray [the landing zone where the desperate fighting was centered] early the next morning for Saigon to brief him and his staff on the battle.' I could not believe I was being ordered out before the battle was over!" Moore, a man who knew where his loyalties belonged, sent his regrets and stayed with his men.[18]

Then, wrote Moore and Galloway, "In Saigon, the American commander in Vietnam, General William C. Westmoreland, and his principal deputy, General William DePuy, looked at the statistics of the thirty-four-day Ia Drang campaign—3,561 North Vietnamese estimated killed versus 305 American dead—and saw a kill ratio of twelve North Vietnamese to one American. What that said to two officers who had learned their trade in the meat-grinder campaigns in World War II was that they could bleed the enemy to death over the long haul with a strategy of attrition."[19]

And, said Moore, "Westmoreland would learn, too late, that he was wrong; that the American people didn't see a kill ratio of 10-1 or even 20-1 as any kind of bargain."[20] A very influential visitor, Senator "Fritz" Hollings from Westmoreland's home state of South Carolina, had warned him about relying on such ratios. Westmoreland told him, "We're killing these people at a ratio of 10 to 1." To that Hollings responded, "Westy, the American people don't care about the ten. They care about the one."[21]

Besides the American people's antipathy to friendly losses of the magnitude experienced in these battles, there was another factor that might have warned Westmoreland and his senior staff about the longer-term viability of the Ia Drang model. The 1st Cavalry Division basically spent the month of December in recovery mode, rebuilding its logistical stocks, putting its overworked helicopter fleet back in shape, and integrating large numbers of replacement personnel. In just two months, October and November 1965, division losses amounted to more than a quarter of total authorized strength. Besides 334 killed and another 736 wounded, there were 364 nonbattle injuries and "2,828 cases of malaria, scrub typhus, and other serious diseases." More than 5,000 replacements had to be brought in.[22] Meanwhile, as of 1 December 1965, 100 of the division's helicopters were inoperable due to parts shortages.[23]

Eric Bergerud wrote, "The engagement provided evidence of what Westmoreland wanted to believe—that the enemy forces would accept battle in dense terrain where helicopter-borne U.S. infantry, backed by artillery and air power, could find them, force them to battle, and defeat them."[24] In those respects, it turned out, the Ia Drang battle had been unusual. Westmoreland's confidence was built on the exception, not the rule. Much more typical were large U.S. formations struggling to find the enemy, then to bring him to battle on favorable terms, with the enemy basically dictating when, where, and for how long such combat would continue, disengaging when it suited his purposes and withdrawing to sanctuaries to rest, refit, and prepare to fight again. David Halberstam observed tellingly that "we underestimated the willingness of these peasants to pay the price. We won every set piece battle. Westy still believes that he never lost a battle. We had absolute military superiority, and they had absolute political superiority, which meant that we would kill 200 and they would replenish them the next day. We were fighting the birth rate of a nation."[25]

Only a few days after the Ia Drang, the 2nd Battalion, 7th Cavalry, was back at the division base camp at An Khe on a cold and rainy Thanksgiving. Lieutenant Colonel Robert McDade, the battalion commander, met a visiting Westmoreland near the mess hall and told him that everyone was just about ready to eat their Thanksgiving dinners. But Westmoreland said, "Get them all together and let me talk to them." The troops had been issued a hot meal, real coffee instead of the powdered stuff that came with C-Rations, turkey, and the trimmings. They were walking back to their squad tents to enjoy this special repast when the order was given to assemble. "There stood General Westmoreland himself," said Sergeant John Setelin. "He made a speech there in the rain and while he talked we watched the rain turn that hot dinner into cold Mulligan stew. Who knew what the hell the man said? Who cared?"[26]

· · ·

BEFORE THE END of that year Westmoreland took time to reassure his West Point classmates that all was going well in Vietnam. "Westy sent greetings to the whole class," reported the Scribe who wrote the 1936 notes for *Assembly*, the West Point alumni magazine. "Until his message," Phil Gage added, "I believe I was not alone in slipping into negative thoughts about the impossibility and meaninglessness of the effort out there. I am thrilled to say that in one short paragraph Westy painted a completely different situation. I am so impressed with what this famous classmate said that I quote him verbatim: 'The American troops over here are doing a tremendous job and have played a major role in keeping this country from becoming completely unglued. I feel there are prospects for success in the future, but the challenge is beyond dimension. Be assured we have our chins up.'" The Scribe then added his own conclusion: "How fortunate we are to have such a great leader in command of our destiny—and he *is* from 1936."[27]

HAVING DECIDED EARLY ON that search and destroy was a dumb way to prosecute the war, General Williamson was equally unimpressed by Westmoreland's direction of it. "Westmoreland was simply present," he said. "A succession of staff and sub-staffs conducted the war. But Westmoreland was very, very responsive to the civilian guidance he was getting from the States. Body count was an element of it. What the newsprint was saying was another. He had back channel communications with a lot of people, some military and some civilian, who were not part of the government."[28]

Williamson was not a Westmoreland admirer on yet other grounds. "General Westmoreland," he said, "had a trait that was repulsive. He *always, invariably*, had someone between himself and anything unpleasant. If it was the slightest bit tenuous, it was always somebody else's fault or responsibility. He was just unbelievably agile at that."[29]

Brigadier General Douglas Kinnard observed that tendency while serving as Chief of the Operations Analysis Branch in MACV J-3. "Westy has more studies, boards, and reorganizations going on than I ever heard of," Kinnard wrote in a letter to his family. "His solution to all problems is study it or form a board of reorganization, none of which, of course, disposes of the problem."[30] Major General Winant Sidle, Westmoreland's long-serving information officer, commented on his related inability to sort out good ideas from bad ones, calling that his "worst trait" and concluding that in consequence "he generated an enormous amount of waste effort."[31] One rather bizarre example, cited by Westmoreland himself in his history notes, was dated 12 September 1965, only weeks after commencement of major U.S. ground force deployments to Vietnam. "Although at this juncture the prospect looks remote," noted Westmoreland, "I asked the J-5 to begin studying how we might proceed to phase down our military effort in Vietnam." One can only imagine how that tasking was received in J-5.

But Westmoreland was very proud of this welter of activity, on one occasion citing in dictation for his history file a rubber plantation study which he "had directed several weeks earlier." That study, he said, was "another example of the initiative that MACV has exercised on other than military matters. Sadly and regrettably, the Embassy and Mission are not organized to do much advanced thinking or planning. It therefore seems that much of this falls to our lot. There have been many other studies outside of my area of interest that should have been made. I am in an awkward position to exercise this initiative; however, if I think these problems bear in any way on the military, I do not hesitate to do so."[32]

IN LATE AUTUMN OF 1965 Secretary of Defense McNamara was briefed in Saigon, then on 28 November 1965 reported his reaction: "I wasn't at all reassured about what I heard yesterday. I have been concerned every time I have been here in the past two years. I don't think we have done a thing we can point to that has been effective in five years. I ask you to show me one area in this country . . . that we have pacified." As we learned much later, McNamara had concluded, possibly—according to his testimony—as early as the latter part of 1965, "that the war could not be won militarily."[33]

Subsequently the futility of massive operations was repeatedly illustrated, as for example by a November 1966 foray by some 25,000 troops from at least five brigades—"in what a U.S. spokesman called the largest American action of the war," reported UPI—looking for the enemy in War Zone C, a notorious communist refuge in the northern half of Tay Ninh Province. "U.S. troops hacked deeper into the jungles today and killed 20 more of the enemy," said the report, "but resistance was scattered and light."[34] The enemy had declined to fight

fair, and 25,000 American GIs had spent a day tracking down 20 of
them.

AN ARRESTING CONVICTION was expressed by General William
DePuy, who after two years as Westmoreland's J-3 then commanded
the 1st Infantry Division for another year. "General DePuy came in to
pay his farewell call," Westmoreland told his history file, "and I enjoyed talking to him about his observations, the primary one of which
concerned the importance of massive firepower against VC units in
populated areas. He feels that military power is the overwhelming
force in influencing the people of the hamlet to abandon the Viet
Cong."

Westmoreland unfortunately shared that view, telling Lieutenant
General William Rosson during the battles of Tet 1968 that "Prov[ince]
Chiefs should tell villagers that if they are going to allow VC to set up
mortars they can expect to get hit with counterfire."

Meanwhile Westmoreland maintained that, "once these main force
units are destroyed," it would be possible for pacification to proceed.[35]
He was never able to destroy those units, and in fact was no closer to
doing so at the end of his four years in command than when he began.
Whatever forces he "destroyed" were simply replaced by the enemy,
leaving the task to be performed over and over again and thus never
actually accomplished.

In consequence, later admitted General DePuy, "we ended up with
no operational plan that had the slightest chance of ending the war favorably."[36] Westmoreland could never bring himself to acknowledge
that reality.

BRUCE PALMER AND WESTMORELAND were West Point classmates
and had known one another for a very long time, but throughout his
assignment to Vietnam, going back to March or so of 1967, Palmer
had been appalled by Westmoreland's way of waging war. "I went over
to Vietnam as Westy's deputy," he said, referring to his eventual assignment as Deputy Commanding General, U.S. Army, Vietnam, "and
it was just a mess. We were losing and trying to put it together, and it
just wasn't working. There wasn't anything that was working."[37] In
the spring, when he was commanding II Field Force, Vietnam, said
Palmer, "it became forcibly obvious to me that we could not achieve

our objectives fighting the way we were. I confided my doubts to General Westmoreland, who really didn't want to talk about it."

When Creighton Abrams came out in May 1967 to become Westmoreland's deputy, Palmer took him aside and poured his heart out about the deficiencies of the Westmoreland approach, about how he had "basic disagreements with Westy on how it was organized and how we were doing it." But Abrams responded, "You know, I'm here to help Westy and, although I privately agree with many things you are saying, I've got to be loyal to him. I'm going to help him."[38] That response may have been conditioned somewhat by Abrams's understanding—not then known to Palmer—that within a matter of weeks he would be taking over from Westmoreland, although ultimately that plan was not carried out, leaving Abrams to serve as deputy for the next thirteen months.

If Westmoreland had problems with fellow Army officers over the viability of his big war approach, those problems were compounded when it came to dealing with Marines. The senior Marine leadership saw the war much differently than did Westmoreland, taking a view entirely compatible with the findings of the PROVN Study (discussed below). In addition—to Westmoreland's disgust—such senior Marines outside Vietnam as Commandant of the Marine Corps General Wallace Greene and Lieutenant General Victor Krulak, Commanding General of Fleet Marine Force, Pacific, maintained close contact with Lieutenant General Lewis Walt, commanding III Marine Amphibious Force in Vietnam, giving him plenty of opinion and advice, much of it contrary to what Westmoreland was advocating. Wrote Allan Millett, a reserve Marine colonel and distinguished historian, "Marine generals like Victor H. Krulak made life miserable for General William C. Westmoreland because of their obsession with pacification and working with the Vietnamese military and paramilitary forces."[39]

Just as the major ground force involvement of U.S. troops was in full swing in October 1965, for example, Krulak followed up a recent visit to Vietnam with a cable to Walt. "I am glad you were able to impress Westy with the magnitude of your activity," he said. "At the same time, I am sure that he has not altered his view that 'find, fix and destroy the big Main Force units' is really the answer, and that patrols, ambushes and civic action are all second class endeavors, more suitable for the ARVN and the paramilitary. I disagree with this, and I know you do, too."[40]

Westmoreland was actually quite unimpressed with what the Marines were doing. "I believed the Marines should have been trying to find the enemy's main forces and bring them to battle," he noted, "thereby putting them on the run and reducing the threat they posed to the population."[41] Later, as input for his memoirs, Westmoreland said he "just did not find the same initiative in I Corps as elsewhere. Marines seem complacent about enemy possibilities. Walt [was] dedicated and sincere but found it hard to grasp the big picture and project into the future." Taking just the opposite tack, Krulak argued that "our effort belonged where the people were, *not* where they weren't. I shared these thoughts with Westmoreland frequently, but made no progress in persuading him."[42]

Lieutenant General Stanley Larsen, observing this situation, rendered a telling judgment. "I don't know what General Westmoreland would say about this," he stated in an oral history interview, "but for all practical purposes the III Marine Force in the I Corps area was not commanded by General Westmoreland, it was commanded by the senior Marine on the CINCPAC staff."

GENERAL JOHNSON WATCHED the proliferation of large-unit sweeps primarily in the deep jungle and was appalled. "I felt in 1962," he stated, "and I still felt in 1968 — with virtually no way to influence it — that what was required was a lot of scouting and patrolling type of activity by quite small units with the capacity to reinforce quickly. We didn't get into very much of that."[43]

Johnson was also dismayed by the incredible expenditures on unobserved artillery fire, known as harassment and interdiction, with nothing much good to show for it and potentially a lot of negative impact on South Vietnamese peasants. During one trip to Vietnam Johnson learned that only some 6 percent of artillery fire was observed, the rest just being fired out into the jungle on the premise that some enemy might happen to be there. "Today," he told Westmoreland, "we are writing checks for a quarter of a billion dollars every month to pay for ammunition, which totals out to three billion dollars a year." Johnson suggested Westmoreland take another look at what was being done. "A reduction of perhaps as much as 50 per cent in the application of unobserved fires as they relate to the destruction of physical facilities in Vietnam, in terms of silencing the battlefield, in terms of scaling down the level of violence, or in terms of reducing the costs of the war, has a lot of attractive features."[44]

Westmoreland was uncomprehending, or at least impervious, to such reasoning. Just over two weeks later he cabled major subordinates to say that "a study of the present rate of fire of artillery currently in the Seven Mountain area reveals that the tempo could be drastically increased in order to effectively harass and interdict enemy movements and actions." He instructed them to "increase rate of fire of existing artillery . . . by orders of magnitude, e.g., 100 to 200 rounds per tube per day."[45]

When in January 1968 Major General Charles Stone went out to command a division under Westmoreland, he had done his homework and already had very strong misgivings about Westmoreland's tactical approach. "Before coming to Vietnam," he said in a later debriefing report, "I studied many after action reports and talked to many persons who had been here. My observations convinced me that the concept of search and destroy operations was not a valid one." As Stone saw it, "I have everything the enemy wants and he has nothing I want. I fight the war on that basis." Thus: "I will not contend him for a triple canopy jungle with no population and no foodstuffs. I am perfectly willing to let him have the border areas where there are no important political or military objectives. He cannot achieve his objectives by staying in the jungle. He has to come to where I am."[46]

As things evolved, Stone's views were shared by a number of other senior Army officers. Douglas Kinnard's survey revealed that 42 percent of Army generals who had commanded in Vietnam thought these large-scale operations were "overdone from the beginning," while 32 percent viewed the search and destroy concept as "not sound," and 51 percent said the execution of search and destroy tactics "left something to be desired," a response that ranked below "adequate." Kinnard observed that "these replies show a noticeable lack of enthusiasm, to put it mildly, by Westmoreland's generals for his tactics and by implication for his strategy in the war."[47]

General Arthur Brown recalled an occasion during his service in Vietnam as an advisor when a colleague had some graphics prepared.

They portrayed what areas of Vietnam were controlled by the Viet Cong and where the government was in control. Then he had flips that depicted where all the resources of the country were—the rice, the fish, and so on. "What it showed was very dramatic," Brown recalled. "The VC were in there with the fish and the rice. What we had was the places where nothing was."[48]

General Stone—and many others—understood the implications of that incongruence, but Westmoreland never did. Perhaps he didn't care since, as South Vietnamese Lieutenant General Ngo Quang Truong observed, Westmoreland had taken for himself much the easier task. "Compared to search-and-destroy operations," said Truong, "territorial security activities were immensely more complex."[49] Those had been left to the South Vietnamese.

Very early on it was clear that the Westmoreland approach was not only saddling the South Vietnamese, more or less unassisted, with that more difficult task—support of pacification—but that Westmoreland personally was determined to remain uninvolved in that aspect of the war. General Earle Wheeler saw the strategic and tactical situations much as Westmoreland did and consistently backed him in implementing his approach. Said the JCS history of the war: "General Wheeler thought that there was too much concentration by 'many Washington agencies' on pacification/RD [Revolutionary Development] as the answer to all problems in RVN."[50]

Despite Westmoreland's many protestations of his abiding interest in and robust support for pacification, there was no substance to the claim. General Phillip Davidson saw it clearly from his post as MACV J-2: "Westmoreland's interest always lay in the big-unit war. Pacification bored him."[51]

Westmoreland was unrepentant. "Pacification was oversold in the United States and oversold to the Johnson administration, where it was the 'end all,'" he insisted in his oral history. "It was never the end all with me," he said, "and I got pressure after pressure after pressure to put emphasis on pacification at the expense of allowing the main forces to have a free rein."[52]

Westmoreland blamed civilian officials for much of that pressure, singling out the Systems Analysis and International Security Affairs staffs in the Pentagon as well as people at State and CIA. "I was fighting them off constantly," he recalled, "just fighting them off. I took steps to demonstrate that I was not against pacification, but I certainly was not going to become so obsessed with pacification that we would give the North Vietnamese Army and the main forces of the Viet Cong a free rein and allow them to attack when and where they chose."[53]

Yet Davidson saw clearly the defects in Westmoreland's approach. "If the United States ground troops could not 'find' the enemy or 'fix' him," he said, "they manifestly could not 'fight' or 'finish' him. Yet that was what the strategy of attrition required—the ability to inflict unacceptable losses on the enemy, and the United States could never do it."[54]

In one of his most disastrous judgments, Westmoreland expressed the conviction—even before large-scale American elements had entered the ground war—that what they would face was "attritional war and we are in a better position than the enemy to fight such a war."[55] Long after the war was over Westmoreland was asked if, looking back, he would do anything differently. "I've thought of that many times," he replied. Apparently the answer was no. "Other courses of action were either infeasible at the time or would have created some unforeseen and some foreseen problems."[56] Thus, he concluded, there was no other option.

A DIFFERENT APPROACH had been advocated by an ad hoc group headed by Lieutenant General Andrew Goodpaster, then Assistant to the Chairman of the Joint Chiefs of Staff. The July 1965 study on winning in Vietnam observed candidly that "the capability of the RVN to handle the problems of pacification, even with the defeat of the VC/DRV main force units, without extensive guidance, is questionable." The reason: "The extensive losses of officials at the village and hamlet levels, coupled with attrition of higher officials, has left the RVN in a weak position." Thus "unless there are substantial desertions from the guerrillas, subversives and sympathizers, extensive pacification programs probably will be required to restore government authority to the countryside and win popular support. Such programs probably will be beyond the RVN capability."[57]

The needed alternative, one wholly compatible with the ad hoc group's approach, was soon developed under the aegis of General Harold K. Johnson, the Army Chief of Staff. Known as PROVN (Program for the Pacification and Long-Term Development of Vietnam), this comprehensive study was not only a direct refutation and dismissal of Westmoreland's way of war but also a detailed articulation of a more availing option.[58] PROVN's summary statement was unequivocal: "The critical actions are those that occur at the village, the district and provincial levels," it held. "This is where the war must be fought; this is where the war and the object which lies beyond it must be won."

The study was, in classic Pentagon fashion, briefed all over the place, including to the Joint Chiefs of Staff. Not much interest was found there, which was not all that surprising with Earle Wheeler as Chairman. His much-quoted summary of his outlook on the war took him in just the opposite direction: "It is fashionable in some quarters to say that the problems in Southeast Asia are primarily political and economic rather than military," he said. "I do not agree. The essence of the problem in Vietnam is military."[59]

In due course Westmoreland was also briefed on PROVN. Since the study emphasized that what he was doing was not working, and could not work, because it was having no effect on the enemy's covert infrastructure, Westmoreland could hardly find favor with it. He suggested it be set aside as a "study document" and also maintained that he was already doing most of what it recommended. Dr. Herbert Schandler, later a well-known author of historical works, was then a staff officer in MACV Headquarters. "I wrote Westmoreland's response to PROVN," he said. "We all thought it was great stuff, but we couldn't say that. We had to write things like 'there are some good ideas here for consideration' and so on. And we said 'we are implementing many of these programs already.' We had to say that—General Westmoreland wanted to show he was ahead of the game."[60] Thus PROVN had to await another day and another commander. Only then, said Lieutenant General Phillip Davidson, as "son of PROVN" would it be implemented, and with gratifying results.

In the papers of Marine General Wallace Greene there is a document, "Force Requirements and Long Range Estimates for I Corps RVN," dated October 1966, when General Greene was Commandant of the Marine Corps. An annotation on the title page in Greene's hand reads: "A vital planning document which was a highlight of General Greene's association with the Joint Chiefs of Staff." Said the document: "The very base of the war is the people of Vietnam. The revolutionary development program is the most critical of all the tasks, with the others being ancillary or contributing."

That language was entirely compatible with the PROVN Study's findings, illustrating a matter of fundamental importance. Not later than the autumn of 1966 the leaders of both services involved in fighting the ground war in Vietnam—Army Chief of Staff General Harold K. Johnson and Marine Commandant General Wallace Greene—saw Westmoreland's approach as fatally flawed and agreed on a viable alternative.

William Colby, who later in the war would head American support for the pacification program, told a postwar audience that "General Westmoreland in Vietnam doesn't seem to have been able to think of anything else but to trade casualties with the North Vietnamese."[61] And, he added, "for several years after we put our troops there we really hadn't figured out that the real nature of the war was at the villages. It wasn't until . . . 1968 that we really began to make progress in the real nature of the war there. The intervening years were just confusion and chaos."[62]

As the record illustrates, almost from the very beginning there was widespread unease about and outright dissent from Westmoreland's conduct of the war. One senior officer after another stated flatly that Westmoreland "does not understand the war," "never understood the war," and so on. But, and this was a critical factor, nobody in the chain of command was really competent to critique his performance. Lyndon Johnson had no understanding of military affairs whatsoever, nor did Robert McNamara. General Earle Wheeler was essentially a staff officer with virtually no troop leading experience, much less combat acumen.

General Harold K. Johnson, the Army Chief of Staff, was an authentic battlefield hero, and he was fundamentally at odds with the Westmoreland approach (expressed over and over again in trip reports after visits to Vietnam, in the PROVN Study, and later in a very extensive oral history), but he was not in the chain of command. As a member of the Joint Chiefs of Staff he theoretically had some influence there, at least to the extent he could shape the collective viewpoint, but even then LBJ and McNamara were famously impervious to advice from the Joint Chiefs. Indeed General Wheeler was not even included in the White House Tuesday Lunches, roughly the functional (although highly dysfunctional) equivalent of the National Security Council, for much of the time LBJ was in office.[63] Thus almost by default Westmoreland was left to go his own way.

EFFECTIVE CRITICAL ANALYSIS of Westmoreland's conduct of the war was provided by Dr. Alain Enthoven's OSD Office of Systems Analysis, an entity much resented by many of the senior military leaders in the Pentagon during the McNamara years. Wrote Enthoven and his coauthor K. Wayne Smith, the way the war was being fought "tended to dissipate its resources on high-cost, low-pay-off operations that happened to be congenial to traditional Service missions in conventional warfare."[64] "More attention to effectiveness in relation to cost," concluded these analysts, "might well have led to reductions in the billions of dollars spent on offensive operations and massive firepower displays—activities yielding small returns." Had even a small part of those resources been expended on more availing aspects of the war—they mention expanding territorial forces to provide better security for the local population, enhancing the effectiveness of the South Vietnamese Army by giving it modern arms, and other measures, "things which were later to be given priority—the course of the U.S. involvement in the war might have been altered sooner."[65] McNamara got much input along these lines but, for whatever reasons, none of it seemed to result in any modification of the field commander's fixation on the large-unit war.

Westmoreland himself was unmoved by such logic, if indeed he was even exposed to it, maintaining even during his final days in Vietnam that the war there had confirmed his "belief that our major advantage in war lies in our superior firepower. . . ."[66]

Meanwhile, in virtually ignoring pacification and the upgrading of South Vietnam's armed forces, Westmoreland failed to advance the security of the populace or the capacity for self-defense of South Vietnam's armed forces. He likewise failed to diminish the enemy's combat forces, despite his near-exclusive focus on that task, as the casualties inflicted were simply replaced. What he *had* done was squander four years of his troops' bravery and support by the public, the Congress, and even most of the news media for the war in Vietnam.

12

★

ATMOSPHERICS

THE FIRST ISSUE of *Time* magazine in 1966 hailed Westmoreland as the 1965 "Man of the Year." This had been in the works for some while, and in Honolulu Westmoreland had devoted considerable time to posing for a sculptor, Robert Berks, who produced a bust of him which was used instead of a photograph on the magazine's cover. A "Letter from the Publisher" said that fifty pounds of clay had gone into the bust and that Westmoreland posed for four sessions of an hour and a half each while on leave with his family. An accompanying photograph showed him in jungle fatigues, standing on the lawn at parade rest. While he posed, *Time*'s Hong Kong bureau chief, Frank McCulloch, quizzed him for the article.[1]

As might be expected, much of the accompanying story was positive, describing Westmoreland as "the sinewy personification of the American fighting man in 1965" and quoting General Richard Stilwell's characterization of him as "just a very straightforward, determined man." Noted *Time*, "in the command he inherited, Westmoreland wears more hats than Hedda Hopper."[2]

Time's story was in many respects also cautionary, noting that communist forces in South Vietnam were increasing by as many as 7,000 men a month, for a total increase in strength during 1965 of 80,000, and observing of allied operations against the enemy that "so far, the results have been less than spectacular." The conclusion was ominous: "As the U.S. troop level climbs toward 400,000 men, as the price of the war begins to crimp Great Society programs and boost taxes, Americans may find it harder than ever to accept the long war predicted by the Administration."[3]

Colonel Spurgeon Neel, an Army doctor who had served with Westmoreland in Vietnam and in fact shared his quarters, wrote to Westmoreland about his reaction to the story: "You are the most important man in the world today." The Scribe writing the Class of 1936 notes for *Assembly* was predictably ecstatic, comparing him to Eisenhower: "Only one other West Pointer has made this honor, and he became President." Dave Palmer, the aide who had accompanied Westmoreland when he first went to Vietnam, wrote that it was "Westmoreland's finest hour."[4]

Dan Cragg, then in Vietnam as a sergeant, found the assignment fascinating. "When the colonel went to lunch I would read the notes dictated by Westmoreland for his 'History,'" he said. "I can recall occasions in the MACV conference room when Westy would doodle the device for a five-star general."[5]

ON HIS BEDSIDE TABLE, said Westmoreland, he kept family pictures and a number of books, including a Bible, a French grammar, Mao Tse-tung's treatise on theories of guerrilla warfare, Jean Lartéguy's novel *The Centurions*, and several works by Bernard Fall. This collection was mostly for atmospherics, however, as Westmoreland admitted that he "was usually too tired in late evening to give the books more than occasional attention."

Sometimes Westmoreland portrayed himself as a student of military history, and in particular of counterinsurgency warfare. It is not clear how he might have come to view himself in that way. He had missed out on all the major military schools, had published no articles relating to military history in professional journals, was not known as much of a reader, and seemed unaware of commonplace matters in even such recent history as World War II.

When Charles MacDonald was ghostwriting his memoirs, for example, Westmoreland told him that "weather forecasting, predicting the monsoons, was a new departure in warfare."[6] MacDonald was a distinguished military historian, as well as a veteran of World War II and the author of acclaimed volumes in the Army Center of Military History's "Green Book" series. He must have wondered what kind of a student of military history had not heard of the critical importance of the weather officers and their forecasts, for example, as Eisenhower pondered the Normandy landings.

WESTMORELAND WAS APPARENTLY determined that sometime during his tenure in command there was going to be an operation in which an American unit would make a "combat" parachute jump. In November 1966 he overrode recommendations from CINCPAC and 7th Air Force that a wing of C-130 aircraft be based somewhere outside South Vietnam for greater security. Westmoreland insisted that they be stationed instead at Cam Ranh Bay. "The thing that influenced me," he dictated for his history notes, "is the need to have the troop carrier wing close by if we are going to exercise our airborne capability, which we intend to do during the coming months."

When the 173rd Airborne Brigade was planning an operation he asked, "Why not include an airborne insertion?" They were not enthusiastic, but said they would look into it. Later he went back for an update on the planning, and again there was no mention of an airborne operation. He asked again, and again they said they would take it under consideration. When he went back for the third time, still no mention was made of an airborne assault. At that point, said his aide Major Larry Budge, who had observed this whole process, "Westmoreland lost patience and ordered that one be made part of the operation."

The jump was eventually programmed as part of Operation Junction City, conducted during February 1967. Jumping in with the airborne troopers was Catherine Leroy, a twenty-one-year-old French photojournalist, who concluded that, at least from her youthful perspective, the operation "clearly had no strategic value."[7]

Westmoreland visited the field headquarters for the operation and found that it had gone according to schedule, but with little to show for it. "I was surprised," he said, "that they had not run into more resistance and enemy installations. A ring of troops had been established around the heavily forested area which we thought was the location of the Viet Cong's Central Office for South Vietnam. But there is a possibility that we have been spoofed by the enemy."[8]

Evaluating the jump itself in his history notes, Westmoreland said that it had gone very well, with very few injuries. "The troops of the battalion that jumped were in very high spirits and proud of the fact that they could now wear a small gold star on their jump wings. General Deane, the brigade commander, was the first man to jump."

Some years later, when *Vietnam* magazine published an article about the operation that took it at face value, Rodney George wrote a heated letter to the editor. "You characterize this airborne insertion of troops as an actual combat operation into a hot LZ," he observed. "Nothing could be further from the truth! It was a publicity stunt that did not even need to happen." George revealed that the airborne troops had landed in a secured LZ that had been staked out and guarded for their arrival, preparations that he himself had taken part in as a member of the 11th Armored Cavalry Regiment. "We watched those airborne troops descend into the LZ in which we were already positioned to ensure their safe arrival."[9]

Another very senior officer viewed the whole matter with some amusement, observing that the jump "lost some of its truly combat quality when it was learned that *Life* photographers had been prepositioned in the LZ so they'd get good shots of the parachutes coming down and be there on the ground to talk to some of the parachutists. It lacked some of the elements of a truly combat assault."

AMONG THE THINGS that troops found to buy, either in Vietnam or elsewhere in the region while on R&R, were very popular ceramic elephants, two or three feet tall and often garishly colored. Thousands and thousands of the things were bought and shipped home. The elephants could be deposited in the Army postal system, without any packing whatever, simply with an address tag, and would miraculously make it to their destination with seldom even a chip in the way of damage. Westmoreland acquired a pair of these artifacts and dispatched them to Lady Bird Johnson in Texas. Diplomatically describing them as "porcelain elephant garden seats," Mrs. Johnson wrote to Kitsy to say "it's so touching that a man with so much on his mind would ever have time to find elephants for us."

During this time Westmoreland also got the first of a number of feelers from people in the publishing industry about his interest in writing a memoir. Responding to Arthur Sylvester, then the Assistant Secretary of Defense for Public Affairs, Westmoreland said, "I don't believe I ever consciously thought of myself as a potential author. If the thought is intriguing now, it is because I am fortunate enough to be serving in the eye of a storm like none other in history."

WESTMORELAND LOVED TENNIS, and in Saigon he played often. An aide-de-camp recalled that one of his duties was to arrange tennis partners for the General. "I did not play with him myself (I was a better player)," said Captain Robert McCue. "I think he played primarily for the exercise." It was clear, though, that Westmoreland took his tennis seriously, even dictating descriptions of various encounters for inclusion in his history notes. One such entry records a game "with the British Ambassador, mostly during a pouring-down rain." An occasional opponent was Major General Joseph McChristian, the MACV J-2, who remembered that Westmoreland would be on his side of the net, you on yours, no handshake, just, "Let's go." They'd play two sets, then shake hands. "Thank you," Westmoreland would say, and off he'd go.

Mishaps on the tennis court were the closest Westmoreland came to being wounded during his four and a half years in Vietnam. In March 1966, attempting to break a fall, he fractured a bone in his wrist and had to wear a cast for six weeks, something he found annoying, but also significant enough to be the subject of a cable he sent to General Wheeler. Columnist Jack Anderson wrote that Westmoreland should have been "at the front" at the time, a comment that infuriated Westmoreland, who observed in his memoirs that "it seems to be an American custom to make anyone in the public eye a ready target for any carping critic. That I sought relief from my fourteen- to sixteen-hour day and seven-day week by an occasional set of tennis, played in the stifling heat of the lunch period, apparently disturbed some people," and he named Anderson specifically.[10]

Westmoreland also noted that "a repatriated prisoner of war later wrote me that North Vietnamese guards taunted the prisoners that 'Westy' would rather play tennis than fight."[11]

Westmoreland's calendar for the years in Vietnam shows tennis games two or three times a week, with the preferred venue the Cercle Sportif, a club centrally located in Saigon. That created certain problems in terms of image and public relations. Westmoreland would show up at noon, just as, observed Robert Komer, all the reporters were having their breakfast. Eventually, harassed (in print) by Peter Arnett and Kelly Smith, Westmoreland gave up his membership in the club, angrily telling Arnett, "You'll be happy to learn that I have quit the Cercle Sportif and tennis." Arnett blurted out that, at his wife's urging, he himself had only recently joined the club. Then: "Westy looked at me with narrowing eyes and said, 'Well, maybe they gave you my slot,' and turned away."[12]

IN MID-AUGUST 1966 Westmoreland's family moved from Honolulu to the Philippines, where Kitsy, Rip, and Margaret took up residence in a duplex at Clark Air Base. Stevie, having graduated the preceding June from the National Cathedral School in Washington, returned to the United States to enter Bradford Junior College in Massachusetts. Westmoreland's attitude toward Hannah the dog as a member of the traveling party had by this time improved sufficiently that she even got mentioned in his memoirs as part of the move to the Philippines, although once again she had to keep a low profile.

Having the family in the Philippines, much closer to Vietnam than they had been in Hawaii, was a great benefit for Westmoreland. In November 1966, for example, he wrote to his mother that he had "been able to get over about every two weeks for a short visit." But these were, inevitably, difficult family years. Kitsy told an interviewer of her particular concern for the children and their relationship with their father. "I don't want him to miss the years in which his children are growing up," she said. "I don't want him to look back when it's all over and say, 'What happened to my children? I never really got to know them.' Yet, deep down, I know that's exactly what's happening." Westmoreland's sister, Margaret, recalled that during these years the children "resented their father's being away in Vietnam."

In the autumn of 1967 Rip, entering eighth grade, was enrolled in the Hawaii Preparatory Academy at Kamuela, the beginning of what proved to be an extended ordeal for him. Soon he dispatched a three-page letter, in pencil on lined paper, to his mother. "It's time to tell the *truth* about this place," he began. "Well . . . I Hate it here. I don't have one friend. This school isn't worth as much as you're paying. I cry every night. I hate everybody here I hate it."

Another letter revealed some of the underlying, and desperately sad, causes of Rip's unhappiness. "People are saying things about Dad now." Rip almost got in a fight with another boy because of it. "This guy says his father told him that's why we're losing the war."

Stevie, now in college, had her own doubts about the war. Westmoreland wrote her a three-page typed letter in which he began: "It appears to be smart in college circles these days to question our government's policy in Vietnam and I gather that you are beginning to follow the same pattern." He sought to reassure her, maintaining that "the senior officials who are making policy are conscientious, dedicated and wise men who are going to weigh all the factors and balance the manifold contributions before arriving at policy decisions. Such has been the case with regard to our national policy in Southeast Asia." There was more along those lines, then some final advice. "I would suggest that you not follow the popular example of criticizing government policy but should acquire the facts and develop an insight that would give you a better appreciation of the basis of the policy which I believe in most cases will intellectually support it. In short, I certainly agree that you should think for yourself but at the same time should refrain from being influenced by emotional currents born of immaturity and ignorance."

Things were better for Margaret, the only child still in the Philippines with Kitsy. "Margaret is very happy with her pony and spends most of her time down at the stables taking care of it," Westmoreland wrote to Rip.

ONE OF WESTMORELAND'S strongest tendencies was to personalize virtually everything. He described Otto Kerner as "a man who served me admirably during the early days of World War II." To Lieutenant General W. R. Peers, about to retire, he wrote that "the job that you did for me in Vietnam was outstanding in every respect." Of Major General Daniel Graham, an intelligence officer, he said, "Danny served me admirably as a colonel in J-2 of MACV." When McNamara came to Saigon, said Westmoreland, he "visited me." The RMK engineering consortium had constructed "my new headquarters." Ambassador Lodge came to say good-bye to "my staff." And when Lieutenant General John Heintges, Westmoreland's West Point classmate, arrived in Vietnam to become the deputy, Westmoreland dictated for his history that "General Heintges was obviously pleased and flattered that he had been selected for the important assignment as my deputy."

The sense that everything somehow revolved around himself comes through most strongly in the history notes, but also in correspondence and cables and in the coaching he gave Charles MacDonald when the memoirs were being put together. At a Mission Council meeting, "the staff gave a special presentation on the number of piasters that have been saved as a result of the aggressive program I initiated several months ago." After an address to some mobile training teams, he observed, "there is every indication that my remarks, which were forcefully delivered, made an impression on the cross-section of Vietnamese officers representing every corps and division." He had to do it all: "With so many things to be done with so few troops, I had to get involved in details that a supreme commander would ordinarily leave to subordinates." Regarding his long tenure, "After three years in-country, my institutional memory was tremendously valuable to subordinate commanders whose experience was less."

Westmoreland also repeatedly micromanaged, getting involved in low-level decisions relating to such matters as minor aspects of the construction program. At one point, for example, he "made a decision to shift the crushed rock and the earth-moving equipment" from an airfield, where work had already begun, to "construction of an Army ammunition depot, an Army warehouse area, and an Air Force bomb dump." Westmoreland viewed this matter as important enough to be recorded in his history file, as was a decision to modify specifications for a planned runway, reducing the width by fifty feet and the thickness by several inches, even—rather bizarrely, or so it seems—referring the airfield decision to Secretary of Defense McNamara. It is virtually impossible to imagine Eisenhower's asking the Secretary of War to approve such a matter during his war, or his even knowing about it.

An area of considerable interest, one often raised and stressed by Westmoreland, was the functioning of an effective inspector general system in MACV. There is great disparity between his own self-congratulatory accounts of the matter and the recollections of others who were centrally involved. Given some of the fairly disastrous occurrences during Westmoreland's command—such as the My Lai massacre, a clubs and messes scandal, and the involvement in racketeering of Westmoreland's own nominee to be Sergeant Major of the Army, none of which was identified by inspectors general before becoming public scandals—the validity of Westmoreland's claims is of considerable importance.

Reflecting on his experience during World War II, Westmoreland concluded that "more emphasis should be placed on the IG, that the IG should be given more emphasis by command." In particular, he stressed, the IG's function should be to concentrate on "detecting trends toward improper conduct rather than totally concentrating on investigating items when evidence was surfaced that something was not right." And, he stated, "I did that in Vietnam. We had the most extensive IG system that we've ever had in a theater of combat."[13] Testifying years later in the trial of a libel suit he brought against CBS (a matter to be examined later in this account), Westmoreland again made such claims, asserting that "when I took command in Vietnam, as I had done in other commands, I . . . put more emphasis on the inspector general than I think has ever been done by a field command."[14]

In August 1967 Colonel Robert Cook arrived in Vietnam to become the new MACV Inspector General. He was welcomed by Major General Walter Kerwin, the MACV Chief of Staff, and also had a brief conversation with the Deputy COMUSMACV. Then, he recalled, he went back to "look at my empire there, and there's me, another colonel, two lieutenant colonels, and four enlisted people."[15] Westmoreland had by then been in command for more than three years, and that was the extent of the "most extensive IG system."[16]

Cook had served with the Deputy COMUSMACV during World War II, and in the IG slot he again worked very closely with him, extending into the years when the Deputy moved up to be COMUSMACV. Within a year of Cook's becoming MACV IG, the handful of staffers he inherited had grown to 96 officers and about 170 enlisted men.[17] Later, at the peak of his tenure, it reached about 90 field grade officers, including 13 full colonels, and a stable of Vietnamese translators—about 400 people in all.[18] In a Military History Institute oral history interview Cook was asked about the apparent discrepancy between Westmoreland's claims of a robust IG system and what Cook found when he arrived in Vietnam. "The point is," he said, "to be quite frank, General Westmoreland wrote glorious papers and philosophical type stuff. But the problem was the implementation afterward. For example, before I got there, he wrote a big paper about setting up a comprehensive Inspector General system and so on. Then when I got there, there were four people sitting around. They didn't have the manpower—they were willing, but they didn't have the assets, didn't have the qualifications. . . . Really nothing going."[19] Reemphasizing the point in a later interview, Cook observed that what "General Westmoreland was prone to do was to have a grandiose idea and put it on paper, and then never do anything more about it."[20]

AS COMUSMACV WESTMORELAND traveled incessantly. In his memoirs he described his typical routine in Vietnam as beginning each day with twenty-five pushups, then visiting units in the field several days a week. On such trips he often took along some member of the media. These visits were a big part of Westmoreland's persona and a source of pride to him as well. "My mobility paid big dividends," he told Charles MacDonald. "My staff had no opportunity to get out as I did. Staff needed the feel of the field which I could provide." And: "Because there were so many things to be done with limited numbers of troops, I personally had to be an action officer and move troops around like checkers." From these and numerous other self-characterizations emerges a portrait of someone who viewed himself as indispensable.

Westmoreland later told military affairs correspondent S.L.A. Marshall that he made "almost daily visits to the field" during his command in Vietnam. The stated purposes of all the travel were to gain information and buck up the troops. This was an enormous expenditure of time and energy, clearly Westmoreland's own personal highest priority among the many claims on his attention, yet it is not clear what he learned or accomplished as a result. Possibly he contributed to good morale on the part of the troops he visited, but he missed a lot of important things that would seem to have been fairly obvious.

One was the malperformance of the M551 Sheridan, a lightly armored vehicle sometimes described as a "light tank." This new system was not yet perfected and quickly manifested many problems. The crews considered it highly vulnerable to mines and rocket-propelled grenades, and the supposedly combustible cases for the 152mm main gun rounds did not perform as advertised, resulting in some sporty moments for crewmen sharing the turret with flaming residuals. Westmoreland was apparently unaware of any of this, later writing with near-perfect inaccuracy to Congressman William Dickinson that "use of the combustible cartridge case with the closed breach scavenger system has proved highly effective" and that "the Sheridan has performed very well in Vietnam where it enjoys a high level of troop confidence."[21] That was total fantasy.

Another important matter that apparently escaped Westmoreland's notice was the very effective use the South Vietnamese armed forces were making of armored vehicles, especially the armored personnel carrier. They devised modifications of the APCs that transformed them almost into light tanks, then used them extensively for convoy security, overwatching fire, even jungle busting. Given Westmoreland's personal antipathy to armored forces, or at least to their employment in Vietnam, he might have learned some useful things had he been more observant. He also apparently missed how woefully underarmed the South Vietnamese were in comparison to the enemy, or didn't think that was important or a problem.

At I Field Force, Vietnam, a corps-level U.S. headquarters in Nha Trang, Lieutenant General Charles P. Graham (then a lieutenant colonel) was the G-3 Plans Officer. In that position he on occasion briefed the visiting Westmoreland. "After the briefing, he would issue some instructions," said Graham. "Each time he did so, I had a strong feeling that he was shooting from the hip, had no overall strategic plan that guided his decisions, and was issuing instructions because he felt as the senior commander he should do something."[22]

Lieutenant General Walter Ulmer recalled, while he was serving as a more junior officer in MACV J-3, going in with a colleague to brief Westmoreland on some matter. "We got into a discussion of mortars versus artillery," he said, "and I observed that General Westmoreland's capacity for handling cognitive complexity was severely limited."[23]

Major General Clay Buckingham was, as a junior officer, a sector advisor in Hau Nghia, a small province directly west of Saigon on the Cambodian border. In those days, he remembered, the province was "flooded with Viet Cong. They owned the night and disappeared during the day." On a given occasion Westmoreland came out by chopper to get a briefing. Buckingham recalled how he "came into our dilapidated hootch and sat in a small circle with my four or five officers and two or three key NCOs." Buckingham introduced his staff members and gave a little background on each. One of the officers was a man named Jim Kernan, who had been football captain at West Point. Buckingham briefed Westmoreland on the situation and told him they were not making much progress. The Viet Cong were everywhere, and the Regional and Popular Forces just could not cope with them. Westmoreland "seemed not to hear. At the end of my short briefing, without any comment to me, he turned to Jim Kernan and talked about Army football for maybe 20 minutes. Then he got up and left."[24]

On another occasion, described by Westmoreland in his memoirs, he visited the barracks of a U.S. unit and observed that the bulletin board was mounted on a wall of female pin-ups. He asked the lieutenant in charge if he required his men to read the bulletin board, and of course the answer was that he did. Westmoreland then "rebuked" him, saying that "somebody might object" to having to view such female pulchritude in the course of reading the daily orders. At that point an accompanying brigadier, Joe Stilwell Jr., stepped in to help out his young subordinate. "By God, sir, you are right," he said. "We'll poll the unit, and if we find somebody who objects, we'll have him transferred."[25]

Brigadier General Edwin Simmons, later the long-serving Chief of Marine Corps History, recalled a commanders' conference convened by Westmoreland at Nha Trang. The room was filled with senior officers who had served in World War II and Korea, and Westmoreland told them: "At the end of World War II I wrote down some principles of war. I have them still today—on a card I carry in my wallet, and I want to share them with you. Whenever possible, feed the troops a hot meal. Make sure they have dry socks, and check their feet. Stress getting the troops their mail." The audience had thought they were about to get the equivalent of Napoleon's maxims or the like, said Simmons, "and instead they got these platitudes of squad leading."[26]

One of the primary stated purposes of all the travel was to boost the morale of the troops, but Westmoreland often seemed oblivious to the human factor. A stunning example occurred during a visit to the 2nd ARVN Division at Quang Ngai for presentation of a Silver Star for valor to Colonel Toan. Westmoreland entered what happened next in his history notes, apparently seeing nothing wrong with the picture presented. "After the ceremony," he said, "we went into Toan's office where he served champagne in celebration of his award. During this period I look[ed] around and observed that we had the entire Vietnamese and American chains of command, from Saigon to Danang, to the 2nd ARVN Division and their counterparts in Quang Ngai and Quang Tin Provinces, and therefore suggested to General Vien that we remove the champagne and have a conference."

Colonel Carl Ulsaker, an advisor, was also at Toan's award ceremony, and he remembered the aftermath: "The champagne turned warm and lost its bubbles as Westmoreland droned on, and he didn't discuss anything worth a damn. He was just so insensitive to people."[27]

At other times in his travels Westmoreland sought to shape how events, especially those in combat situations, were understood and portrayed. Visiting an evacuation hospital where casualties of a recent engagement were being treated, he encountered 1st Sergeant "Bud" Barrow of Delta Company, 2/28 Infantry. Westmoreland asked him how he had been wounded. "Well, sir," he responded, "we walked into one of the damnedest ambushes you have ever seen." Those words were anathema to Westmoreland. "Oh, no, no, no," he protested. "That was no ambush." "Call it what you want to," said Barrow. "I don't know what happened to the rest of the people, but, by God, I was ambushed."[28]

13

★

BODY COUNT

Westmoreland often maintained that "our purpose was to defeat the enemy and pacify the country, and the country couldn't be pacified until the enemy was defeated." In his terms, defeated meant killed, which in turn established body count as the measure of merit in this war.

Inevitably controversy arose over the reliability of the body count and—even in those cases where the count matched the number of bodies left on the battlefield—whether all those killed had in fact been enemy rather than innocent South Vietnamese civilians. Westmoreland was very sensitive to such charges and for years publicly maintained faith in the accuracy of the body counts reported by his forces. "I believe one of the great distortions of the war has been the allegation that casualties inflicted on the enemy are padded," he asserted. "I can categorically state that such is not the case."[1] That was not, however, his private belief, for as early as August 1965 he had cabled General Wheeler that in MACV "we have taken the position that KIA figures are inflated and are sufficiently accurate only to indicate trends in battle casualties."[2]

A large majority of Westmoreland's generals were basically appalled by body count, as documented by Douglas Kinnard's survey. In fact, 61 percent of the respondents thought body count was "often inflated," and that question, reported Kinnard, "brought forth a torrent of comment from the generals." Among the examples he cited were these: "They were grossly exaggerated by many units primarily because of the incredible interest shown by people like McNamara and Westmoreland." "Often blatant lies." "A fake—totally worthless." It should be emphasized that these were not the views of antiwar critics, or even rank-and-file observers, but the conclusions of Army general officers who had commanded in Vietnam.[3]

Westmoreland's fallback position was that, even if the reports of body count were inflated, they were ultimately probably accurate, perhaps even understated, because of enemy deaths caused by unobserved effects of bombing and artillery and those who later died of wounds. While he was Chief of Staff, Westmoreland spoke to this issue during an appearance on the CBS television program *Face the Nation*. Asked whether he believed the body counts, Westmoreland said, "I was aware of the fact that there was some double counting from time to time, no doubt from time to time some inaccurate reports." But, he added, it was the assessment of his staff, an assessment with which he agreed, that "offsetting any exaggeration" would be those other losses.[4]

Undoubtedly there were such uncounted losses, but Westmoreland was apparently not troubled by the ethical question of false reporting by his chain of command. There is substantial evidence that in some places, perhaps many, unit commanders were made to compete with one another on the basis of reported body count. Josiah Bunting's 1973 roman à clef, *The Lionheads*, describes such pressures and the agonizing dilemmas they created in a thinly disguised 9th Infantry Division.

There was yet another problem with Westmoreland's metric. If the enemy replaced his real losses, then those reported losses had little or no significance in terms of "progress" in the war or improved prospects for its successful conclusion. The fighting simply continued at the same level against an equal, or perhaps even larger, number of enemy forces.

Later Robert McNamara described his skepticism regarding the reported body count. "The point is that it didn't add up," he recalled.

"If you took the strength figures and the body count, the defections, the infiltration and what was happening to us, the whole thing didn't make—didn't add up."[5] Ambassador Ellsworth Bunker, who took up his post in Saigon in March 1967, was even less impressed, questioning whether body count, even if accurate, had any utility as a measure of progress in the war. "Well," he said, "I think body count is not a very good way of calculating progress in the war."[6]

By the time the war was over Westmoreland had reverted to his original position, that body counts were reliable and accurate. "All I can say," he told the BBC, "is they were the most honest estimates that we could make."[7] Later he reconsidered yet again, saying during a news conference at Vanderbilt University that "great emphasis was placed on accuracy of reporting," but that he was sure "in many cases [the body counts] were exaggerated."

DURING A FEBRUARY 1966 conference in Honolulu a very important marker was laid down by McNamara, a set of six statistical indicators of progress that Westmoreland was tasked to achieve by the end of the year. One involving body count was central: "Attrite, by year's end (the communist) forces at a rate as high as their capability to put men into the field."[8] This meant, in other words, reaching the much-discussed "crossover point," where the enemy's losses exceeded his ability to replace them through infiltration from the north or recruitment in the south. This tasking would assume enormous importance during the 1967 "Progress Offensive," and again after the war when Westmoreland launched a libel suit. Richard Holbrooke observed that "this crossover point was terribly important to Westmoreland and to his command. He kept waiting for it. He kept looking for it. It had a profound influence on the well advertised, much debated question of the enemy order of battle."[9]

LBJ also attended that Honolulu conference. Westmoreland recalled that "in private talks with me, President Johnson seemed intense, perturbed, uncertain how to proceed with the Vietnam problem, torn by the apparent magnitude of it."[10] It was during this conference that the President made two famous comments to Westmoreland. "General, I have a lot riding on you," he stressed at one point, and then, alluding to the problems President Truman had with

General Douglas MacArthur during the Korean War (problems that, as both Johnson and Westmoreland knew very well, led to Truman's relieving MacArthur from his post), "I hope you don't pull a MacArthur on me." In his memoirs, recalling that rather dramatic moment with the President, Westmoreland wrote that "since I had no intention of crossing him in any way, I chose to make no response."[11]

The *Pentagon Papers*, describing this conference, quoted Ambassador Henry Cabot Lodge's remarks in a plenary session: "We can beat up North Vietnamese regiments in the high plateau for the next twenty years and it will not end the war—unless we and the Vietnamese are able to build simple but solid political institutions under which a proper police can function and a climate [can be] created in which economic and social revolution, in freedom, are possible."[12] Among those who heard Lodge express these convictions were General Nguyen Van Thieu and Air Vice Marshal Nguyen Cao Ky of South Vietnam. Westmoreland's later dismissive comment was that "Ambassador Lodge does not have a deep feel of military tactics and strategy."

Summing up the results of the conference, the *Pentagon Papers* noted that, from that time forward, "it was open and unmistakable U.S. policy to support pacification and the 'other war,' and those who saw these activities as unimportant or secondary had to submerge their sentiments under a cloud of rhetoric."[13] Westmoreland acted accordingly.

Before returning to Vietnam Westmoreland held a press conference in Honolulu. Asked how the war was going, he responded: "I am very optimistic at this time," citing body count achieved as the basis. Westmoreland also blithely stated that enemy guerrillas could be dealt with and local security established "once these main force units are destroyed." That fatal miscalculation persisted throughout his tenure in command.

Another major conference was held in Honolulu in July of 1966, during which various briefings were presented, including one by Admiral Sharp's staff on progress made in reaching the six goals set at the February conference, especially prospects for reaching the crossover point by the end of the year. "This goal is unlikely to be achieved," said the briefer, "because of the enemy's demonstrated ability to increase his forces despite losses."

Ambassador Lodge had commented on precisely that point in an 8 June 1966 cable to LBJ: "The best estimate is that 20,000 men of the Army of North Vietnam have come into South Vietnam since January and, as far as I can learn, we can't find them."[14] Frustrated, Westmoreland sent a memorandum to his major subordinate commanders stating: "I am increasingly concerned about the fact that we are not engaging the VC with sufficient frequency or effectiveness to win the war in Vietnam. The VC/PAVN buildup is continuing and our attrition of their forces in South Vietnam is insufficient to offset this buildup."[15]

Yet another in this year of conferences, this one a seven-nation summit, was convened in Manila in late October 1966. Afterward President Johnson headed for Vietnam, where he spent, as all the press reports carefully noted, two hours and twenty-four minutes at Cam Ranh Bay. During the stop he awarded Westmoreland a Distinguished Service Medal for his leadership of U.S. forces in the war.

Despite the cruel math and ineffective strategy, Westmoreland was at that point riding high. Later he told Charles MacDonald of his pleasure at how he had been treated by the President in Manila: "Nice touch for LBJ to make me the senior military officer at the conference. He did not bring Wheeler, or Sharp, said he wanted me, put me in the center of the military stage, to be the one who spoke." Back in the United States, LBJ underwent his celebrated gallbladder operation and from Bethesda Naval Hospital wrote to Westmoreland: "When it [victory in Southeast Asia] comes, you will stand first in the ranks of those your countrymen will celebrate. You already stand first in your President's affection and esteem."

WESTMORELAND'S RELATIONS WITH the media form an important subtext in any account of his experience in Vietnam and, indeed, of his career in general. He courted reporters and journalists assiduously, but seemed baffled by lack of the good press for which he thirsted. Said his aide Dave Palmer, "I had instructions from Westmoreland that when he was traveling I was to have a reporter or a USAID person or Sir Robert Thompson [a British counterinsurgency expert] or the like with him." In such encounters, said Palmer of Westmoreland, "he was very compelling." Sometimes, instead of the single media target of Westmoreland's attention, there would be a large entourage. In a letter to his wife Palmer described a trip of that type, a flight on an Army Caribou aircraft in which "the cargo compartment was jam-packed with Mr. Lodge, Gen. Westmoreland, a colonel, me, and 27 reporters of all nationalities and descriptions." Another officer recalled how Westmoreland stage-managed interviews at his headquarters, changing into jungle fatigues (field uniforms) beforehand if he had been wearing something less picturesque.

Sensitivity to press coverage and public relations was one of Westmoreland's most pronounced traits. Even when a matter involved serious ethical or disciplinary issues, Westmoreland's first—and often only—concern was how it would play in the press. The jump fatalities soon after he took command at Fort Campbell were one dramatic example. Another took place soon after major American ground forces entered the war in Vietnam. Westmoreland was shown a number of pictures, taken by an Associated Press photographer, depicting South Vietnamese using torture to interrogate prisoners of war. In several of the photos U.S. Marines were also shown, although they were not actively participating in the mistreatment of captives. Westmoreland cabled his guidance to Major General Lewis Walt, then the senior Marine in Vietnam: "We should attempt to avoid photographs being taken of these incidents of torture and most certainly in any case try to keep Americans out of the picture."[16]

From his earliest days in Vietnam Westmoreland had been critical of how the war was being reported. "In my opinion, the press has done a poor job in relaying to the American public a realistic feel of the situation here," he wrote to Robert Stevens in June 1964. "They seem to accept the bad and unhappy events with very little attention given to the other side of the ledger." At various times Westmoreland spoke of the press as negative, cynical, vicious, and immature.

Westmoreland's key associate, General William DePuy, located the problem somewhat closer to home. "Neither the public nor the media had the slightest conception of the scope or intensity of the war," he concluded. "We in the military failed miserably in portraying the war for what it was."[17]

Barry Zorthian, who headed the Joint U.S. Public Affairs Office in Saigon, maintained that "when Westmoreland took over he was under very strong pressure from Washington to straighten out the media." But, he said, in his dealings with reporters Westmoreland was "quite pedantic. He tended to lecture the newsmen."[18]

George Wilson, longtime military affairs correspondent for the *Washington Post* and less hostile to Westmoreland than most reporters, identified Westmoreland's demeanor as part of the problem, recalling that "Westy was the stiff, ceremonial-looking general who sounded platitudinous in the formal press conferences he tolerated."[19] And there were a lot of formal press conferences. The press, said General "Dutch" Kerwin, for over a year Westmoreland's MACV Chief of Staff, "were alienated by General Westmoreland, there is no doubt about it."[20]

Nevertheless journalistic assessments of Westmoreland continued to be mostly favorable through 1966. "In this modern and political conflict," wrote Don Moser for *Life*, "Westmoreland is so strikingly uncontemporary that meeting him is a little like stumbling across a live dinosaur." As evidence of that judgment, Moser said that Westmoreland was "methodical rather than clever, organized rather than intellectual, and outside of military affairs no one has ever called him sophisticated," a man "no more inclined to profanity than, say, Tarzan."

Letting Westmoreland speak for himself, Moser's piece quoted his recent remarks to some of the troops: "We're going to out-guerrilla the guerrilla and out-ambush the ambush," he told them. "And we're going to learn better than he ever did because we're smarter, we have greater mobility and firepower, we have more endurance and more to fight for." "And we've got more guts."[21] That these anticipatory achievements were being cited some two and a half years into Westmoreland's command tended to undercut their impact considerably, however, as did Moser's report of "nervously" sensing "a kind of resemblance between Westmoreland and the focal character of Graham Greene's best-selling novel *The Quiet American*, that wholesome and well-intentioned type who, out of his naïve idealism, generates a tragedy for those whom he would help."[22]

Westmoreland explained how a perceived negative media outlook drove him and other government officials to overstate their case in the opposite direction. They "realized that some of these negative stories were giving an improper perspective, and it was only natural that they and the Ambassador and myself were very anxious to balance this picture by putting forth the positive when the press became the proponents of the negative."[23]

Several years after his Vietnam service Westmoreland was even more candid, explaining how, "during 1967 and 1968 a polarization situation occurred in Saigon. The media were inclined to accentuate the negative. To balance the picture, the U.S. authorities in Saigon were prone and encouraged by Washington to accentuate the positive."[24] Perhaps this situation led to such occasions as a Westmoreland visit to the 1st Battalion, 28th Infantry, recalled with distaste by James Parker. When the unit was advised of an impending Westmoreland visit, word went out that "the COMUSMACV is coming to walk among his people" and to hand out some Silver Stars. Since officers had no pending recommendations for such awards, that sounded rather strange, but they picked some soldiers and lined them up as instructed. Westmoreland arrived on schedule, as did a second ship full of photographers. "They swarmed around him like gnats and took pictures of his every move," said Parker, who remembered that, despite having his arm in a sling (later revealed as due to a tennis injury), Westmoreland "had a regal manner." He handed out the medals, got back in his helicopter, and headed for the next stop. As his ship lifted off, the battalion operations officer went down the line of troops who had been decorated, taking back the medals, "an awards ceremony in reverse." Said Parker, "The visit turned out to be pure public relations for General Westmoreland. We were props."[25]

A more serious case, a strenuous effort to discredit certain reporting after the fact, took place in early December 1967. Westmoreland was shown the draft of a Jonathan Schell article, scheduled for publication in the *New Yorker*, which was highly critical of a U.S. operation in two northern provinces, especially in terms of alleged callousness of the troops toward the civilian populace. Describing Schell as "an avowed pacifist," Westmoreland said the article was highly exaggerated and would create major problems if published. He urged Robert Komer to see Ambassador Bunker about the article, "which I felt could get involved in the war crimes arena if taken at face value." The point of seeing Bunker was to urge that he send a message to Washington which, said Westmoreland, "would hopefully bring about some pressures to withhold publication until the matter could be given further study."[26] As it happened, there was some further study, with most interesting results. Said Bunker, "We sent an embassy officer to investigate. He reported that much of what Schell described was true."[27]

Westmoreland's efforts to control what was printed about the war also extended to trying to limit what news was available to the troops in Vietnam. In particular he had an aversion to a tabloid known as *Overseas Weekly*, a publication often called *Oversexed Weekly* by the troops, who delighted in its spectacular cheesecake photographs and exposés of the misdeeds (actual or alleged) of those in authority, especially senior officers. That publication was widely available wherever American GIs were stationed, most notably in Germany, but Westmoreland sought to limit its availability in Vietnam. He denied it space in the post exchange distribution system on the premise that "since it did not fall into the category of a magazine, and we do not allow tabloid-type papers (with the exception of the service newspapers) to be sold because of their bulk, we could not find a place for it on our stands."[28] That was, of course, a totally bogus rationale, as was demonstrated conclusively when Westmoreland's successor, soon after taking command, authorized *Overseas Weekly* to be sold at *Stars and Stripes* newsstands throughout Vietnam.

EARLY IN HIS VIETNAM assignment Westmoreland had conceived a very unfavorable impression of many in the press corps, writing to his father in April 1964 that "this war has been very badly reported to the American people through the press, and I might say the *New York Times* is perhaps the best example of what I mean." That paper, he said, had not sent its best reporters to the war zone, "and the results speak for themselves. Many of the reporters have been young, immature, impetuous men who have been unprepared to report the situation objectively." After Tet 1968, said Westmoreland in a later oral history interview, "I was very disgusted with the media, particularly CBS and Walter Cronkite. . . . I think they deceived the American people. . . ." Of course, he took it personally: "This was an effort to lift the onus of the adverse public reaction toward the Vietnam War, following the Tet Offensive, lift it from their backs and put it on my back. That was the objective."

The targets of such criticism had a different view. One who said so very strongly, and one of the more controversial Vietnam War report-ers, was Peter Arnett. "When Westy took command in 1964," he said, "I was thirty years of age. I had been in Southeast Asia for eight years, and had been all over Vietnam. I was married to a Vietnamese woman. My brother-in-law was a colonel in the Vietnamese army. I knew John Paul Vann and most of the American advisors. What did he [West-moreland] mean that we were too young and didn't know anything? Westy was wrong."[29]

Another experienced Vietnam-era journalist rebutted a Westmore-land criticism of reporters. Westmoreland had told Charles MacDon-ald, in preparation for his work on the memoirs, that "ARVN had an unfair and unfortunate image in the US, thanks to a vicious press." In the book Westmoreland then wrote that "the American news media contributed to a false image of the ARVN's performance." Attempting to shore that up, he continued, "I convinced the South Vietnamese to conduct a daily press briefing similar to that held at MACV, but only two or three American reporters showed up the first day and seldom any thereafter."[30]

Not so, wrote Phil Jordan to Westmoreland after reading that al-legation in the memoirs. "I was in 'Nam as a reporter—freelance and on the staff of *Overseas Weekly* for more than two years, 1966–68." And "for most of the time I was in Vietnam, I had a standing assignment from USIS to cover, when I was in Saigon in the afternoon, the ARVN daily press briefing and to prepare from it one or more stories for USIS distribution. I can assure you, Sir, the ARVN briefings were, for the time I was attending them, usually quite well attended by Ameri-can and other non-Vietnamese reporters; anywhere from 25 to 40 or more reporters were common, as I recall, and on occasion a lot more." Jordan closed by giving Westmoreland the names and addresses of both American and South Vietnamese officials who could confirm his account.[31]

IN A FINAL definitive contribution to the credibility gap, at a postwar symposium where reporting from the combat zone was being dis-cussed, Westmoreland asserted: "I could care less. My job is on the battlefield."[32]

14

★

M-16 RIFLES

N O SINGLE FACTOR more definitively illustrates Westmoreland's neglect of the South Vietnamese armed forces than the M-16 rifle, then a new, lightweight, automatic weapon considered ideally suited for the Vietnam environment.

When improved weaponry and other materiel became available, U.S. forces got first call on the M-16 rifle, the M-60 machine gun, the M-79 grenade launcher, and better radios. For much of the war the South Vietnamese were armed with castoff U.S. equipment of World War II vintage, such as the M-1 rifle (not well suited in weight or configuration for the relatively slight Vietnamese, let alone in terms of its limited firepower) and the carbine. Meanwhile the North Vietnamese and Viet Cong were getting the most modern weaponry their communist patrons could provide, including the famous AK-47 assault rifle. As a consequence, during the Westmoreland years the South Vietnamese were consistently outgunned, with predictable results in battlefield outcomes and morale, not to mention reputation. "Great emphasis was placed on improving the ARVN constantly," claimed Westmoreland,[1] but that was simply not the case.

Air Vice Marshal Nguyen Cao Ky, South Vietnam's sometime Vice

President, later recalled bitterly how "the big, strapping American GI carried a light, fully automatic Colt M-16 rifle into combat with hundreds of rounds of ammunition, a match for the enemy's AK-47 assault rifle. Until after the Tet Offensive of 1968, our small soldiers carried heavy, eight-shot American M-1 rifles so obsolete that the U.S. National Guard did not want them."[2] Lieutenant General Ngo Quang Truong, widely viewed as South Vietnam's best field commander, agreed. "In general," he noted of ARVN units during these years, "they were inadequately equipped to respond effectively to operational requirements."[3]

The disparity in weapons would become particularly apparent during the 1968 Tet Offensive. Remembered Colonel Hoang Ngoc Lung, South Vietnam's chief intelligence officer, "The RVNAF was equipped with modern weapons only after comparable ones had been employed by the enemy. M-16 rifles were supplied to all RVNAF units only after the 1968 *Tet* Offensive when the enemy employed Communist AK-47s in large numbers."[4] The recollections of Lieutenant General Dong Van Khuyen, South Vietnam's chief logistician, were particularly poignant: "During the enemy Tet offensive of 1968, the crisp, rattling sounds of AK-47's echoing in Saigon and some other cities seemed to make a mockery of the weaker, single shots of Garands and carbines fired by stupefied friendly troops."[5] The M-1 rifle weighed over 11 pounds loaded and was 43 inches long. According to one calculation, the average South Vietnamese soldier stood five feet tall and weighed 90 pounds.

Westmoreland later stated that in 1964 he had asked his deputy, Lieutenant General John Throckmorton, specifically to look into "the feasibility of my asking for M-16's for the Vietnamese forces," and that when Throckmorton recommended against it because of the cost, he had approved that recommendation, although "with some reluctance." A year later, after the initial large-scale battles in the Ia Drang, said Westmoreland, "I decided that the M-16 was essential, not only for the American troops but for the Vietnamese. I made such a request in December 1965." But, he had to admit, "upon my departure in the summer of 1968, only a fraction of the Vietnamese forces had been equipped."[6] Of course they had been relegated to last priority for the new weapons, and there had been influential officials who opposed giving them modern weapons at all, so that result was predictable.[7]

In April 1968 *Time* magazine reported that the new Secretary of Defense, Clark Clifford, had announced "a dramatic increase in the U.S. production of the M-16 so as to equip all ARVN units by mid-summer."[8] That was something McNamara had never agreed to and that Westmoreland made only sporadic and at best halfhearted efforts to advocate, further evidence of just how pervasive was his belief that U.S. forces could come in and do the job for the Vietnamese without the necessity of ever equipping them to do it for themselves.

Charles MacDonald interviewed many other people to obtain background for Westmoreland's memoirs. One of them was General Harold K. Johnson, Army Chief of Staff during 1964–1968. MacDonald laid out his problem: "In talking with General Westmoreland from time to time, I've gotten his story of having recommended the M-16 for the ARVN as early as the [word(s) inaudible: Ia Drang?] fight back in the fall of 1965. And I have also asked him on occasion, 'Well, why has this only recently been fulfilled?' And the only answer he could give me was—said probably production difficulties in the United States. Can you shed any light on that at all?"[9]

General Johnson could and did. He did not remember exactly when Westmoreland's initial request for the rifles came in, he said. "But you will find—this is a personal view, and one in which I am perhaps being too candid—General Westmoreland has a request to cover every contingency. He has a magnificent file as far as Vietnam is concerned."[10]

General Frank Besson was at that time commanding the Army Materiel Command, and he remembered Westmoreland's request for 100,000 M-16 rifles. "I also recommended that we give it to the Vietnamese, the South Vietnamese," recalled Besson, "because I felt we ought to give our allies the best we could. But they said, 'No. We can't give it to the South Vietnamese because it will undoubtedly be captured by the Viet Cong and the North Vietnamese and will be used against us.' The honest-to-god fact—that is what they said."[11]

It was not until March 1967 that an allocation of M-16 rifles for the South Vietnamese was reinstated, the first shipments arriving the following month. "But," said Brigadier General James Lawton Collins Jr., "until 1968 there were only enough to equip the airborne and Marine battalions of the General Reserve."[12]

In his debriefing report upon leaving Vietnam in August 1968

General Fred Weyand emphasized the effects: "The long delay in furnishing ARVN modern weapons and equipment, at least on a par with that furnished the enemy by Russia and China, has been a major contributing factor to ARVN ineffectiveness."[13]

To the last, Westmoreland sought to evade responsibility for the longstanding failure to properly arm the South Vietnamese, and he appears never to have even considered giving them the M-16 rifle and other advanced weaponry *before* similarly equipping U.S. forces. He was reinforced in this view by the similarly uncomprehending Wheeler, who at one point cabled to tell him: "You will readily perceive the sensitive public relations issues which would be raised if we provide M-16 rifles to non-U.S. units while U.S. combat units are issued less preferred rifles." Thus: "I must request that you defer equipping non-U.S. units with the M-16 rifle until we can sort out the rifle situation."[14]

A TELEPHONIC EXCHANGE with Colonel "Hap" Argo about the preparation of Westmoreland's *Report on the War in Vietnam* is also revealing. Argo: "On equipping of ARVN, on 22 Jan[uary 19]66 you requested M-16s for them and this was turned down because of failure of US to go to wartime production. Do you want to use this language?" Westmoreland: "No. Don't explain why request denied, just say due to reasons beyond my purview they were not immediately available." As he knew very well at the time, "not immediately" was going to mean not for the next two years, but his report was not going to say that.

Another phone conversation reveals Westmoreland's feeble efforts to get M-16s for the South Vietnamese. In July 1971 Westmoreland spoke with Walt Rostow, who asked him what the problem had been with acquiring M-16s. The memorandum of the phone conversation reads like this: "General Westmoreland replied that this was a long story, but that he had sent a message urgently requesting that all US and ARVN troops be equipped with the M-16 in December 1965." However, "it was two whole years before things really got moving and it wasn't until June 1968 that this December 1965 request was actually filled. CSA [Chief of Staff, Army, Westmoreland's position at the time of this conversation] noted that he had mentioned this problem to

President Johnson on at least one occasion, and the President had been amazed to hear about it."[15]

IN A THREE-PAGE paper headed "M-16 Rifle" dictated as material for his memoirs, Westmoreland related a tale about White House concern over the adequacy of the M-16 rifle. "President Johnson sent a trusted old friend from Temple, Texas, a Mr. Frank Mayborn, to discreetly investigate the adequacy of the M-16 and the satisfaction of the troops," said Westmoreland. "I received word that General Bruce Clark[e], U.S. Army, Retired, had requested permission to come to Vietnam accompanied by Mr. Mayborn. I later learned that the principal in the party was Mayborn and that General Clark[e] had been used as a cover for Mayborn, which was neither known at the time by me or General Clark[e]."[16]

"When they left and asked what they could do for me," continued Westmoreland, "I pointed out that I had asked for the M-16 rifle for my troops and the ARVN in December 1965. Now over two years later, I had them for my troops and only certain selected ARVN units. It was extremely important that I get the full complement of weapons soonest because the Tet Offensive clearly demonstrated the wisdom of the decision. This message was taken back to the United States and presumably reported to Mr. McNamara."[17]

"Later, when I came back to the United States in May [1968], I pointed out to the President that the M-16's were not arriving as rapidly as should be the case. I reminded him that I asked for them almost two and a half years before. The President acted surprised," said Westmoreland, "as if he had not heard about this before, turned red in the face, and said he would do something about it soonest. After that, things really began to happen — orders went to Mr. McNamara to increase the production base and in due time this was done."[18]

There are serious problems with the entirety of Westmoreland's account.

Westmoreland repeated the claim of having spoken about this matter directly with the President in a subsequent letter to the President's executive assistant, saying that "the President and I discussed in some detail the importance of modernizing the Vietnamese forces, and I put in a strong plea for accelerating the delivery of M-16 rifles to the Vietnamese. The President was," said Westmoreland, "unhappy with the fact that this had been a slow process and gave immediate orders to speed it up."[19]

Then, in his memoirs, Westmoreland returned to this matter. "President Johnson later sent an old friend from Texas, Frank W. Mayborn, to investigate the M-16 for him personally," he wrote. "Mayborn was ostensibly merely accompanying a retired World War II general, Bruce Clarke, on an inspection trip to Vietnam, and it was only after their departure that I learned the true purpose of Mayborn's visit."[20]

Actually, as General Clarke later wrote to Brigadier General Hal Pattison, the Army's Chief of Military History, "At Christmas 1967, Gen. Westmoreland called me and asked if I would come to VN to visit his troops in the field. We set 1 Feb 1968 as the target date." Clarke then explained to Pattison how he had asked Mayborn to go along, how they had drafted Clarke's report on the aircraft returning home, how General Wheeler had passed a copy of the report to President Johnson, how LBJ had Clarke and Mayborn come to the White House to talk about the lack of equipment for ARVN, and how they had then discussed that matter for three hours with two presidential staff members. "Within a few days of our visit to the White House," said Clarke, "a presidential aide called me to say the President had released 100,000 M-16 rifles to ARVN."[21]

Westmoreland's contrary account in the memoirs prompted a letter from Clarke to Westmoreland setting the record straight. "It was I who invited Frank Mayborn to come with me," he began. "I do not believe [President] Johnson knew about it until after we returned when Pres. Johnson sent for me over my report on the poor weapons of the V.N. soldier. I took Mayborn with me. They were and had been on a 1st name basis for many years in Texas near Ft. Hood. It was at this meeting when Johnson released 100,000 rifles for the VN army."[22]

After the trip to Vietnam Clarke drafted a report that included this observation: "The Vietnamese units are still on a very austere priority for equipment, to include weapons. This affects their morale, effectiveness, and their ability to supply the first ingredient of success—*security*. This should be corrected as a matter of urgency. Troops know and feel it when they are poorly equipped."[23]

Westmoreland had every reason to know the facts of Clarke's involvement and the key role he played in breaking loose M-16s for the South Vietnamese, since Clarke had written to him on 15 March 1968: "Tuesday we were called to the White House, where the President thanked us for the report and then turned us over to two members of the staff—Col. Cross and Bob Fleming—to be debriefed."[24]

Another source, Ambassador Ellsworth Bunker, had also informed Westmoreland of Clarke's role in dealing with the M-16 matter. Cryptic notes of a telephone conversation between Bunker and Westmoreland on 16 February 1968, shortly after the end of Clarke's visit, include this from Bunker: "Amb. reported Gen. Clarke's remarks on too bad ARVN not armed with M-16 sooner." Then Westmoreland's efforts to explain the problem: "Gen. relayed that he urged this in 1965, ran into peacetime methods in Washington—bureaucratic snafu, compounded by Colt patent on weapon. Colt wouldn't let anyone else produce, they couldn't fast enough by selves; result: never did get to ARVN like should have."[25]

Later Clarke received a fine letter from Admiral John S. McCain Jr., who told him: "I am proud of your role in getting new weapons to the Vietnamese Army."[26]

SOON AFTER BECOMING Chief of Staff Westmoreland tasked the Deputy Chief of Staff for Logistics to provide him the file of messages exchanged about equipping ARVN forces with the M-16 rifle. The tabulation included these pivotal items:

> 18 November 1966: JCS informed CINCPAC that "conditions require that the allocation of XM16E1 rifles be revised on the basis of giving priority to first equipping US Forces, followed by non-US Forces engaged in combat in Vietnam."
>
> Joint Chiefs of Staff, in a memorandum for the Service Chiefs, recognized a need for a re-evaluation of the allocation plans for M-16 rifles. This was based upon a significant change which occurred in the fall of 1966 that planned greater use of ARVN Forces in a pacification role while continuing the use of US Forces and some elements of non-US Forces in an almost exclusive combat role.

• • •

WHILE HE WAS Army Chief of Staff Westmoreland responded in writing to certain questions posed by Townsend Hoopes, a former civilian defense official who was writing a book. Westmoreland told him that his strategy had had two parts, to put maximum pressure on the enemy and "to build up the Vietnamese forces both quantitatively and qualitatively, to include the provision of modern equipment to the point where they could assume progressively more of the battlefield burden."[27]

If that latter task was indeed a part of his strategy, Westmoreland totally failed to achieve it, or even in any real sense attempt it, during his long years in Vietnam. In 1965, noted Herbert Schandler, when Westmoreland asked for more U.S. troops he was "recommending a virtual American takeover of the war. There was little or no mention in General Westmoreland's request of South Vietnamese forces, or any program to utilize those forces, or to make them more effective."[28]

So relentless was Westmoreland in giving first consideration to American forces and last to the South Vietnamese that those priorities even extended to munitions. "For a while during 1966 ammunition stocks were low, forcing me to limit ARVN artillery to two rounds per day per gun," said Westmoreland in his memoirs, "but no American unit ever wanted for necessary ammunition."[29]

Meanwhile the enemy was energetically and effectively improving virtually the entire arsenal of his forces. "During 1966," stated a communist history of the war, "many new types of weapons and implements of war were sent to the battlefield, increasing the equipment of main force and local force units." The upgrades included both more weapons and improved ones, to include B-40 and B-41 rocket launchers. "AK assault rifles were issued to units down to the local force level." And new and better models of recoilless rifles, mortars, and anti-aircraft machine guns were issued down through provincial local force battalions.[30]

General John Galvin expressed a strong opinion on the matter: "Westmoreland firmly believed that any help to the ARVN would be a disadvantage to U.S. forces. He thought the money all came from the same pot, and he was damned if he wasn't going to get it all."[31]

South Vietnam's armed forces were criticized by many during these years, including some Americans who served in Vietnam. But General

Fred Weyand tied the admitted deficiencies of those forces to the paltry support Westmoreland was giving them. "The reason why some ARVN battalions, as well as RF and PF units, never operated at any distance from their fortified bases in 1965 and 1966," he said, "was that they were quite literally surrounded by a strong, but well-hidden enemy and these lightly armed, under-strength units simply did not have the capability to deal with them."[32]

General Norman Schwarzkopf, later prominent in the first Gulf War, was an Army major advising South Vietnamese airborne troops at the time American ground forces were committed. He welcomed the added firepower, wrote Schwarzkopf, but "the Americanization of the war disturbed me. We were suddenly going in the wrong direction with the South Vietnamese. It was their country, their battle: eventually they would have to sustain it. I thought we should give them the skills, the confidence, and the equipment they needed, and encourage them to fight. Yet while our official position was that we were sending forces to help South Vietnam fight, the truth was that more and more battles were being fought *exclusively* by Americans." And, he added, for the South Vietnamese "supplies and equipment were harder to come by because American units had priority."[33]

This disparity in resources, especially weapons, persisted throughout Westmoreland's tenure in Vietnam. Ambassador Bunker noted it in a reporting cable to the President only weeks before Westmoreland's departure. "The enemy has also been able to equip his troops with increasingly sophisticated weapons; they are in general better equipped than the ARVN forces, a fact which has an adverse bearing on ARVN morale," said Bunker in a 29 February 1968 message.

THESE WASTED YEARS—when the South Vietnamese could have been developing in terms of leadership, combat operations experience, and skill in the use of more modern weaponry—had cascading effects in the years of American withdrawal. Many of Westmoreland's senior associates understood, at least in retrospect, the negative consequences of ignoring the South Vietnamese armed forces during these years, and they said so. In his survey of Army general officers who commanded in Vietnam Douglas Kinnard included a list of actions that, given another chance, they would most like to see given more emphasis. Ninety-one percent of the respondents (the highest percentage opting for any of the eight items on Kinnard's list) selected "improving the ARVN."[34]

Ambassador Maxwell Taylor put it succinctly: "We never really paid attention to the ARVN army," he said. "We didn't give a damn about them."[35]

LATER, WHEN THE Nixon administration took office in January 1969, the White House settled on a policy of "Vietnamization" for ending the war. That entailed, among a number of other steps, upgrading the South Vietnamese military establishment, especially in terms of providing arms comparable to those the enemy had long possessed, and progressive unilateral withdrawal of U.S. forces from the war zone. That was a new idea.

When Creighton Abrams had been in Vietnam for only a month, in early June 1967, he sent a significant message to General Harold K. Johnson, the Army Chief of Staff. "It is quite clear to me," said Abrams, "that the US military here and at home have thought largely in terms of US operations and support of US forces." Thus: "ARVN and RF/PF are left to the advisors." Added Abrams, "I fully appreciate that I have been as guilty as anyone. The result has been that shortages of essential equipment or supplies in an already austere authorization has not been handled with the urgency and vigor that characterize what we do for US needs. Yet the responsibility we bear to ARVN is clear." And finally: "The ground work must begin here. I am working at it."[36]

Westmoreland would later claim that Vietnamization had been his idea, albeit one he had been unable to sell to Lyndon Johnson. "I tried to get the President to adopt this as policy," he told a group of Marine Corps historians, "but the Johnson administration did not do it. They did not want to pay the price of giving the South Vietnamese the equipment that they needed if they were going to take over progressively more and more of the war."[37]

Westmoreland complimented himself for discerning, early on, that the South Vietnamese were deficient in communications equipment, in his judgment "a major contributor to deficiencies in combat." He came up with a solution, diverting "sufficient funds to provide every squad and platoon leader with a brass whistle and every company with a bugle. That was no final answer to the communications problem," he admitted, "but it helped. Aside from practical use, the whistle became a kind of prestige symbol, and the bugle had a second use for ceremonies."[38]

Long after the war, Westmoreland continued to claim that Vietnamization had been his idea. "I became particularly disenchanted with the political strategy in 1967," he told interviewers. "And it was during that time frame that I came up with the strategy of withdrawing our troops and turning over to the Vietnamese."[39] In an oral history for the Military History Institute Westmoreland noted that "Vietnamization was not my term," acknowledging that it was coined later by Secretary of Defense Melvin Laird. "But," he asserted, "I was the one that initiated the concept."

The cupboard of countervailing evidence is richly stocked. Westmoreland's late requests for large numbers of additional U.S. forces for Vietnam—200,000 in 1967 and 206,000 in 1968—head the list. All along Westmoreland had given priority to U.S. forces for modern weapons and combat force multipliers such as close air support, B-52 missions, artillery, and helicopters. Bringing in large numbers of additional U.S. forces to compete for such assets would inevitably move the Vietnamese armed forces backward, not ahead, in their ability to take over the primary role in defending their nation against the communists.

Given Westmoreland's single-minded focus on the main force war and on his personal conduct of it using American units, it apparently never occurred to him that the wiser course of action would have been to give all the good modern gear to the South Vietnamese first, then to U.S. units if there was any left over.

At the very end of his tour in Vietnam, half a year after Westmoreland insisted he had come up with the scheme to Vietnamize the war, an approved priority list for providing equipment to the South Vietnamese refuted such a claim. Priority 1 was U.S. forces in Southeast Asia. Priority 2 was U.S. forces deploying to Southeast Asia. Priority 3 was the training base in the United States. Priority 4 was U.S. forces in Korea. Priority 5 was U.S. reserve forces called to active duty. Not until Priority 6 were South Vietnamese forces so much as mentioned, and even then it was to be materiel dribbled out over time: "Equipment for time-phased modernization of the RVNAF. Equipment for the unprogrammed Civilian Irregular Defense Group improvement and modernization." There were three lower priorities on the list, and last place was also instructive. Priority 9: "Equipment for RVNAF expansion in FY 1969."[40]

In his memoirs, which appeared in 1976, Westmoreland maintained that "only with the advent of the Nixon administration and new officials in the Pentagon was *my strategy* of strengthening the ARVN to stand to the total Vietnamese Communist enemy fully implemented."[41]

General Richard Stilwell provided a definitive view of the actual situation: "I think one of the most significant differences between the Westmoreland tenure and that of Abrams is that, under the former, overriding priority was given to the buildup and sustainment of US forces. And the training, equipment, mothering, helping the ARVN forces took a relative back seat—until Abrams got there."[42] Asked in Douglas Kinnard's survey about Vietnamization efforts "beginning in 1969," 73 percent of responding Army generals who had commanded in Vietnam said that program "should have been emphasized years before."[43] Said South Vietnam's top soldier, General Cao Van Vien, after the war: "The U.S. and the RVN wasted seven valuable years since 1961 by developing the RVNAF in a half-hearted way."[44]

Jeffrey Clarke, later the Army's Chief of Military History, commented in an official history: "Despite all the public relations hoopla, Westmoreland had not yet planned any new role for the South Vietnamese military. The combined campaign plan prepared in late 1967 differed little from its predecessor regarding the employment of South Vietnamese troops. The division of missions between American and South Vietnamese forces remained unchanged."[45]

Another authoritative source, the JCS History, is confirmatory: "The United States had included the strengthening of the Republic of Vietnam Armed Forces (RVNAF) among its objectives since the beginning of its involvement in South Vietnam, but in the period 1965 through early 1968, major US attention was devoted primarily to the conduct of combat operations. It was only after the 1968 Tet offensive, when President Johnson ruled out a further US troop increase in South Vietnam, that the United States undertook serious preparations for eventual South Vietnamese assumption of the combat effort."[46] By that time Westmoreland was on his way home.

15

★

PROGRESS OFFENSIVE

D URING MOST OF 1967 the Johnson administration orchestrated
what came to be known as the "Progress Offensive," a systematic
effort to convince the American people that the war in Vietnam was
being won. Westmoreland became an important part of the campaign,
making several trips back to the United States for speaking engagements and briefings of political leaders.

Even before those trips began, a most revealing flap on the implications and integrity of reporting came to light. Westmoreland had
submitted to General Wheeler, in an early 1967 cable, statistics that
showed the enemy, not allied forces, holding and indeed increasing
the tactical initiative. Wheeler was distraught. "If these figures should
reach the public domain," he wailed, "they would, literally, blow the lid
off Washington." First, an interim solution: "Please do whatever is
necessary to insure these figures are not repeat not released to news
media or otherwise exposed to public knowledge."[1]

Two days later Wheeler followed up with a longer and even more
anguished message. He had been informed that the MACV Periodic
Intelligence Report for January 1967 showed that, of about 385 enemy battalions making contact with friendly forces during the period
1 February 1966 through 31 January 1967, approximately two-thirds

had occurred as a result of enemy initiative. "I must say I find this difficult to believe and certainly contrary to my own impression of how the war has been going during the past six to eight months," he observed.[2]

"The implications are major and serious," Wheeler continued. "Large-scale enemy initiatives have been used as a major element in assessing the status of the war for the President, Secretary of Defense, Secretary of State, Congress, and, to some degree, the press here in Washington." Moreover, "these figures have been used to illustrate the success of our current strategy as well as over-all progress in Vietnam. Considerable emphasis has been placed on these particular statistics, since they provide a relatively straightforward means of measuring the tempo of organized enemy combat initiative. (In cold fact, we have no other persuasive yardstick.) Your new figures change the picture drastically." Thus: "I can only interpret the new figures to mean that, despite the force buildup, despite our many successful spoiling attacks and base area searches, and despite the heavy interdiction campaign in North Vietnam and Laos, VC/NVA combat capability and offensive activity throughout 1966 and now in 1967 has been increasing steadily, with the January level some two and one-half times above the average of the first three months in 1966."[3]

"I cannot go to the President," Wheeler maintained, "and tell him that, contrary to my reports and those of the other Chiefs as to the progress of the war in which we have laid great stress upon the thesis that you have seized the initiative from the enemy, the situation is such that we are not sure who has the initiative in South Vietnam."[4] Of course Wheeler *could* have done that, could have told the President the truth, could have provided him with the information he needed to make informed decisions about the future course of the war. But he did not. Instead he sent his Special Assistant, a general officer, out to Vietnam to confer with Westmoreland about how to make the problem go away.

By 22 March 1967 Westmoreland could report by cable to Admiral Sharp the fixes that had been made. Quoting a memorandum sent to General Wheeler: "Lieutenant General Brown's team and members of my staff have developed terms of reference in the form of new definitions, criteria, formats and procedures related to the reporting of enemy activity which can be used to assess effectively significant trends in the organized enemy combat initiative."[5] General Wheeler could rest easy. They had redefined the problem out of existence.

MEANWHILE THE JOINT Chiefs of Staff were unhappy on other fronts as well, including the most basic of all, the way Westmoreland was fighting the war. This discontent led to establishment as early as October 1966 of a working group charged to develop an alternate course of action for Vietnam. One product, sent to McNamara by the JCS on 23 December 1966, was a draft NSAM on "Strategic Guidelines for Vietnam." The basic observation was that, "while military accomplishments of the past two years provide encouragement, we have not reached a turning point in the over-all effort where successful accomplishment of our objectives is clearly in view." Worse yet, concluded the JCS, "unless the present course is altered, either the U.S. may never attain its objectives or the effort to reach them will be excessively long and costly."[6]

At about this time a novel idea was floated: Westmoreland might be appointed U.S. Ambassador to the Republic of Vietnam. General Wheeler cabled the news to Westmoreland, reporting that "at luncheon today Secretary McNamara told me that over the weekend discussions were renewed concerning the merits of your being designated as Ambassador to Vietnam." Two options had been considered. One involved Westmoreland's taking on the post as a civilian, the other as Commander, U.S. Forces, Vietnam. "I told Secretary McNamara that . . . you personally would be unhappy to give up your military rank," said Wheeler.

McNamara had not reacted well to the "military ambassador" proposal, arguing that because of Westmoreland's personal involvement in the war he would inevitably give too much time to that while neglecting his political and economic responsibilities. Now Wheeler told Westmoreland, "I personally consider what we need in Vietnam is an operation of the MacArthur type wherein you would plan the strategy, both on the military and non-military fronts, insure coordination between the two, and devote your principal energies to the weakest segment of your operation."

Secretary of State Dean Rusk didn't oppose the appointment in principle, but balked at what he viewed as the complete "militarization" of the senior structure in South Vietnam. McNamara, having been lobbied by Wheeler on Westmoreland's reluctance to take off his uniform, counseled against putting Westmoreland in the job as a civilian. The next morning McNamara talked with Wheeler and came away with the understanding that both Wheeler and Westmoreland would prefer to see another civilian ambassador installed, with Westmoreland retaining the military command. That stance on the part of the two officers torpedoed the scheme. A week later Wheeler again cabled Westmoreland: "I learned this morning from Secretary McNamara that the proposal to give you overall control in South Vietnam by making you ambassador as well as commander-in-chief is almost surely dead." There would be no Ambassador Westmoreland.

For a moment, it appears, there was the further prospect that there would be no more COMUSMACV Westmoreland, either. According to columnist Drew Pearson, he met with LBJ a few days after the ambassadorial issue had been resolved, at least insofar as Westmoreland's candidacy was concerned, and was told by the President that "Lodge has got to go" and also that "probably Westmoreland will go."[7]

IN APRIL 1967 Westmoreland was brought back to the United States to help with the Progress Offensive. In New York City he addressed Associated Press Managing Editors at the Waldorf-Astoria while, outside the hotel, protestors burned him in effigy. "We will have to grind him down," Westmoreland told the newsmen, describing how to deal with the enemy in the war of attrition they were fighting in Vietnam. Even so, he admitted, "I do not see any end of the war in sight."[8]

Before going to Manhattan, Westmoreland and his entourage had spent several days at West Point. There he had rehearsed his AP Editors talk, using a television camera with instant playback. General Donald Bennett, the Superintendent, put together a "group of young upstarts, all Vietnam vets." Their mission was to listen to the talk, then ask the hardest questions they could come up with. At the first question, Bennett remembered, Westmoreland "got all insulted and really became angry. We calmed him down, showed him how he looked on the television screen." Then some more questions. "I told him at one point," said Bennett, "that his face was saying one thing and his words another. Showed him that on the screen, too." Afterward the conclusion was that they hadn't succeeded in changing the content of what

Westmoreland had to say, but at least they had helped him polish his delivery.[9]

Moving on to Washington for talks at the White House, Westmoreland reported having reached the point at which his forces were killing enemy troops faster than they could be replaced. "It appears that last month we reached the crossover point," he told the President on 27 April 1967.[10]

Next came Westmoreland's most cherished moment. On 28 April 1967 he addressed a Joint Session of Congress on progress in the war. He was very warmly received, rendered a confident and optimistic report, and was trailed by applause all the way from the Capitol chamber out to his waiting limousine. But that, as Westmoreland's biographer Samuel Zaffiri put it, was all the applause there was: "When it finally died, it died for good."[11]

Westmoreland had told the legislators that the allies were winning the war militarily, citing various statistical indicators of progress, and suggested that withdrawal of U.S. forces could begin within two years. "Given the nature of the enemy," he said, "it seems to me that the strategy we are following at this time is the proper one, and that it is producing results." He assured them, "I command the most professional, competent, dedicated and courageous servicemen and women in our military experience."[12]

Afterward Westmoreland was extremely pleased with his performance, dictating for his history file that "the speech was well received and I was overwhelmed by the ovation. After it had continued for some time I turned around and saluted the Vice President, the Speaker, and both sides of the aisles. My spontaneous reaction brought forth further response." The speech also elicited a charming message from Westmoreland's older daughter, a telegram sent care of Secretary of State Dean Rusk that said simply: "Maximum I love you. Stevie."

Even the gossip columnists, however, were surprised by Westmoreland's return from the combat zone for such a speaking engagement. Wrote Lloyd Shearer, "Westy Westmoreland is the only general in U.S. history who ever left a war he was leading to address a joint session of Congress." But Westmoreland treasured the occasion, later telling a correspondent that "my most memorable moment in my military career was the occasion of my address to a joint session of Congress in April 1967."[13] It seems both ironic and sad that, as he saw it, the most memorable moment in a famous general's military career was a political event.

For some reason Westmoreland later insisted that he had not known beforehand that he would be speaking to a Joint Session of Congress, a claim he made in later press interviews, speeches, correspondence, his memoirs, and his official oral history, saying in the latter, " [A]fter I got back [to the United States] I learned that I was going to talk to a joint session of Congress." But that was incorrect. Westmoreland had advance notice of the impending speech and prepared for it beforehand.

On 24 March 1967, a full month before the trip to the United States, General Wheeler cabled Westmoreland about concerns that during his visit various Congressional elements would want to drag him in for hearings on the war. "The problem," he said, "is to keep you away from all Congressional committees. I believe that the solution will be for you to address a joint session and then depart the country rather promptly."[14]

In Saigon the staff had worked on Westmoreland's speech. Douglas Kinnard, then Chief of the Operations Analysis Branch of MACV J-3, recalled it. "I became aware of this journey when called by Fly Flanagan, working for Westmoreland and drafting the talks. What he needed from me was some ammunition from the Measurement of Progress report. Material that I gave him over the telephone appeared in both speeches," said Kinnard, referring to "the April 24 luncheon of the Associated Press executives in New York and, on April 28, a joint session of Congress."[15]

WESTMORELAND MAINTAINED THAT on these occasions he had not been told what to say and that there was no pressure on him to report good news.[16] Whether or not the pressure was overt, the record does reveal other cases of what might be termed "directed optimism." On 1 March 1967, for example, Wheeler cabled Westmoreland to say that "Highest Authority" wanted him to hold a press conference as soon as possible and to state the results of air operations in "positive terms."

Later, acknowledging such a reality in the course of his libel suit against CBS, Westmoreland would write to his attorney that "at the same time [referring to June 1967], and continuing until the end of the year, there was cable traffic between the State Dept. and the White House and the Ambassador's office to attempt to emphasize the positive side of the war and to make known progress that we in Vietnam thought was being made in the field of pacification and in our efforts to attrite the enemy in the South."[17]

Addressing a university audience in 1972, Westmoreland stated that "during 1967 and 1968 a polarization situation occurred in Saigon. The media were inclined to accentuate the negative. To balance the picture, the U.S. authorities in Saigon were prone and encouraged by Washington to accentuate the positive."[18] Over a decade later, during his libel suit against CBS, Westmoreland reverted to his denial stance. "Was there pressure on you, General Westmoreland, to come up with better news about Vietnam than you were capable of giving at that point?" he was asked. "Well I—I was unaware of the pressure, and I—if I had discerned pressure, I would've—would've really resented it," he maintained.[19]

Westmoreland's accomplishments at this time were, as always, reported by the Class of 1936 Scribe in the pages of *Assembly:* "Westy is developing a very particular image in our history which may stamp him as one of our great wartime commanders. His record: perfect!"

BACK IN SAIGON Westmoreland gave a background briefing for the press and was quickly challenged on his optimistic pronouncements. R. W. Apple of the *New York Times* led the charge, pointing out large increases in enemy strength over the past two years and asking Westmoreland, "How do you make that fit with the general picture of success you've painted?" "Well," ventured Westmoreland, "it is a relative thing. As we entered troops, so did he." Challenged on his claims of having reached the crossover point, Westmoreland took several stabs at it, finally reduced to this: "This numbers game is a very vexing thing—for us all." And: "Frankly, our intelligence is just not that good." And finally: "You have to make certain extrapolations and certain estimates."[20]

IN EARLY JULY Secretary McNamara made another of his periodic visits to Saigon, this one to be marked by two dramatic episodes. In the first, after Westmoreland had given an overview of the current situation, the J-2 briefer addressed the estimated enemy personnel situation. "We hear frequently of the so-called cross-over point," he said. "This is a nebulous figure, composed as you have seen of several tenuous variables. We may have reached the crossover point in March and May of this year, but we will not know for some months."[21] Westmoreland's intelligence staff thereby undercut, in the presence of McNamara, Westmoreland's statement during his last visit to Washington that the crossover point had been reached. McNamara of course perceived the discrepancy, later writing that between 1965 and 1967 Westmoreland had "intensified his pursuit of an attrition strategy aimed at inflicting more casualties on the Vietcong and North Vietnamese than they could replace. But the facts proved otherwise. However much Westy, I, and many others wished differently, the evidence showed that our adversaries . . . expanded their combat numbers substantially. Vietcong and North Vietnamese forces increased in size throughout 1966 and into 1967."[22]

The second notable episode was even more dramatic because it was public. After the deliberations were finished, McNamara gave his usual press conference before he departed. When someone asked about possible further troop increases, his response was newsworthy. We will, as in the past, he assured the press, "provide the troops which our commanders consider necessary." But, he added, "I want to emphasize a corollary, that what is necessary depends on the extent to which we are using effectively the resources we have available to us." That was breaking news, and McNamara added emphasis: "We have over a million men here under arms and there are many, many opportunities open to us to increase the effective use of those men, and we will set our minds and hearts to doing that."[23] This constituted a stunning revelation that the administration's patience with Westmoreland was finally running out.

Westmoreland had returned to the United States before McNamara's visit ended, a trip precipitated by his mother's death. After attending her funeral, Westmoreland made his way to Washington and the inevitable press conference. "Do you agree with Secretary McNamara's statement that some parts of your forces are not being used efficiently?" he was asked. Westmoreland refused to validate the premise of the question. "I don't think Secretary McNamara made that statement," he responded.

But they had McNamara's statement on tape, including his observation that "more effective use" could be made of the forces already in Vietnam. Westmoreland reverted to a fallback position: "At least he has not discussed it with me." The reporter switched to Westmoreland's own views: "Do you feel, sir, that your force is being used as efficiently as possible?" "I do, indeed," he said, launching into a lengthy dissertation on logistical buildup and base development. "Are you annoyed, sir, that this feeling has gotten so widespread that you weren't using your troops as efficiently as possible?" "Well, I'm—I don't accept that this has been interpreted. Because he did not discuss it with me."

Regardless of the semantics, the days of automatic approval of Westmoreland's troop requests had clearly come to an end.

WHEN CREIGHTON ABRAMS was sent out in May 1967 to be Westmoreland's deputy, it was planned that he would succeed to the top command within a matter of weeks. In anticipation of that, he was allowed to choose his own chief of staff and to bring that officer with him when he traveled to Saigon. Abrams chose Major General Walter T. "Dutch" Kerwin Jr., who did become MACV chief of staff almost right away.[24] Meanwhile Abrams languished in the deputy's slot for over a year.

There were difficult challenges for Kerwin in the assignment. One was that, as he later stated it, "Westy was a man of *very* strong opinions. Once he glommed onto something, it was difficult for him—once committed—to modify it or drop it." Veteran journalist George McArthur noted the same trait. "Westy was so convinced he was right on everything," he said. "Westy really thought he was God's gift to the military."[25] Even more awkward for Kerwin was that there were things Westmoreland told him not to reveal to General Abrams, even though Abrams was the Deputy COMUSMACV. Kerwin found it a long, hard year.

Initially it had not looked as though there would be such a lengthy delay in the change of command. Brigadier General John Chaisson had written to his wife on 20 May 1967 that "I have no doubt whatsoever that Westy will leave this summer. The air is filled with the feeling of change."[26] Former Deputy Secretary of Defense Cyrus Vance confirmed that "when Abe went out to Vietnam as deputy it was planned that he would succeed Westmoreland within two or three months. I remember talking with Bob McNamara and saying we had to get Abrams out there."[27]

Veteran reporter Keyes Beech said, "I can confirm that we expected Abrams to take over soon after arrival. I did a profile on Westmoreland after Tet, and I remember his saying: 'You know, I was supposed to leave before this. Now I'm going to be looked on as the guy who lost the war.'"[28]

When the succession did not happen as anticipated, Major General Phillip Davidson, Westmoreland's J-2, thought he knew why and described it to some of his staff officers during a briefing rehearsal. Recalling McNamara's July 1967 trip to Saigon, Davidson suggested the issue of effective utilization of troops had been the hidden agenda item for the whole visit. The central question was, he said, "What are you doing with your own troops?" That was the theme. "Mr. McNamara bad-mouthing General Westmoreland's manpower conservation and use, I'm confident, kept poor General Westmoreland over one more year than he was supposed to have been kept over. And, in effect, led him down the Tet path, et cetera, et cetera. I'm absolutely convinced of it."[29]

Columnist Jack Anderson reported similarly, noting that earlier "the White House had every intention of relieving Gen. Westmoreland of the South Vietnam command," and that accordingly "Gen. Creighton Abrams was sent to Saigon to be on hand to relieve him. After the widely publicized reports of differences between Westmoreland and McNamara, however, the general will not be relieved. It would be interpreted as a slap."[30]

The President may also have realized that, while Westmoreland was willing to play his assigned role in the Progress Offensive, it was not assured that a successor would be similarly amenable.

There is one more possible reason for LBJ's delay in replacing Westmoreland, which, while entirely speculative, has some plausibility given the source. "I got to know LBJ pretty well," Westmoreland told John Raughter, editor of the *American Legion Magazine*. "He had good personal instincts and, early on, thought I was after his job. Once I assured him that I wasn't, we got along fine."[31]

In 1967 the South Carolina General Assembly had adopted a resolution to "prevail upon" their native son Westmoreland to become a candidate for the presidency. A cautious and canny politician, Johnson may well have decided to take no chances. By leaving Westmoreland 12,000 miles away and under his authority as commander-in-chief, he nullified any chance of having to confront him in the political arena.

MORE PROGRESS OFFENSIVE markers had been laid down by Westmoreland during his July 1967 visit to Washington, where at a press conference he asserted that "the statement that we are in a stalemate is complete fiction. It is completely unrealistic. During the past year tremendous progress has been made."[32]

In contrast, just two months earlier LBJ had stated publicly his conclusion that the war was a "bloody impasse."[33] During that same month Westmoreland himself laid out a radically more pessimistic view. At a MACV commanders' conference he presented an assessment including the admission that "the main force war is accelerating at a rapid, almost alarming, rate. The enemy is reinforcing his four main force fronts with people and weapons."[34] The conclusion is inescapable that Westmoreland had not believed all those positive things he proclaimed back in the United States and thus deserved the subsequent loss of credibility and collapse of his reputation after the Tet Offensive.

By late summer Wheeler was cabling Westmoreland and Sharp with a warning. "Both of you should know," he cautioned, "that there is deep concern here in Washington because of the eroding support for our war effort."[35] Despite all the rosy reports and forecasts, the Progress Offensive was, like the allied offensive in Vietnam, failing to gain its objectives. Westmoreland again got on the bandwagon. "General Westmoreland has become increasingly concerned with the fact that the U.S. press is painting a pessimistic, stalemated situation in RVN," said General Bruce Palmer in a 19 August 1967 message to General Johnson. "To counteract this distorted impression of the true situation, he is launching a local campaign to portray and articulate the very real progress underway in the Vietnamese War."[36] Far from being the reluctant participant he sometimes claimed to be, Westmoreland was opening his own branch office of the Progress Offensive. He himself reported these plans to General Wheeler and others, noting in early August 1967 that "of course we must make haste carefully in order to avoid charges that the military establishment is conducting an organized propaganda campaign, either overt or covert."

Thus, "while we work on the nerve endings here we hope that careful attention will be paid to the roots there—the confused or unknowledgeable pundits who serve as sources for each other."[37]

MORE OF THE DEEP concern reported by Wheeler surfaced in an August 1967 *New York Times* article by R. W. Apple, "Vietnam: The Signs of Stalemate." Murray Fromson, then based in Saigon as a CBS correspondent, had met an officer—identified only as "an American general"—who told him, "Westy just doesn't get it. The war is unwinnable. We've reached a stalemate and we should find a dignified way out." Subsequently that officer met for further discussion with Fromson and one other reporter of his choice. Fromson selected Apple, and later both reporters filed stories based on that discussion. "I've destroyed a single division three times," said their host. "I've chased main-force units all over the country and the impact was zilch. It meant nothing to the people. Unless a more positive and more stirring theme than simple anti-communism can be found, the war appears likely to go on until someone gets tired and quits, which could take generations."[38]

Among those outraged by these anonymous views were LBJ, Wheeler, and of course Westmoreland. Wheeler sent out an anguished cable, to which Westmoreland responded that "no general of mine would ever have said that." One had, however, a very senior, experienced, and highly regarded general. His identity remained secret for nearly four decades, even though earlier the reporters had asked to be released from their pledge of confidentiality. "Westy is an old friend," said their informant, "and I would not want to hurt or embarrass him. Let's wait until he's no longer with us." Wait they did until Westmoreland and Apple were both gone. Then Fromson again contacted the source, and this time General Fred Weyand agreed to be publicly identified.[39]

Meanwhile, also in August 1967, Ambassador Bunker submitted a paper entitled "Blueprint for Viet-Nam" which included this assessment: "We still have a long way to go. Much of the country is still in VC hands, the enemy can still shell our bases and commit acts of terrorism in the securest areas, VC units still mount large scale attacks, most of the populace has still not actively committed itself to the Government, and a VC infrastructure still exists throughout the country."[40] That was what Westmoreland had to show for his more than three years in command, and that was the reality as opposed to the version being retailed to the public.

WESTMORELAND MADE YET another trip to the United States, his third of the year, in November 1967. He got off to an optimistic start at a press conference on his arrival at Andrews Air Force Base, where he was asked how he saw the situation in South Vietnam. "Very, very encouraged," he responded. "I've never been more encouraged during my entire almost four years in country. I think we are making real progress."

He appeared on NBC's *Meet the Press*, where he claimed that "we are winning a war of attrition now," adding that the enemy was having "very serious" manpower problems in South Vietnam and that "this manpower cannot be replaced." That last was a particularly unfortunate attempt to claim the crossover point, but in no way accurate. During this broadcast Westmoreland also held out the hope of diminished American involvement in the war, saying, "I believe it is quite conceivable that within two years or less we can progressively phase-down the level of our commitment."[41]

Another crucially important appearance that November was a widely reported speech at the National Press Club in Washington. "In that address," said Westmoreland, "I permitted myself the most optimistic appraisal of the way the war was going that I had yet made."[42] The Press Club talk contained a number of dramatic assertions, including this: "I am absolutely certain that whereas in 1965 the enemy was winning, today he is certainly losing. There are indications that the Vietcong and even Hanoi may know this." And: "The enemy's hopes are bankrupt." And: "We have reached an important point where the end begins to come into view."[43]

In response to a question, Westmoreland made his most extreme claim yet, arguing that "this body count figure which we've reported is, in my opinion, very, very conservative. Probably represents, I would say, 50% or even less, of the enemy that has been killed."[44]

PETER BRAESTRUP NOTED not only how Westmoreland had been co-opted, but how historically unique that was. "For the first time in American history," he wrote, "a field commander, Westmoreland, had allowed himself to be snookered into becoming a political spokesman. It was his vanity. He loved being on TV and he came home twice at Johnson's behest to speak, and it tainted him not only in the eyes of the press but in the eyes of a lot of military men. Westmoreland had in effect taken the king's shilling and become a propagandist—a soldier for the administration."[45]

Braestrup was right about the reaction of other military men. Some of Westmoreland's senior colleagues were quite uneasy about his being used in this way. General Harold K. Johnson, the Army Chief of Staff, cabled General Abrams in Vietnam. "Westy's trip has gone extremely well," he reported, "and I only hope that he has not dug a hole for himself with regard to his prognostications. The platform of false prophets is crowded!"[46] General Bruce Palmer was equally concerned. "Since it was obvious that Westmoreland was being used for political purposes," he wrote, "many of us in Vietnam at the time resented having our field commander put on the spot in this manner. Westmoreland enjoyed these occasions, however, and would return to Saigon still 'up on cloud nine.'"[47] General "Dutch" Kerwin saw it the same way. "I think he enjoyed it," he said of Westmoreland's trips and talks. "He liked to be a part of things at the higher level. He was in the spotlight. He seemed to be rejuvenated each time he came back."[48]

Westmoreland's participation in the Progress Offensive was costly to him, both personally and professionally. "By playing a prominent role in [the optimism campaign], General Westmoreland, who previously had enjoyed much respect as a nonpolitical, professional military leader, became in the eyes of the press simply another pitchman for the administration line," wrote Army historian Graham Cosmas. "From then on, his command's assessments, no matter how valid they might be, would be received at best with skepticism."[49]

COLONEL (LATER GENERAL) Donn Starry served in Vietnam in several posts, first in the headquarters known as U.S. Army, Vietnam. There he had some fascinating experiences, including being part of a study group charged in 1966 to calculate what it would take to win in Vietnam. Westmoreland had asked for this following McNamara's visit in July of that year. They concluded it would take a million and a half men—500,000 U.S. and 1,000,000 South Vietnamese—and ten years. That infuriated Westmoreland, who refused to send in the estimate. "He said it was politically unacceptable," recalled Starry. "But the Secretary of Defense's staff kept urging MACV to respond to the Secretary's question, so finally Westmoreland sent it, but with a disclaimer. He said the war was going to be over in the summer of 1967."[50]

Later Westmoreland challenged Starry's assertion that he had said he could end the war by 1967. "I'm sorry, General," Starry replied, "I was there and heard you say it several times."[51]

That prediction turned out to be another of the positions Westmoreland worked hard to distance himself from in later years. In his memoirs he cites numerous documents and conversations in which he predicted a long war. Be that as it may, in 1965 he had promised otherwise. "At this writing," indicated one of the *Pentagon Papers* authors, referring to 1967, "the U.S. has reached the end of the time frame estimated by General Westmoreland in 1965 to be required to defeat the enemy." Thus "the strategy remains search and destroy, but victory is not yet in sight." Fox Butterfield commented, in the *New York Times* compilation of the Pentagon Papers, that "according to the Pentagon study, General Westmoreland's plan shows that 'with enough force to seize the initiative from the VC' sometime in 1966, General Westmoreland expected to take the offensive and, with appropriate additional reinforcements, to have defeated the enemy by the end of 1967."[52]

According to Alain Enthoven, Assistant Secretary of Defense for Systems Analysis, when those months had elapsed it was clear that Westmoreland's approach had failed. "We know that despite a massive influx of 500,000 US troops, 1.2m tons of bombs a year, 400,000 attack sorties a year, 200,000 enemy killed in three years, 20,000 US killed, our control of the countryside and the defense of urban areas is now essentially at pre-August 1965 levels," Enthoven wrote. "We have reached a stalemate at a high commitment."[53]

"THE CULT OF OPTIMISM fostered in Saigon and Washington," observed Army historian Jeffrey Clarke, "was self-defeating and, in the end, only encouraged the continuation of policies and practices that had little hope of success."[54]

As 1967 neared an end David Halberstam wrote a powerful essay, published in *Harper's Magazine*, as an antidote to the official optimism to which Westmoreland had lent himself throughout the year. Halberstam had talked with a rural pacification official, described by him as "a competent American professional" who had been working in Vietnam for four years, and quoted him: "We are losing. We are going to lose. We deserve to lose."[55]

As the year of the Progress Offensive drew to an end, President Johnson visited the troops at Cam Ranh Bay in the Republic of Vietnam. There he observed that "all the challenges have been met. The enemy is not beaten, but he knows that he has met his master in the field. For what you and your team have done, General Westmoreland, I award you today an oak leaf cluster."[56] Back in Washington, meeting at the White House with General Wheeler and others, the President observed: "I like Westmoreland. . . . Westmoreland has played on the team to help me."[57]

16

★

ORDER OF BATTLE

DURING MOST OF 1967, in ironic parallel with the Progress Offensive, there raged a fierce struggle within the Intelligence Community, spilling over into the decision-making apparatus on the war, a struggle over enemy order of battle. "Order of Battle" is a military term of art having to do with the enemy's estimated strength and organization.

The disagreement centered primarily on the number of enemy forces, an argument of great significance to Westmoreland in terms of being able to demonstrate that he was making "progress" in his prosecution of the war and that he was nearing the point of attriting enemy forces faster than they could be replaced.

MACV J-2, the intelligence staff, was responsible for estimating and reporting enemy order of battle.[1] Besides the numbers themselves, the other most pertinent, and contentious, element was how the enemy was organized and how the troops were distributed in the constituent categories. Intelligence on the various categories was considered of decreasing reliability: best on main forces, then descending to least on the shadowy irregulars. In August 1966 a CIA memorandum entitled "The Vietnamese Communists' Will to Persist" stated that "recently acquired documentary evidence, now being studied in detail, suggests that our holdings on the numerical strength of these irregulars (now carried at around 110,000) may require drastic upward revision."[2]

In a retrospective analysis of the year 1966, Westmoreland acknowledged that "the enemy continued to build up his forces by recruiting in the South and infiltrating from the North." As a result, "his total strength exceeds 280,000." That was despite enemy losses of "at least 50,000 killed" during the year and compared with an estimated strength of approximately 240,000 at the beginning of the year. The analysis also gave estimates of monthly enemy infiltration into South Vietnam from the north. It was clear that the crossover point had not yet been reached. In fact, enemy strength was going in the other direction, increasing rather than dropping.

Shortly after the MACV analysis was forwarded to Washington, Walt Rostow wrote a note to the President: "I do not for one minute believe the infiltration rate is 8,400 per month. I believe it is a MACV balancing figure to give them what I strongly suspect is an inflated order of battle. They are being excessively conservative both as an insurance policy and to protect themselves against what they regard as excessive pressure to allocate more forces to pacification."[3] Since Rostow of course had no independent knowledge of the validity of the figures he was questioning, what he was really saying was that Westmoreland's reporting could not be trusted.

IN EARLY FEBRUARY 1967 elements of the Intelligence Community met in Honolulu for a week-long conference on enemy order of battle methodologies and estimates. Among the very important conclusions was "acknowledgement that the irregulars—guerrilla, SD [Self-Defense], and SSD [Secret Self-Defense]—and political categories carried unchanged at, respectively, 112,000 and 39,000 since May 1966, had been too low and would have to be revised upward."[4]

Much else was accomplished at this important conference, including unanimous agreement that Self-Defense and Secret Self-Defense forces, along with the political order of battle, should continue to be included in the enemy order of battle. Describing these results later, Major General Joseph McChristian, at that time the MACV J-2, was asked for his professional judgment on whether the Self-Defense and Secret Self-Defense forces belonged in the order of battle. "It was my strong conviction from the beginning that they were definitely a force who could and who did adversely affect the accomplishment of the commander's mission and should be in the order of battle," he replied.[5]

In the spring of 1967 work began in Washington on what is called a SNIE, or Special National Intelligence Estimate, a document to be developed collegially by the various members of the Intelligence Community. MACV and CINCPAC were also participants. The topic was "Capabilities of the Vietnamese Communists for Fighting in South Vietnam," or in other words the enemy order of battle. Early in the process it became clear that unanimity was going to be hard to come by. Led by CIA, civilian agencies had begun to believe that the true enemy numbers were greater, perhaps much greater, than MACV was reporting. MACV clung tenaciously to its much lower figures, arguing that, as the people on the ground, they were entitled to specify the counting rules. CIA's George Allen later observed that "the higher numbers [in the draft SNIE] proposed for the enemy's order of battle in South Vietnam sparked one of the most heated and prolonged controversies in the history of American intelligence."[6]

Simultaneously MACV, in the person of Westmoreland himself, was trying to change the longstanding basis for the order of battle calculations. Although they had been a component of the order of battle since before Westmoreland arrived in Vietnam, and MACV had continued to carry them for some three years under Westmoreland's command, he now decided that the enemy's Self-Defense Forces and Secret Self-Defense Forces no longer belonged in the tabulations. In earlier statements he had described these forces as dangerous, only the year before observing that "you have your political cadres that are non-military, but play a very important role in their war of insurgency."[7] But now he said otherwise.

There was a larger and indeed fundamental issue involved in defining, assessing, and enumerating various elements of the enemy's apparatus, and it went to the very essence of the conflict. Ronald Smith, a senior officer in CIA's South Vietnam analytical branch, was one who understood this, maintaining that MACV's manipulation of the order of battle was so serious that it "misrepresented the very nature of the war we were fighting."[8] That essential point was later emphasized in a comprehensive analysis of the war prepared on contract by the BDM

Corporation with a view to identifying strategic lessons learned: "MACV's preoccupation with viewing the OB in classic military terms prevented the command from assessing the enemy in the context of a much broader people's war, in which the enemy mobilized civilians to assist his efforts."[9]

And, concluded BDM's analysts, "many military intelligence personnel and commanders reflected an unfortunate lack of appreciation of the importance of the VCI [Viet Cong infrastructure] in the communists' scheme of things. Further, White House insistence on showing an enemy OB under 300,000 contributed to the obdurate position on OB taken by MACV J-2 and by senior DIA officials."[10]

WESTMORELAND WAS VERY clear that it was his own decision to remove the Self-Defense Forces and Secret Self-Defense Forces from the order of battle. In an apparent effort to justify the excisions, he now began to refer to the "military" order of battle. Later still he took to calling the document the "so-called" order of battle. Colonel (later Major General) George Godding, who was then in MACV J-2, confirmed what Westmoreland himself had said: "It was General Westmoreland's decision to take the Self-Defense Forces and Secret Self-Defense Forces out of the enemy order of battle. We basically accepted this."[11]

Removing these two categories from the order of battle had a huge and immediately beneficial impact—on paper—on reaching the crossover point. David Maraniss described the process aptly as an exercise in "political mathematics." In a postmortem assessment conducted in the aftermath of the 1968 Tet Offensive, the President's Foreign Intelligence Advisory Board reported to the President that "MACV's method of bookkeeping on enemy strength, unfortunately, had been designed more to maximize the appearance of progress than to give a complete picture of total enemy resources."[12]

This new Westmoreland outlook emerged at just the time special studies undertaken by MACV J-2 under General McChristian's guidance had come up with the most accurate and best estimates yet of enemy forces in the very categories Westmoreland wanted to eliminate. Begun in the autumn of 1966, a study known as "Ritz" concentrated on the enemy's irregular forces, while a companion assessment called "Corral" dealt with political cadres. At the end of May 1967 Colonel

Gains Hawkins, MACV's top order of battle specialist, briefed Westmoreland on the results of these two studies, significantly greater
numbers in both categories than had previously been carried.

According to Hawkins, Westmoreland reacted with alarm. "What
will I tell the President?" he reportedly asked. "What will I tell the
Congress? What will be the reaction of the press to these high figures?"[13] Later Westmoreland admitted in a deposition that he had reacted to the briefing by observing, "We've got a public relations problem here." When McChristian brought Westmoreland a proposed
cable reporting the new figures to higher headquarters, he recalled,
Westmoreland told him, "If I send this cable to Washington it will create a political bombshell." Westmoreland declined to approve it, instructing McChristian to leave the cable with him. This was, said Mc-
Christian, "the first time he had ever questioned my intelligence."[14]
Nevertheless McChristian was not surprised, saying of Westmoreland: "I don't think he ever knew anything about intelligence. I don't
think he ever got interested in it. I don't think he appreciated it."[15]

With the new numbers—198,000 in the irregular category and
88,000 for the political order of battle—the order of battle total
would have reached 429,000. That compared to just under 300,000
then being reported by MACV.[16]

In a later interview Westmoreland said of the new estimate: "I did
not accept his recommendation. I did not accept it. And I didn't accept
it because of political reasons—that was—I may have mentioned this,
I guess I did—but that was not the fundamental thing: I just didn't accept it." Asked to explain the political reason, Westmoreland said, "Because the people in Washington were not sophisticated enough to understand and evaluate this thing, and neither was the media."
Westmoreland later said of the new estimate, "I was not about to send
to Washington something that was specious. And in my opinion, it was
specious."[17]

CIA in particular resisted the idea of taking whole categories out of
the order of battle, arguing that doing so "could give consumers a misleading impression that the enemy forces had been reduced." Wrote
George Allen, "This fundamental controversy posed an extraordinary
dilemma for the intelligence community, whose embarrassment was
compounded when word of the dispute was leaked to the press. This
development confounded top levels of the administration, then fully

engaged in the campaign to demonstrate progress in the war and thereby eliminate Vietnam as a divisive issue in the coming presidential election."[18]

Westmoreland sought to blame a single CIA analyst, Sam Adams, for the new numbers. But that charge was both inaccurate and unfair. "In 1967–1968," wrote veteran senior Agency official Hal Ford in a detailed scholarly analysis, "available evidence convinced virtually all CIA officers that the enemy had additional tens of thousands of irregular troops that were militarily significant, but which MACV would not count. CIA's concerns on these scores were validated when the enemy employed such irregulars in great numbers in his 1968 Tet Offensive, and despite very heavy casualties was able simultaneously to conduct major operations at Khe Sanh, Hue, and elsewhere."[19]

AT A CRUCIAL JUNCTURE in this unfolding drama, Brigadier General Phillip B. Davidson Jr. arrived in Saigon to replace Major General McChristian as MACV J-2. Davidson moved quickly to take personal control of the relevant order of battle figures, stating in a 15 August 1967 memorandum to his staff that they have got to "attrite main forces, local forces, and particularly guerrillas. We must cease immediately using the assumption that these units replace themselves. We should go on the assumption that they do not, unless we have firm evidence to the contrary. The figure of combat strength, and particularly of guerrillas, must take a steady and significant downward trend, as I am convinced this reflects true enemy status." And, he added, "due to the sensitivity of this project, weekly strength figures will hereafter be cleared personally by me."[20] Those instructions were a sure sign of trouble ahead.

Later General McChristian remembered that fateful development. "DePuy and Chaisson asked me to change the reporting to indicate that we had reached the crossover point," he recalled. "We did not reach it. Phil Davidson came in after me and said that we had."[21]

Eventually CIA sent a senior officer, George Carver, out to Saigon to try to resolve the impasse on what enemy strength figures would be included in the SNIE. Carver soon cabled Director Richard Helms to report his mission as "frustratingly unproductive since MACV stonewalling obviously under orders," leading to the "inescapable conclusion that [MACV] has been given instruction tantamount to direct order that VC total strength will not exceed 300,000 ceiling. Rationale seems to be that any higher figure will not be sufficiently optimistic and would generate unacceptable level of criticism from the press."[22]

Helms later characterized Carver's mission to Saigon as a "vitally important" order of battle conference and the conflict he faced as "mean and nasty," but before it was over CIA had given in completely to Westmoreland's demands. Even though CIA's position, according to Helms, was a calculated enemy strength "of about 500,000—with some . . . analysts holding out for an even more substantial figure," after meeting one-on-one with Westmoreland Carver reported "circle now squared," which turned out to mean that CIA accepted the MACV numbers. Helms could accommodate that. Carver, he said later, "already knew my basic views: that because of broader considerations we had to come up with agreed figures, that we had to get this O/B question off the board, and that it didn't mean a damn what particular figures were agreed to."[23]

In his memoir, published posthumously, Helms acknowledged that there was "a significant political problem," one he attributed to the likelihood that, "in view of the continuing increase in U.S. personnel and armaments in South Vietnam, any admission that the Viet Cong were actually gaining strength would obviously have stirred a severe public reaction on the home front."[24]

Helms was sympathetic to the plight of MACV's intelligence officers dealing with the order of battle controversy. "In effect," he concluded, "the commanding general MACV had taken a 'Command Decision' as to the facts bearing on the O/B problem, and his subordinates had no choice but to fall in line."[25]

Later Hal Ford published a case study in which he concluded that "the most important regulator of the MACV O/B estimates was the fact that General Westmoreland and his immediate staff were under a strong obligation to keep demonstrating 'progress' against the Communist forces in Vietnam. After years of escalating US investments of lives, equipment, and money," said Ford, and "of monthly increases in MACV's tally of enemy casualties, and of vague but constant predictions of impending victory, it would be politically disastrous, they felt, suddenly to admit, even on the basis of new or better evidence, that the enemy's strength was in fact substantially greater than MACV's original or current estimates."[26]

GEORGE GODDING FROM MACV J-2 represented the headquarters at the Washington conference convened to develop the new SNIE. In mid-August 1967 General Davidson cabled him from Saigon to report that "the figure of about 420,000, which includes all forces including SD and SSD, has already surfaced out here. This figure has stunned the embassy and this headquarters and has resulted in a scream of protests and denials."[27] The 420,000 represented the order of battle total that would have resulted from including the new figures for SD and SSD derived from the Ritz and Corral studies. Clearly any claims of having reached the crossover point were going to be cataclysmically undermined if the new estimates in those categories were to be accepted and those categories themselves continued to be part of the order of battle.

Thus Davidson put it to Godding pretty straight: "In view of this reaction, and in view of General Westmoreland's conversations, all of which you have heard, I am sure this headquarters will not accept a figure in excess of the current strength figure carried by the press."[28] That figure was 292,000, a number that had recently appeared in the *Wall Street Journal,* and Godding was being given a ceiling he was not to let the new SNIE total exceed. Much later shilly-shallying would be devoted to denying that fact, but the language of Davidson's instructions to Godding was unambiguous.

Davidson took the precaution of protecting Westmoreland. "Let me make it clear," he added, "that this is my view of General Westmoreland's sentiments. I have not discussed this directly with him but I am 100 per cent sure of his reaction."[29]

Colonel Gains Hawkins was also part of the MACV delegation to the August 1967 SNIE conference at Langley. Later, under oath, he answered two crucial related questions. First, when he went to that conference, "Did you have any understanding as to whether at that time there was a MACV command position with respect to enemy strength estimates?" Hawkins: "Yes, sir, I did." And second, "What was that understanding?" Hawkins: "That there was a ceiling of 300,000 and we would not exceed that ceiling."[30]

Hawkins also addressed that central issue in a deposition: "I knew that when we went on that trip that there was a ceiling that we weren't going to get above and General Godding was perfectly aware that the ceiling was there, and he corroborated that when he called me at my home after the CBS documentary. That there was a ceiling that we had to stay under."[31]

Eventually, of course, Helms and the CIA conceded the argument to MACV, and SNIE 14.3-67, *Capabilities of the Vietnamese Communists for Fighting in South Vietnam*, was published in November 1967. It was apparent that MACV had prevailed, not only in the numbers and the categories included in the order of battle, but in the central conclusion: "Manpower is a major problem confronting the Communists. Losses have been increasing and recruitment in South Vietnam is becoming more difficult. Despite heavy infiltration from North Vietnam, the strength of the Communist military forces and political organizations in South Vietnam declined in the last year." The crossover point had been reached, or so it had now been decreed by the Intelligence Community.

EN ROUTE BACK TO Vietnam from his final Progress Offensive foray of the year, Westmoreland sent a cable to his deputy distilling the message he had been promulgating in various forums. His presentation, said Westmoreland, had been "compatible with the evolution of the war since our initial commitment and portrays to the American people 'some light at the end of the tunnel.'"[32]

17

★

KHE SANH

IN MID-DECEMBER 1967 Westmoreland confided in his history notes: "My analysis is that the enemy's next major effort will be in I Corps, and I believe he will be prepared to initiate this by next January." In late January 1968 he cabled Washington: "I believe that the enemy will attempt a country-wide show of strength just prior to Tet, with Khe Sanh being the main event."[1]

To Westmoreland I Corps meant preeminently Khe Sanh, the remote and primitive base near South Vietnam's western border with Laos, a place of consuming interest to him. Later he wrote that "it was important to avoid trying to hold positions too close to the Laotian and Cambodian borders, for in view of the proximity of large enemy forces just across the border, that would have been inviting trouble."[2] Yet Khe Sanh, and nearby Lang Vei, were exactly such places. Westmoreland told General Wheeler that Khe Sanh was important as a base for clandestine teams operating cross-border in Laos and to provide what he termed "flank security" for the strongpoint obstacle system, but that it was "even more critical from a psychological viewpoint. To relinquish this area would be a major propaganda victory for the enemy."[3]

Since mid-autumn Lyndon Johnson had had a sense of foreboding about Khe Sanh, fearing it could turn out to be an American version of the French defeat at Dien Bien Phu. He had a replica of the Khe Sanh plateau and the base situated there built and placed in the White House Situation Room, and reportedly demanded from each member of the Joint Chiefs of Staff a signed affirmation of his belief that the base could successfully be defended. The base was garrisoned primarily by the 26th Marines, a regimental-sized unit commanded by Colonel David Lownds, with other U.S. forces and a South Vietnamese Ranger group alongside. For weeks the position had been cut off by road, meaning all support had to be delivered by air drop or by aircraft landing there.

At MACV Westmoreland established a task force to monitor what was going on at Khe Sanh, then directed that someone from the task force go up there every day. "I went first," said Colonel "Hap" Argo. "The Marines were not dug in. Colonel Lownds was there. What a loser, what a loser! A complete loser. They were afraid to relieve him. The Marines were afraid they'd lose face if they relieved him. 'Prop him up!' was their approach."[4]

Westmoreland, though outwardly confident of Khe Sanh's ability to defend itself, asked Argo, the MACV historian, to study and compare Dien Bien Phu and Khe Sanh. When Argo reported his findings, said Westmoreland, "he gave a gloomy presentation" that "stunned" the staff. Afterward Westmoreland, he later wrote of himself, took charge, "deliberately getting the attention of all," speaking in a "firm voice," and insisting that the allies would not be defeated at Khe Sanh. Then he "strode deliberately from the room." The tenor of this account makes it clear that Westmoreland was proud of how he had handled the matter.[5]

The Marines did not want to be at Khe Sanh, and senior Marine officers such as Lieutenant General Victor Krulak were frustrated and angry about Westmoreland's fixation on the place. "Like the people who were bearing the brunt of it," wrote Krulak later, "I hated the bad choice that put them there."[6] To Krulak, Khe Sanh was "a tactical albatross."[7]

Westmoreland sent a tutorial cable to Marine Lieutenant General Robert Cushman, Commanding General of III Marine Amphibious Force, expressing his deep concerns about Khe Sanh. "I keep getting recurring reports that the Marines are not digging in at Khe Sanh.

Most of these reports are from the press and I am sure that there is the usual exaggeration, but I want to be sure that you are aware of this." Continuing, "I cannot overemphasize the importance of continuous improvement of your position at Khe Sanh, and that this requires strong and constant command attention. The chain of command must constantly emphasize this important matter to insure that troops are given the maximum protection and capability to fight under strenuous conditions."[8]

The enemy's buildup around the position eventually amounted to most of two divisions, positioned on overlooking hill masses that gave their artillery and rockets excellent vantage points for pouring near-continuous fire into Khe Sanh. General William DePuy wrote after the war that the communist forces at Khe Sanh "were invisible except to Marine patrols in contact and, eventually, aerial observation of encircling trench lines." The enemy, he said, "fought by stealth and silent encroachment." DePuy also calculated the extraordinary costs of achieving the claimed casualties inflicted on the enemy. "If 50 percent of all enemy combat deaths are attributed to air attack (a generous allocation)," he wrote, "the return on investment in air power was one kill for every three sorties, including those of B-52s."[9]

On 20 January 1968 the Battle of Khe Sanh commenced with strong enemy attacks by fire against the combat base. Three days later Westmoreland sent a message to Sharp and Wheeler, trying to head off any suggestion that Khe Sanh should be evacuated. "I unreservedly maintain that Khe Sanh is of significance, strategic, tactical,.and most importantly, psychological," he insisted.

Westmoreland understood that the key to successfully holding off enemy forces at Khe Sanh was allied air power. He instructed General William Momyer, his Air Force deputy, to plan such defenses. "I gave it the code name Niagara," said Westmoreland, "to invoke an image of cascading shells and bombs." There was plenty of cascading, all right, with by one account over 24,000 fighter-bomber sorties and 2,700 by B-52s dropping a total of 110,000 tons of bombs during the enemy's seventy-seven-day siege.[10] Westmoreland was gratified by the results, predicting in his memoirs that "Khe Sanh will stand in history, I am convinced, as a classic example of how to defeat a numerically superior besieging force by co-ordinated application of firepower."[11] For his part General Momyer simply observed that "enemy attacks never developed to more than a large scale probing maneuver."

Westmoreland added one more thing: "There was another possibility at Khe Sanh: tactical nuclear weapons."[12]

He was serious about it. "If Washington officials were so intent on 'sending a message' to Hanoi," he wrote, "surely small tactical nuclear weapons would be a way to tell Hanoi something." Washington was having no part of it, however. "Although I established a small secret group to study the subject," Westmoreland reported, "Washington so feared that some word of it might reach the press that I was told to desist. I felt at the time and even more so now that to fail to consider this alternative was a mistake."[13]

"No single battle of the Viet Nam war has held Washington—and the nation—in such complete thrall as has the impending struggle for Khe Sanh," offered *Time* magazine in mid-February.[14] Actually, though, Khe Sanh had by then more or less ebbed away, not producing the climactic set piece battle apparently envisioned, perhaps even hoped for, by Westmoreland, and certainly not turning out to be determinative in any way for either side. That had to be disappointing, for Westmoreland had earlier told Sharp and Wheeler that in his opinion "the confrontation in Quang Tri may well be the decisive phase of the war."[15] Even so, Westmoreland viewed himself as having bested his opponent, writing in his memoirs that Khe Sanh "evolved as one of the most damaging, one-sided defeats among many that the North Vietnamese incurred, and the myth of General Giap's military genius was discredited."[16]

Afterward Westmoreland stated that "no single battlefield event in Vietnam elicited more public disparagement of my conduct of the Vietnam war than did my decision to stand and fight at Khe Sanh."[17] He sounded rather pleased about it. In an interview published two decades later he was asked which of the decisions he made as commander in Vietnam he was most proud of. "The decision to hold Khe Sanh," he responded.[18]

When it came time to assess what had been accomplished, though, Westmoreland's nominal superior, Admiral Ulysses S. Grant Sharp, expressed a dramatically different outlook. "The Communist strategy continued to reflect an effort to draw Allied forces into remote areas,"

he observed in a 1969 report, "especially those areas adjacent to border sanctuaries, leaving populated areas unprotected. This enabled enemy local and guerrilla forces to harass, attack, and generally impede
government efforts. Through these means the Viet Cong continued to
exert a significant influence over large portions of the population."[19]

The Americans did hold on to their base at Khe Sanh. Westmoreland later added up the cost: "My staff estimated that the North Vietnamese lost 10,000 to 15,000 men in their vain attempt to restage
Dien Bien Phu. The Americans lost 205."[20] Defense of the base had
required massive commitment of air power and the shifting of large
numbers of ground forces to the northern provinces, but an all-out
enemy ground attack (if one was ever intended) had been deterred and
Khe Sanh was held.

THERE WAS ONE FINAL bit of the Khe Sanh drama to be played out
a few months later, shortly before Westmoreland left Vietnam for the
final time. He was away on some preliminary business in the United
States. Abrams, left in charge, convened a meeting at Phu Bai, where
he announced plans to move out of Khe Sanh. Marine Major General
Rathvon Tompkins, commanding 3rd Marine Division, was "charged
with the phaseout of the Marines and the dismantling of the base." But
when Westmoreland returned and reasserted control, stated Tompkins, he "was quite adamant that Khe Sanh would not be abandoned
until such time that General Abrams was in the chair; that he, Westmoreland, felt it was necessary to retain Khe Sanh. It was just that simple. And he said it, too, in front of a group of staff officers. So the decision was actually Abrams'."[21] For the time being the withdrawal was
put on hold.

Marine Brigadier General John Chaisson reported these matters in
a letter to his wife. "Sunday we went up to Phu Bai for a meeting on
Khe Sanh," he wrote. "I never saw Westy so mad. They were making
plans . . . to pull out. Westy lowered the boom. He was so mad he
wouldn't stay around and talk with them. Instead he told me what he
wanted and left me to push it with Rosson and Cushman." That account was dated 17 April 1968.[22]

Later, though, Westmoreland would claim (at planeside as he arrived at Andrews Air Force Base in Maryland) that in April he had had
a study done, then "made a decision in principle that we would change

our tactics in the area of the Khe Sanh plateau." In a 257-word disquisition in response to a reporter's question, Westmoreland sought to imply that the withdrawal from Khe Sanh (which had been effected while Westmoreland was en route home for reassignment) had been his idea and that all Abrams had done was decide on the date of its execution.

When Abrams did take command, the first thing he did was get on the intercom with Major General Walter Kerwin, still the MACV Chief of Staff. "Dutch," he asked, "what about Khe Sanh?"[23] Soon the withdrawal plans were back on track. "I agreed with Abe," said Kerwin. "Khe Sanh just sat there and was a blister, a boil."[24] So did Admiral Sharp, the CINCPAC. "I concur completely" with Abrams's outline of the situation involving Khe Sanh, he cabled General Wheeler. "The advantages of the move . . . override any possible psychological or political disadvantage."[25]

Abrams, seeking to spare Westmoreland the appearance of having had his judgment reversed when Khe Sanh was evacuated, had suggested some language for use in a MACV press release. Lieutenant General Charles Corcoran remembered a meeting Abrams convened in his office. "The topic was Khe Sanh. The consensus of his staff was that Khe Sanh didn't do anything for us, and we could use the troops elsewhere. There was a big discussion. Then Abe said: 'We've got to handle this very carefully so it won't be interpreted as a repudiation of Westy.' I don't think," observed Corcoran, "Westy ever appreciated Abe's consideration."[26]

Westmoreland countered with his own proposed version of a press release: "We are taking military action to reinforce the successes won by General Westmoreland at Khe Sanh earlier this year." General Wheeler sent that suggestion out to Abrams, where it elicited a lukewarm response. "Paragraph 1," said Abrams, referring to the "reinforce the successes" element, "has a trace of the 'hard sell' and, furthermore, it is not quite true, since basically we are not reinforcing success." Abrams suggested instead a modified statement that he described as "low key" and containing "a minimum of future booby traps."[27]

In the event, Khe Sanh proved to be merely a distraction for American and South Vietnamese forces. The real battle — and the terminating factor for Westmoreland's tenure in Vietnam — lay just ahead.

18

★

TET 1968

On 23 JANUARY 1968, getting ready for the annual celebration, Westmoreland cabled Sharp and Wheeler to report that "we are developing a plan to broadcast from ground and air PA systems sentimental Vietnamese music to the NVA during Tet."[1] The new year observance about to commence was, as things turned out, going to be marked in a very different way.

Shortly before the Tet Offensive began Westmoreland decided to send Abrams north to the I Corps region to establish and run a tactical headquarters that he designated MACV Forward. From there Abrams was to control the operations of all U.S. forces in the area, Army and Marine alike.

General Phil Davidson had returned from a visit to Khe Sanh on 20 January 1968 and briefed Westmoreland and Abrams on the situation there. "The description of the unprotected installations at Khe Sanh and the general lack of preparation to withstand heavy concentrations of artillery and mortar fire agitated General Westmoreland," said Davidson. "Finally, he turned to Abrams and said something to the effect that he (Westmoreland) had lost confidence in Cushman's ability to handle the increasingly threatening situation in his (Cushman's) area. Westmoreland concluded his remarks by saying, 'Abe, you're going to have to go up there and take over.'"[2]

Marine reaction was predictable. Major General Rathvon Tompkins, commanding the 3rd Marine Division, was one of the most vocal. Marines viewed this development with "shock and astonishment," he said. "The most unpardonable thing that Saigon did!"[3] Davidson later wrote that "General Westmoreland's establishment of MACV Forward with authority over General Cushman and the marines in the two northern provinces raised a storm of protest within the Marine Corps and a flurry of hostile and speculative comment by the news media. Westmoreland promptly held a press conference in which he denied that he had lost confidence in Cushman and for that reason had placed Abrams over him."[4]

Westmoreland also cabled Cushman to say "there has been extensive backgrounding here [in Saigon] with the various news bureau chiefs to point out that the establishment of MACV Forward carried no stigma whatsoever with respect to the Marines, that it was merely a normal military practice of establishing a forward headquarters near the scene of impending critical combat, and that it was only temporary."[5] Only the "temporary" was true.

The other denials were false, as evidenced not only by Davidson's eyewitness account but by a lengthy and anguished cable Westmoreland sent contemporaneously to General Wheeler. "As you perhaps appreciate," he began, "the military professionalism of the Marines falls far short of the standards that should be demanded by our armed forces. Indeed, they are brave and proud, but their standards, tactics, and lack of command supervision throughout their ranks requires improvement in the national interest."

There was more: "I would be less than frank if I did not say that I feel somewhat insecure with the situation in Quang Tri province, in view of my knowledge of their shortcomings. Without question, many lives would be saved if their tactical professionalism were enhanced." Westmoreland added that he was sending this message only because he was "concerned as a military man, and feel it is my duty to give you the benefit of my views."[6] He also sent an emissary, Army Lieutenant Colonel Richard Cavazos (later a full general), to Washington to describe these problems in more detail, although it is not clear that he actually got in to see Wheeler.

General Kerwin, as MACV's Chief of Staff, also had some insight into Westmoreland's evaluation of the Marines, described in an 8 February 1968 letter to his wife: "Westy just came back from the I Corps, up with the Marines. Surprisingly, he was very despondent, more than I have ever seen him. He says the Marines are inflexible—good for over-the-beach operations, storming hills, and holding onto positions. In this type of warfare they don't have the ability to change pace quickly. Also their organization and equipment is not configured for this type of war."[7]

Westmoreland was also very candid in discussing the Marines in his memoir, even after a trusted former aide who had seen the manuscript in draft counseled him to tone it down, which he apparently did, at least to some extent. During early 1968, said Westmoreland, he had significantly reinforced the Marines in I Corps, "yet General Cushman and his staff appeared complacent, seemingly reluctant to use the Army forces I had put at their disposal." In a conference with Marine leaders in Danang, he said, he listened to various reports for more than two hours, "becoming more and more shocked at things that virtually begged to be done yet remained undone. Local decisions were urgently needed. I ended up giving direct orders myself to General Cushman's subordinate units, an unusual and normally undesirable procedure."[8]

After the war, when the Marines were writing their history of the conflict, they sent a draft of the 1968 volume to Westmoreland for comment. He marked it up so extensively, and took issue with so many of the judgments rendered, that he was invited to discuss the whole matter in person. He accepted and, in a session with a number of Marine Corps historians, again insisted with regard to establishment of MACV Forward that "that particular action had not a damned thing to do with my confidence in General Cushman or the Marines, not a damned thing."[9] This was not only false, but reckless, given the existing paper trail.

LATE ON THE NIGHT of 29–30 January 1968 the enemy attacked —prematurely—at a half-dozen or more different locations, mostly in the II Corps Tactical Zone, the apparent result of a mix-up in instructions.[10] The next night the full panoply of the Tet Offensive erupted throughout South Vietnam, with thirty-six of the country's forty-four provincial capitals and five of the six autonomous cities assaulted. In Saigon the American Embassy, the Vietnamese Joint General Staff compound, the radio station, Independence Palace, and Tan Son Nhut Air Base were among the facilities targeted.

In November 1967 Westmoreland had been quoted in *Time* magazine as saying, "I hope they try something because we are looking for a fight."[11] Now he had his wish.

The Tet Offensive caused Westmoreland many problems, perhaps the least of which was helping the South Vietnamese turn back the attacks. In the succeeding weeks he was deeply engaged in attempting to refute charges that he had been surprised by the enemy offensive, in claiming that he had caused that offensive—a "desperation" move by the enemy, he styled it—by the successes he had achieved during the previous year, in defending his decision to hold at Khe Sanh and in seeing that as the enemy's primary objective, in denying that he had made a request for a large number of additional troops while simultaneously claiming a great victory, and then in rejecting any idea that his subsequent reassignment to Washington was in any sense a criticism of his performance as the U.S. commander in Vietnam.

Westmoreland was in his quarters in downtown Saigon when the attacks began, then found himself pinned down there, unable to get to his headquarters until morning. But, he claimed, "by my Marine aide talking to the Marine guard inside the embassy and by my numerous telephone conversations with the US Army MP command, I was able to follow the course of the battle and direct action."[12]

At mid-morning on 31 January Westmoreland was driven to the U.S. Embassy in Saigon, where during the night some nineteen enemy sappers had blasted through the wall and entered the grounds, where all were eventually killed. There Westmoreland gave an impromptu press conference. Television reporter Don North was there. "I couldn't believe it," he later wrote. "Westy was still saying everything was just fine. He said the Tet attacks throughout the country were 'very deceitfully' calculated to create maximum consternation in Vietnam and that they were 'diversionary' to the main enemy effort still to come at Khe Sanh." North was also present later that day when Westmoreland appeared at the daily MACV press briefing to "emphasize the huge body counts" that allied forces were racking up. Then, "to add to Westy's growing credibility gap," recalled North, "it was also reported at his press briefing that the city of Hue . . . had been cleared of enemy troops. That false report had to be retracted, as the enemy held parts of Hue for the next 24 days."[13]

Once he reached MACV Headquarters, Westmoreland found himself isolated, cut off from his residence, which was in any case insecure. "It was humiliating for Westmoreland. He couldn't get out of his own headquarters," said Brigadier General Zeb Bradford, then a field grade officer and aide-de-camp. That first day, he said, the senior generals assembled in a makeshift dining room for the evening meal. "The mood was grim, even despondent," he recalled. "It appeared that all that had been so painfully achieved over the years had been for nothing." Then "the gloom was made complete when a stray bullet smashed through a window in the room where the generals were eating. With as much dignity as possible these senior officers had to evacuate themselves to a safer part of the building."[14]

"Our HQ has been under continuous attack all day," Brigadier General Chaisson wrote to his wife on 1 February. "We are all holed up here—including Westy." The next day Chaisson wrote home that "Gen Westmoreland had to shift his office. He was getting sniper fire." The following week he reported on another aspect of the situation: "Chow is intermittent. Gen West—— is living in the building, also. He has a rotating mess. I get an invite to dinner or breakfast every other day. In the interim, I have 'C' rations." Brigadier General Larry Caruthers was another sometime beneficiary of that makeshift hospitality. "We slept in the headquarters at MACV during Tet," he recalled, and "every third night we would be detailed to have drinks and dinner with Westy."

Captain Ted Kanamine was then an aide-de-camp to General Abrams, and he too remembered the situation at MACV Headquarters. "You know, my god, they were in that building and Tet was raging all about them, and bullets were going through the windows and there was a superhuman effort made for those two gentlemen [Westmoreland and Abrams] to be in a certain configuration in that building so that they wouldn't be exposed to direct bullet fire and all that kind of stuff."[15]

MACV Headquarters was not a war headquarters, recalled Bradford. "There were no bunkers. No defenses. It was like being in a department store." Some 4,000 people who worked in the headquarters were in a billeting area at Newport or somewhere about four miles away and could not get to their place of duty. "It didn't seem to make any difference," he observed. "The war went on without them."[16]

And, said Bradford, "I remember Westmoreland saying: 'Everything I have worked for is lost. It's all been a failure.'"[17]

In his history notes for 7 February 1968 Westmoreland recorded that "in the course of the next several weeks, I put in extraordinarily long hours." He cited receiving numerous briefings and approving B-52 strikes as examples of what he was doing. "In other words, I was living with the situation minute-by-minute, and making the appropriate decisions."[18]

ABOUT THREE WEEKS before the Tet Offensive began, Lieutenant General Fred Weyand, then commanding II Field Force, Vietnam, had called MACV and asked to meet with Westmoreland. He was given an appointment for the following day. He made his way to Saigon, where he first met with General Abrams. Weyand described recent traffic analysis of enemy radio transmissions which seemed to indicate major formations were moving toward the Saigon area. "General Westmoreland had ordered us to move many of our major units northward toward the border," said Weyand, "apparently to relieve the pressure on Khe Sanh. I went to Saigon and talked with Abe, then with his backing explained to Westmoreland what was happening and got the orders to move to the border rescinded. We then set up to protect Saigon."[19]

That was highly fortuitous, for when the enemy attacks began Weyand had forces in position to interdict their routes into the Saigon area and then to assist South Vietnamese forces in dealing with those that made it into the city. Those same assaults demonstrated the utter futility of the large-scale search and destroy sweeps of the previous year, including Operation Cedar Falls in the enemy stronghold known as the Iron Triangle. During Tet, observed the authors of *The Tunnels of Cu Chi*, "the most damaging thrust—that against Saigon itself—would come straight from the Iron Triangle."[20]

It was noteworthy that Weyand had had to dissuade Westmoreland from his fixation on deployments in the border regions, far from most of the indigenous population. The 1967 MACV Command History, contrariwise, had favored the tactic of disposing forces to protect the population. "The enemy's strategy continues to reflect an effort to draw Allied forces into remote areas of his choosing," read that account, "especially those areas adjacent to border sanctuaries, enabling his local and guerrilla forces to harass, attack, and generally impede the GVN nation building effort."[21] Apparently Westmoreland was not aware of, or chose to discount, these implied lessons of recent history.

In his memoirs Westmoreland sought to take some credit for these pre-Tet precautionary troop dispositions, claiming that "Weyand's information reinforced doubts that had already begun in my own mind."[22] Since Weyand had needed the backing of Abrams to persuade Westmoreland to cancel sending Weyand's forces out to the border areas, there is some reason to doubt Westmoreland's claim.

Weyand also had a more realistic outlook on the larger sweep of the war than Westmoreland's optimistic accounts, noting in his debriefing report after lengthy service in Vietnam that "the infiltration of large North Vietnam forces and the concomitant dominant control of the conflict taken by Hanoi in 1966, the massive infusion of the most modern Soviet weaponry down to include the guerrilla squads in 1967 and the all-out country-wide assaults coupled with introduction into South Vietnam of multi-division forces in 1968, each brought the conflict to a point of great crisis for the allies."[23]

OF COURSE THE issue of surprise at Tet became the centerpiece of a raging controversy. Westmoreland tried hard to make the case that he hadn't been surprised, or at least not all *that* surprised. It was a difficult case to make. On 26 December 1967 Westmoreland had said in a Mutual Broadcasting System interview that 1967 had been a year of "great progress" in the war and that he saw "no evidence that communist strategy will change in the coming year."[24] Thirty-five days later the Tet Offensive began, sending Westmoreland scrambling to make a case that the enemy had been forced to change his strategy because of allied successes in the year just past.

Inevitably Westmoreland's year-end report for 1967 came to light and was excerpted in the *New York Times* on 21 March 1968, just as the Tet fighting was winding down. That report predicted, wrote another journalist, "'more military victories in 1968' and contained not a hint of any possible enemy drive such as the Tet offensive."[25] In fact, stated Westmoreland's report, "In many areas the enemy has been driven away from the population centers; in others he has been compelled to disperse and evade contact, thus nullifying much of his potential. The year ended with the enemy resorting to desperation tactics in attempting to achieve military/psychological victory; and he has experienced only failure in these attempts."[26] Summarizing, the report stated that "1967 was characterized by accelerating efforts and growing success in all phases of MACV endeavors."[27] No wonder the Tet Offensive, erupting only days after this report reached Washington, created such widespread consternation. General Wheeler reacted by cabling Westmoreland to report the "pernicious leak" of the optimistic assessment.

In rationalizing Tet, Westmoreland was all over the map. "We knew preci—almost exactly when he was gonna attack," he stated in a CBS television documentary interview. "We—we thought he would attack before Tet or after Tet." On another occasion he maintained that "we had full warning that the offensive was coming."[28]

But Ambassador Ellsworth Bunker acknowledged, in an unpublished manuscript, that "there was an intelligence failure. We had no inkling of the scope, the timing, or the targets of the offensive." Brigadier General John Chaisson briefed the press early in the offensive, observing afterward that "Westy thought I pounded down too hard on surprise. He pouted."

Air Vice Marshal Nguyen Cao Ky expressed the South Vietnamese outlook on the offensive: "Savage, brilliantly executed, it caught all of us off balance. Almost before we knew what happened, we were fighting for our lives in the streets of Saigon, Hue, and a dozen other cities."[29] ARVN Major General Nguyen Duy Hinh wrote that "the surprise achieved by the enemy was absolute."[30] And Colonel Hoang Ngoc Lung, South Vietnam's top intelligence officer, was also unequivocal: "One thing was certain. The enemy had really achieved the element of surprise."[31]

IN WASHINGTON THERE was also plenty of surprise to go around. In a White House meeting soon after the offensive began, Secretary of Defense McNamara stated that "Westmoreland did not expect the strength of attacks throughout the cities."[32] George Christian, the White House press secretary, observed that "the Tet offensive came as a brutal surprise to President Johnson and all of his advisors. We had been led to believe that the Viet Cong were pretty well defanged by that period, that the pacification program had worked very well, that most of the villages in South Vietnam were secure, and that it was virtually impossible for the Viet Cong to rise to the heights that they did in 1968."[33]

The reaction to these events was costly to Westmoreland's reputation. *Newsweek* cited a devastating measure of the fall: "In November, when he was conjuring up the light at the end of the tunnel, he was affectionately called 'Westy.' But by last week he was 'General Westmoreland' in most official and unofficial briefings."[34]

Army Chief of Staff General Harold K. Johnson accompanied the President when he flew down to Fort Bragg to see off troops being sent to Vietnam. "On the flight," recalled Colonel Fred Schoomaker, "LBJ was on the telephone with General Westmoreland in Vietnam, and he was just tearing into him: 'What the hell is going on out there?!! You just told Congress light at the end of the tunnel and all that!!!'"[35]

Still, Westmoreland sought to shed the disgrace of having been surprised, writing to Congressman Dale Milford: "I want to lay the canard to rest that the Tet offensive represented an intelligence failure. The large-scale attacks that occurred were not only anticipated, but I personally directed each commander to place his forces in a maximum alert posture, in anticipation of the attack I knew was coming, 36 hours in advance." And: "The only surprise was its rashness. The enemy assumed risks, inviting great casualties, due to attacks on heavily defended areas where superior firepower could be brought against them."

In later years Westmoreland became more and more expansive on this point, claiming for example that the Tet Offensive "was a surprise to the American people but not to us on the battlefield."[36]

Westmoreland did concede one element of surprise discovered at Tet: "The offensive revealed an unexpected determination by North Vietnam to pursue the war regardless of the price."[37]

· · ·

WESTMORELAND'S EXPLANATION FOR the enemy's offensive, one he clung to until the end, was that "in 1967 we hurt the enemy to the point where he changed his strategy from that of a protracted war, a war of attrition, so to speak, to a general offensive."[38]

In reality, however, all Westmoreland's attrition strategy had been doing was killing large numbers of the enemy, losses the enemy was able and willing to replace. Meanwhile the enemy's covert infrastructure in South Vietnam's hamlets and villages, totally untouched by the war of the big battalions in the remote jungle, continued to maintain control of the rural populace through coercion and terror. Far from seeing this as a losing situation, the enemy leadership seemed quite satisfied with engaging the Americans in useless bloody battles that did nothing to undermine their control of the war's real object, the people. General DePuy, among many other senior officers, saw this clearly. Before Tet 1968, he said in his oral history, "It was a stalemate."

Despite Westmoreland's insistence that the enemy's new departure at Tet was forced on him by the reverses that he, Westmoreland, had been inflicting during 1967, the MACV Historian, Colonel Argo, suggested a different interpretation. In a memorandum to Westmoreland dated 11 May 1968 he commented on "the enemy's current willingness to do battle under conditions where success not only is not assured, but is extremely improbable." That may be explained, he said, by the fact that "he no longer needs to win the battle to attain his basic political-psychological goals. He merely has to demonstrate his presence by fighting. The tactical defeat is inconsequential."[39]

Robert McNamara was also among those who never accepted the Westmoreland viewpoint. "MACV interpreted Tet as a great victory for the United States," he said. "I always felt this was a preposterous claim."[40] He had a point. Westmoreland's claims of victory were based on the familiar criterion of body count. He had indeed stacked up plenty of enemy bodies in years gone by, but had nothing to show for it in terms of a favorable outcome to the war. More bodies were probably not going to change that.

While Westmoreland was busy trying to portray the Tet Offensive as forced by his successes, his boss had a different outlook. In his 1967 Memorial Day proclamation Lyndon Johnson called the war a "bloody impasse," then at a White House meeting on 4 November 1967 complained that, in prosecuting the war, "We've been on dead center for the last year."[41]

ON THE NIGHT OF 6–7 February 1968, the Special Forces camp at Lang Vei, near Khe Sanh, was attacked and eventually overrun by communist infantry and sappers reinforced by nine PT-76 tanks. In such circumstances the Marines at Khe Sanh were supposed to reinforce Lang Vei, as they admitted in their after action report: "The mission of the 26th Marine Regiment was to . . . be prepared to conduct operation in relief of or reinforcement of the CIDG Camp at Lang Vei."[42] But when the crunch came, the Marines sat tight and let the outnumbered defenders of Lang Vei fend for themselves. "No marines came down from the big base in Khe Sanh to help out," wrote Charles Simpson bitterly in a history of Special Forces in the war.[43] The defenders of Lang Vei also asked the Marines at Khe Sanh to fire preplanned artillery in their support. That didn't work out very well, either. "The mission initially landed on top of the camp, but was adjusted on target," said the Special Forces after action report.[44]

There were 24 U.S. Special Forces soldiers with the indigenous troops at the base. John Prados reported that "Lang Vei resisted heroically but was overrun." Of the 24 Americans, 13 were wounded and 10 more reported missing and presumed dead.[45]

This failure of the Marines to carry out an assigned mission immediately became a matter of furious controversy. "When he learned the Marines would not carry out the relief plan," wrote Robert Pisor in his account of Khe Sanh, "the Special Forces commander in Vietnam, Colonel Ladd, called Saigon and demanded to be put through to COMUSMACV. Westmoreland was awakened, but he said he would not second-guess a commander at the scene. He rang off."[46]

Westmoreland addressed the issue in his memoirs, astonishingly siding with the Marines. Recalling that he had twice during the night been awakened by General Chaisson, he acknowledged that "under established plans, the marines at Khe Sanh were to send a relief force if Lang Vei got into trouble, but twice the marines had turned down Lang Vei's call for help. They reasonably considered that a relief force moving by road was bound to be ambushed and that a helicopter assault in the darkness against a force known to have armor was too hazardous." Thus, said Westmoreland, "honoring the prerogative of the field commander on the scene, I declined to intervene until I could ascertain more of the situation."[47] That was, as things played out, going to take quite a long while, and in the interim things at Lang Vei became very bloody.

Westmoreland maintained that the next day he "called a conference of Marine Corps and Army commanders operating in the I Corps Zone and flew to Danang. There were a number of topics to be discussed . . . , but of first priority was the crisis at Lang Vei. I directed General Cushman to provide helicopters for a relief force of CIDG troops with Special Forces advisers to bring out American and South Vietnamese survivors."[48]

Colonel Jonathan Ladd, the Special Forces commander, remembered it differently. "I called General Abrams on the phone and told him what was happening, and I said, 'I just can't get Westmoreland's attention long enough to do anything.'" Abrams called the Marine air chief, Major General Norman Anderson, said Ladd, "and told him that he was telling me to go ahead and evacuate the camp at the earliest opportunity and for the Marines to help us, which they did." Further, per Ladd: "Now in Westmoreland's book, he said that he did all that, but he didn't. Abrams did it, because he, Westmoreland, kept going back to these civic action meetings. Abrams was the one who just said go ahead and do it." Also: "Anderson was a great help." "And old Lownds, you know, he just kept saying that it was a suicide mission."[49]

Westmoreland appeared to have some later regrets about the matter. "I will say in all candor that I did have two disillusioning experiences with the Marines, and I may as well tell you about them," he said in his meeting with the Marine historians. "One was . . . the question of that outpost that was overrun, not too far from Khe Sanh. Nothing was being done about that."[50] In Robert Pisor's view, "the fall of Lang Vei evaporated Westmoreland's last reserves of confidence in the Marines."[51]

General Cushman spoke about Lang Vei in his oral history. "We didn't relieve Lang Vei because we didn't want to loose [sic] a lot of men unnecessarily," he said. "It would have been a grave risk to send Marines from Khe Sanh to Lang Vei in the hours of darkness."[52] Of course, it was also something of a grave risk to be defending Lang Vei when it was attacked by tanks.

Subsequently, even before succeeding Westmoreland and taking formal command of MACV, General Abrams cabled General Cushman with some pertinent observations. "I have been reflecting on the actions at Lang Vei, Ngok Tavak and Kham Duc," he began. "Based on the scant detail which is available to me through reports, I have assessed each of these actions as minor disasters. I do not believe that we would have had this string of failures if our plans for the support, or relief, of these camps had been carefully prepared." Abrams provided some suggestions for reviewing and improving command and control, communications, planning, reaction procedures, and reporting. "Through this process I would expect that when your command is confronted with a similar imminent problem, appropriate actions would be taken so that we would not lose another camp."[53]

ON 6 FEBRUARY 1968 the President sent Westmoreland a reassuring cable professing continued confidence in him, pretty close to the kiss of death. On 22 March 1968 the White House announced that Westmoreland was coming home to be Army Chief of Staff, a post he would assume at the beginning of July. General Chaisson wrote that "we had a birthday dinner for [Westmoreland] Tuesday night at the HQ. He gave a rather bitter speech in which, half facetiously, he said he didn't think he could get confirmed for his new post, because the Senate should hold him responsible for all of our failures—such as 'search and destroy,' ARVN corruption, lack of fight in ARVN, the Tet offensive, the Khe Sanh trap, and others."

That outlook was shared, and not in jest, by at least some of Westmoreland's senior associates. "I felt very strongly that, with Tet having occurred, General Westmoreland's credibility as a commander had suffered damage that could not be repaired," said Lieutenant General Charles Corcoran. "I think it was important to the country and important to the military forces there that General Westmoreland leave quickly."[54]

Two months or so after the Tet Offensive erupted the *Washington Post* ran a devastating piece describing the progress of the Vietnam War as charted in statements by Westmoreland. They began the compilation with this 20 June 1964 comment: "I don't see any reason for expansion of the U.S. role in Vietnam. I am optimistic and we are making good progress." On 8 July 1964: "At the present time I believe the whole operation is moving in our favor." Then on 14 April 1967, one of the most infamous: "We'll just go on bleeding them until Hanoi wakes up to the fact that they have bled their country to the point of national disaster for several generations." And from his address of 28 April 1967 to the Joint Session of Congress: "Backed at home by resolve, confidence, patience, determination and continued support, we will prevail in Vietnam over the Communist aggressor." In Honolulu en route back to Vietnam he had said, on 1 May 1967: "Based on what I saw and heard [while in the United States], I'd say 95 per cent of Americans are behind us 100 per cent."[55]

There was more. Again visiting the United States in the summer, Westmoreland said in Washington on 13 July 1967: "We have achieved all our objectives, while the enemy has failed dismally." And again in Washington, at the National Press Club on 23 November 1967: "The end begins to come into view. The enemy's hopes are bankrupt. I am absolutely certain that whereas in 1965 the enemy was winning, today he is certainly losing. We are winning the war of attrition." For that appearance, said Westmoreland, "I wrote the speech." Then came the Tet Offensive, leading him to comment on 1 February 1968: "That was a treacherous and deceitful act by the enemy."[56]

Sir Robert Thompson, who had followed the war closely for a number of years, wrote a powerful assessment published in the *London Sunday Times* as the Tet battles raged. Noting that General Giap's objectives had "been achieved only at high cost to the Vietcong and North Vietnamese in manpower—the one resource in which they have a surplus," he stressed that the initiative was with the enemy and that they had "calculated the price so that, at a cost acceptable to themselves, they are imposing a cost which cannot indefinitely be acceptable to the United States." Then, in a direct swipe at the position Westmoreland had staked out, Thompson observed that "if some American reactions are to be believed that the battles are acts of desperation or a 'go for broke' attempt to improve Hanoi's negotiating position, then there is still a complete lack of understanding of the strategy of the war and of the stage it has now reached."[57]

THERE WAS ONE further consequence of the Tet Offensive, one having to do with the order of battle controversy of the previous year in which Westmoreland, and MACV, had prevailed in arguing for a far lower estimate of enemy strength than the number much of the rest of the Intelligence Community advocated. That matter now surfaced again. Confirmed an officer of the Defense Intelligence Agency: "DIA did join CIA in publishing a report reintroducing the self defense militia and upping enemy strength to 500,000."[58] The battlefield realities had induced second thoughts about going along with Westmoreland's insistence on reporting an artificially constrained number.

Ronald Smith, then CIA's South Vietnam Branch Chief, verified that after Tet the Agency had "argued for quantification and inclusion of all categories—including self-defense and secret self-defense militia forces and political cadres—in estimates of total enemy strength because all of these categories worked together and contributed to the enemy's war effort." In April 1968, at the request of CIA Director Richard Helms, a new order of battle conference was convened at CIA Headquarters "as a result of the increasing and widespread concern in the wake of the Tet Offensive that MACV's official strength estimates were understated." CIA estimated total enemy strength as of 31 March 1968 at 440,000 to 590,000, while MACV "held to its familiar estimate of about 300,000."[59]

Concluded General DePuy, "[T]he North Vietnamese lost the battle, but they won the war as a result of Tet. It terrified and horrified the people in Washington."[60]

Westmoreland himself, however, professed no second thoughts in the wake of the Tet Offensive and three years of attrition warfare. "Basically," he said in a contemporaneous interview with the Associated Press, "I see no requirement to change our strategy."[61]

★

19

TROOP REQUEST

WESTMORELAND DENIED THAT he ever made a request for 206,000 more troops in the wake of the Tet Offensive. It was a hard case to make.

When the offensive erupted, in several places the night of 29–30 January and then pretty much all over the country the following night, Washington had been stunned and—some people said—panicked. MACV was a good bit stunned as well. Almost at once there began a fascinating exchange of messages between Westmoreland and General Earle Wheeler, Chairman of the Joint Chiefs of Staff.[1]

As early as 3 February 1968 Wheeler was cabling Westmoreland, "at the President's request," to ask whether "there is any reinforcement or help that we can give you."[2] Westmoreland replied the following day that "the enemy has dealt the GVN [Government of Vietnam] a severe blow," but made no mention of any need for additional forces.

On 8 February Wheeler cabled again, and again he raised the issue of more troops. "Do you need reinforcements?" he asked Westmoreland. This is a new situation and the old criteria need not apply, he said reassuringly. "If you consider reinforcements imperative you should not be bound by earlier agreements." Just to be sure Westmoreland got the message, Wheeler drove it home one more time: "In summary, if you need troops, ask for them."[3]

The following day, 9 February, Wheeler was back again: "Please understand that I am not trying to sell you on the deployment of additional forces. . . . However, my sensing is that the critical phase of the war is upon us, and I do not believe that you should refrain from asking for what you believe is required under the circumstances."[4] Westmoreland finally got the word, replying within hours to this latest invitation: "Needless to say, I would welcome reinforcements at any time they can be made available."[5]

Then, in the same way that Wheeler had become increasingly insistent in his elicitations, Westmoreland escalated the specificity and urgency of his replies. On 11 February he told Wheeler: "I am expressing a firm request for additional troops. . . . A set back is fully possible if I am not reinforced."[6]

Westmoreland commented on the response to that message in a later White Paper: "According to General Wheeler, the group [Rusk, McNamara, Helms, Taylor, Clifford, Rostow, and Wheeler, meeting at the White House on 12 February] interpreted my message as expressing the following thoughts: 'You could use additional U.S. troop units, but you are not expressing a firm demand for them; in sum, you do not fear defeat if you are not reinforced.'"[7]

On that same date Westmoreland fired back his most urgent depiction of the situation yet: "I desperately need" reinforcements. "Time is of the essence."[8]

Subsequently Wheeler scheduled a trip to Vietnam to discuss troop augmentation at first hand. General Chaisson of course wrote home about it: "Westy seemed kind of blue tonight," he told his wife. "Wheeler is coming out tomorrow and his visit is not causing much glee. He is here on a fact-finding deal I guess and Westy is miffed. He had me cross 'Warm Regards' off the bottom of a couple of messages I wrote for him today to Wheeler!"

The Chairman arrived in Saigon on 23 February 1968. It was the eleventh time he had made the long haul to visit the war, and Westmoreland saw in him an "exhausted and ill man." Wheeler and Westmoreland almost immediately plunged into a series of conferences in which it was ultimately agreed that they would seek a large number of additional troops, some 206,000 in all. It was in many ways a rehash of the request for 200,000 troops that Westmoreland had submitted, without success, in March 1967.

Westmoreland characterized the proposal as not a request but merely a "contingency plan" premised on a major change in the over-all strategy for conduct of the war. This additional force increment (not all necessarily for deployment to Vietnam, at least initially) would, maintained Westmoreland, enable him to go into Laos and Cambodia to clear out the enemy sanctuaries there and possibly to make some forays into North Vietnam as well.

The remainder of the forces sought could be used to reconstitute the strategic reserve — the active forces available to respond to contingencies anywhere in the world. These had been severely depleted to meet Vietnam requirements and the Joint Chiefs of Staff were very uneasy about the resultant strategic vulnerability. Of course additional forces of this magnitude could not possibly be provided without mobilizing the reserves. It is clear that Wheeler saw in the cataclysmic events of the Tet Offensive a new and possibly better opportunity to convince, or perhaps coerce, the President to take that long-avoided step.

"General Wheeler came over there and he was actually begging me to ask for more troops," Westmoreland told historian Mark Perry. "Really, just begging me. And he told me the president was ready to call up the reserves, and if that were to happen how many men would I need, how many men would I use. And that was when I said well, we could use another 200,000."[9]

Brigadier General Douglas Kinnard wrote of this visit that, "with Wheeler's encouragement, Westmoreland developed a plan for approximately 206,000 additional troops. Of this total, Wheeler would recommend that 100,000 go to Vietnam and the remainder be held in strategic reserve."[10] On 26 February 1968 McNamara discussed the Wheeler report with the Joint Chiefs: "They are stunned by the 206,000 request." McNamara later revealed that, "in my last official act on Vietnam, on February 27, 1968, I opposed Westy's renewed appeal for 200,000 additional troops on economic, political, and moral grounds."[11]

STOPPING IN HONOLULU on his return to Washington, on 26 February Wheeler cabled ahead with the results of his meetings in Saigon, along with the request for 206,000 troops. No mention was made of any change in strategy. Rather the request was presented as an essential measure necessitated by battlefield developments in Vietnam. Two days later, just off the plane and having gone directly to the White House for a breakfast meeting with the President and his advisors, Wheeler repeated the briefing, and the troop request, in person.

If indeed "Washington panicked," Wheeler had most assuredly helped them do so. Again he made no mention of any contingencies such as changes in strategy, but instead stressed the parlous situation of the forces in Vietnam. Incoming Secretary of Defense Clark Clifford and others understood that he was stating a firm and unqualified requirement for 206,000 more troops. The White House presentation, Clifford remembered, was "so somber, so discouraging, to the point where it was really shocking." And it had an impact on the President, although perhaps not the one Wheeler had intended. LBJ was, said Clifford, "as worried as I have ever seen him."[12]

Westmoreland was comfortable with this Wheeler ploy, then and later. General Wheeler, said Westmoreland in his memoirs, "saw no possibility at the moment of selling reinforcements in terms of future operations." Thus, dealing with the "civilians," it was "better to exploit their belief in crisis to get the troops, then argue new strategy later."[13] Of course that was a belief induced by Wheeler.

Robert Komer confirmed that no longer-term issues, such as calling up reserves and using the additional troops to go into the sanctuaries, were "ever mentioned by Wheeler to either the President or McNamara. By God they would have thrown him out on his ear." Komer was bitter about it: "The Chiefs never tell anybody anything. The goddamn Chiefs of Staff. Wheeler's the evil genius of the Vietnam war in my judgment."[14]

As this drama was playing out, Clark Clifford arrived at some clear judgments of the respective roles of Westmoreland and Wheeler. Clifford recalled Westmoreland's blaming Wheeler "for presenting his request for reinforcements in a tone so pessimistic" as to exploit the sense of crisis in Washington.[15] Clifford, a highly regarded attorney, understood the implications. "Westmoreland's charge was a serious one," he wrote. "If true, the Chairman of the Joint Chiefs was guilty of deliberately misleading the President in order to get additional troops authorized for Vietnam." Added Clifford, of Wheeler: "He was an honorable man, though, and I do not believe the accusation."[16]

ON SUNDAY, 10 March 1968, the *New York Times* front-page headline read: "Westmoreland Requests 206,000 More Men." That was sensational but not determinative. Two days earlier Wheeler had cabled Westmoreland to alert him to "strong resistance in all quarters to putting more ground force units in South Vietnam. . . . You should not count on an affirmative decision for such additional forces."[17] And a McNamara chronology of the period included this entry for 8 March: "LBJ nixes the 206,000 troop request."[18] The request was dead even before it became known publicly.

Westmoreland had in fact helped to undermine his own request by describing a venturesome plan to take the war north. LBJ recalled that report, rendered in early March 1968, in his memoirs. "I had just received a message from Westmoreland describing his plan for a major offensive against the North Vietnamese and the Viet Cong in northern I Corps on or about April 1," he wrote. "I thought that if Westmoreland had enough confidence to launch an offensive with the forces he had in Vietnam, it would be wise to limit additional commitments."[19]

IN MID-MARCH LBJ met with Dwight Eisenhower in Palm Springs, where he "asked the general how long a President should back a military commander. Eisenhower replied that a President must back his general as long as he had confidence in him—if he lost this confidence, he should replace him."[20] Ten days later the President announced his decision to appoint Westmoreland Army Chief of Staff.

SOON WESTMORELAND AND Wheeler met again, at Clark Air Base in the Philippines on 24 March 1968. Said Westmoreland, Wheeler "told me that a significant change in our military strategy for the Vietnam War was extremely remote, and that the administration had decided against a large call-up of reserves."[21] What happened instead was an emergency deployment of 11,000 men, plus another 13,500 to support and sustain them, for a total of 24,500, far short of the 206,000 requested. That brought the overall authorization to 549,500, the high-water mark for the war (and even that was never fully achieved, deployments eventually peaking at 543,400).

The deliberations leading to that decision had been agonizing. Robert McNamara, who for months (or years, as was later revealed) had been disenchanted with the war and had belatedly let LBJ know it, was leaving to head the World Bank and Clifford was replacing him, effective 1 March 1968. LBJ thought that in his longtime friend Clifford he would again have in the post of Defense Secretary a strong supporter of the war and his policies there. Perhaps Clifford approached the job from such a position, although as long ago as the previous August, when Westmoreland's *first* request for some 200,000 troops was still under consideration, he had predicted in a conversation with the President that "a year from now we again will be taking stock. We may be no closer a year from now than we are now."

Now the arrival of the 206,000 request sent a jolt through everyone, and it wasn't going to be a year before they took stock again. On the same day Wheeler told LBJ in person about the troop request, the President asked Clifford to head up a task force to examine the request, specifying that he report back by 4 March. Clifford began his task that same afternoon, then spent three full days huddled with the Joint Chiefs of Staff in the Pentagon. Those three days permanently changed his outlook on the war and on U.S. policy for the prosecution of it.

Clifford remembered: "How long will it take? They didn't know. How many more troops would it take? They didn't know. Would 206,000 answer the demand? They didn't know. Might there be more? Yes, there might be more. So when it was all over, I said, 'What is the plan to win the war in Vietnam?' Well, the only plan is that ultimately the attrition will wear down the North Vietnamese and they will have had enough. Is there any indication that we have reached that point? No, there isn't." By the end of this extended interview, said Clifford, now just days into his tenure as Secretary of Defense, "I had turned against the war."[22]

On 26 March 1968, in a meeting at the White House, LBJ told Rusk, Wheeler, and Abrams how damaging he thought the request had been: "We have no support for the war. This is caused by the 206,000 troop request, leaks, Ted Kennedy and Bobby Kennedy."[23]

Later, in a book review of Westmoreland's memoirs, noted military affairs commentator Hanson Baldwin scoffed at any such prospects for additional forces and a wider war as postulated by Westmoreland. "In the wake of Tet and the battle of Khe Sanh which followed," he wrote, "it would have been politically and psychologically impossible to follow up the abortive enemy offensive with still another major escalation of troop strength."[24] Stanley Karnow wrote that "the chances of Johnson endorsing all or any of these moves were remote, but the gullible Westmoreland took Wheeler at his word."[25]

Many years later, at a Vietnam roundtable held by the LBJ Library, Westmoreland assigned the blame to Wheeler. "The 206,000-man troop request was after the Tet offensive, when 'Bus' Wheeler came over," he said. "We had long discussions, and I went along." But, he added, taking a legalistic approach, "There's no paper you can find signed by me requesting 200,000 troops."[26]

Les Gelb wrote that Westmoreland later said privately he had been "conned" by Wheeler into making the request and, added Gelb, after he came up with the figure of 206,000 "Westmoreland was shocked and bitter when he later discovered that Wheeler had portrayed his request" as based on battlefield necessities.[27] Whatever their respective shares of blame, it is depressing to think that Wheeler and Westmoreland colluded in seeking to manipulate and deceive the President. As soon became clear, the results of their machinations would be catastrophically different than what they had sought.

PERHAPS BOLSTERED BY Wheeler's visit, Westmoreland now showed a new outlook, at least as perceived by General Chaisson: "Westy is getting tougher with the Washington crowd," he wrote on 26 February 1968, just after Wheeler had left for Washington. "His attitude is that he is holding things together with minimum assets. The enemy is making major efforts and cannot be defeated with patchwork. He (Westy) wants what he thinks he needs; no more discount jobs."[28]

Westmoreland had discussed his troop aspirations with Ambassador Bunker, but not fully. "At that time he asked for a moderate increase . . . amounting to about twenty or twenty-five thousand troops," recalled Bunker, "and I did support that. I did not support his later request for a much larger increment of troops." Westmoreland later maintained that he "had the close support of the ambassador" for his post-Tet troop requests, but Bunker's account contradicts that assertion. In fact, when Secretary of State Rusk asked Bunker to comment on the larger troop request, Bunker replied that "an increase of the size contemplated might well nullify the purposes that had brought the United States into the war in the first place by destroying what was left of South Vietnamese initiative."[29]

CLARK CLIFFORD RECALLED in his memoirs the effects in Washington when Wheeler returned from Vietnam and reported the need for more troops, many more. "Wheeler's report contained an assessment of the situation so bleak, and a request for additional troops so large," he said, "that it had a profound effect on the course of the war and American politics." Clifford also said of Wheeler's report: "It was clear that, although he avoided criticizing Westmoreland directly, he had lost some degree of confidence in his theater commander and no longer fully accepted Westmoreland's judgment of the war." And, he went on: "His [Wheeler's] report damaged Westmoreland because in a disagreement between the two men, President Johnson and his advisers would unquestionably give far greater weight to the views of Wheeler." Finally, and decisively: "The request was *never* presented as a 'contingency' [plan]."[30]

Perhaps even more significant at this time, again as described by Clifford, was that in a 4 March 1968 White House meeting with the President and others Wheeler "surprised me by backing away from the Wheeler-Westmoreland request in order to keep alive *his* primary objective, a call-up of the Reserves." Thus "Wheeler disassociated himself from the request, referring to it throughout as 'Westmoreland's request': 'If we could provide Westy the troops he wants, I would recommend they be sent — but they clearly cannot be provided.'"[31]

General Maxwell Taylor, who knew both Westmoreland and his approach to conduct of the war intimately, was a member of the working group convened at the President's direction and under Clifford's chairmanship to consider how to proceed. "In fairness to him [Westmoreland]," Taylor wrote, "I thought he should be told that, in this phase of the conflict, remote terrain along the frontiers of South Vietnam meant nothing in itself insofar as Washington was concerned, that the President and his advisers looked with favor on the avoidance of combat close to the cross-border sanctuaries of the enemy where he had the advantage of short lines of communications, and that they saw no advantage in paying a high price to hold exposed outposts like Khe Sanh." That was a pretty thorough indictment of Westmoreland's way of war, one to which Taylor added this: "Apart from other disadvantages, such border operations offered the enemy the opportunity of gaining cheap, minor successes which propaganda and our own media would then blow up into a major victory for them and a disaster for us."[32]

In the Pentagon there was little doubt a request had been made. Lieutenant General Elmer Almquist, then the Southeast Asia program coordinator in Army DCSOPS, remembered it well. "All I know is that I was working my ass off on a message requesting 206,000 troops," he said. "I never thought Westy knew what the hell was going on over there. Then I was told to recover the copies of the message. They denied there was ever a request. They had to get all the copies of the messages."[33]

LBJ, RUSK, CLIFFORD, McNamara, Ambassador Bunker, Admiral Sharp, the Joint Chiefs of Staff, the JCS and OSD historians, CIA Director Helms, General Bruce Palmer Jr., even Wheeler, so he said — all viewed the 206,000 as a request for more troops. Only Westmoreland argued — in vain — that it was merely a "contingency" plan.[34] Finally even he admitted the obvious, telling reporter Sam Donaldson in 1985 that "General Wheeler persuaded me to ask for 200,000 reinforcements."[35]

Charles MacDonald wrote that the Westmoreland-Wheeler request, "which would require a large-scale call-up of reserves and major additional appropriations, stunned the president's civilian advisers and convinced many that a time had come for a change in policy."[36] And, as Graham Cosmas put it in an official Army history: "The troop request all but nullified General Westmoreland's claims of victory."[37]

"The Vietnam War was not unwinnable," said General Fred Weyand. "It was just not winnable Westmoreland's way."

20

★

HEADING HOME

HAVING SERVED FIFTY-FOUR months in Vietnam, in addition to thirty-nine months abroad during World War II and another sixteen in the Korean War, Westmoreland had in the aggregate spent more than nine years overseas, over a quarter of his thirty-two years' service. Now he headed home for the final time.

Immediately there was much press speculation that Westmoreland was being relieved because of the Tet Offensive, viewed by many in the United States as a victory for the enemy, at least in terms of achieving surprise and disruption on a grand scale. Westmoreland argued strenuously that he was not being relieved for cause, that his reassignment had been in the works for some time, and that it was in fact a promotion. He had stated during an April 1968 visit to Washington that the President revealed to him as far back as January 1967 that he wished to make him Chief of Staff. When the new appointment was announced, he told Brigadier General Chaisson that he had hoped to be given one of two other posts, CINCPAC or Chairman of the Joint Chiefs of Staff, and would have preferred CINCPAC, "but supposed that the Navy fought it."[1] General Phillip Davidson scoffed at even the possibility of such an appointment, saying that Westmoreland had "no chance" of becoming CINCPAC and that for General Wheeler to have broached such a prospect was "exploitation of Westmoreland's naiveté."[2]

There is confirmatory evidence for Westmoreland's contention that his reassignment from Vietnam had been in the works for some time, although (had he been aware of it) Westmoreland would not have been pleased with the underlying premise. In 1967 Dr. Edwin Deagle, then a captain, was studying at Harvard under Army auspices when McGeorge Bundy came to deliver a series of lectures. Richard Neustadt of the Harvard faculty told Deagle to be sure he attended Bundy's talks and instructed him to wear his uniform. It was then arranged for Deagle and Bundy to meet privately during one of the receptions. Bundy began by saying, "Captain Deagle, some of my former colleagues here at Harvard tell me you can help me. We must soon bring General Westmoreland back from Vietnam and replace him with someone with a more agile, creative mind. What should we do with him and who should be his replacement?"

Naturally this staggered the young captain, but he did his best to respond. "You should not make General Westmoreland Chief of Staff of the Army, as he will have to go to great lengths to justify his record. Make him the new NATO commander, where he will have no responsibility for Vietnam policy. The Chief of Staff should be General Andrew Goodpaster, who is brilliant and wise in the ways of Washington. Replace Westmoreland with General Creighton Abrams, who is a gifted combat commander." Bundy thanked him and left, whereupon Deagle "walked out with rubbery knees and got a martini."[3]

WESTMORELAND'S TENURE IN Vietnam had gravely damaged the war effort. Ambassador Bunker summarized the evidence, observing that "because this kind of war was new to the American experience, it is clear that we have made mistakes. We did not in the beginning, I think, fully understand the complexities of this kind of warfare. Prior to Tet 1968 we underestimated the capabilities of the enemy. And we were slow in equipping our Vietnamese allies while the enemy was being equipped by the Soviets and Chinese with a wide range of the most sophisticated weapons."[4]

General Bruce Palmer offered a similar assessment. "South Vietnam needed more time," he concluded, "and the real tragedy was that if we had concentrated more of the time we had on building the South

Vietnamese forces rather than trying to win it ourselves, we might have done better in the long run. I think it could have been done. We misjudged how tough it was."[5]

Bunker and Westmoreland also had divergent views on a subsidiary but important matter, as became clear at Westmoreland's departure press conference. The issue was enemy shelling and rocketing of cities, especially Saigon. Said Westmoreland, "The Viet Cong shelling is being blown out of proportion. . . . The shelling has not done major damage. There have been a few people killed, there's been some destruction, but it's been minor. There have been a number of innocent people wounded, but it has not disrupted life in Saigon. Life goes on as usual."

Westmoreland did, however, deplore the public relations aspects of the shelling. "I was recently in the United States," he told the assembled reporters, "and I was shocked to see how a few rockets were ballooned in the public eye."[6] And, he claimed, "It's of really no military significance. It does make headlines, I must say."[7]

That outlook was very much at variance with Ambassador Bunker's views.[8] Only a month before Westmoreland's departure Ambassador Bunker had told the President, in his periodic reporting cable, that "the indiscriminate attacks on civilians in Saigon had had an impact in psychological terms as well as added to the list of dead, wounded, and homeless." He reinforced that view in his next cable, only six days later, saying that "the terror attacks on the cities in the South, which are essentially attacks against the civilian population, are obviously designed to destroy popular morale, to create loss of confidence in the government, and eventually to bring about its downfall." Thus: "It is not so much the amount of damage inflicted, but the fact that this is a continuing pattern which is important."[9]

Those diametrically opposed views on the significance of enemy rocket attacks on Saigon illustrated, as dramatically as any other single aspect of the conflict, how Westmoreland simply never understood the war he was assigned to fight.

S.L.A. MARSHALL HAD spent considerable time in Vietnam during Westmoreland's tenure there, traveling extensively and producing books describing major battles. Now, as Westmoreland prepared to leave Vietnam for Washington, Marshall wrote a devastating critique of the way he had fought the war. "There is little payoff in any of the large sweeps where we crank up a division and more to beat out the countryside," he stated. "Least of all are these ventures advisable when thrown against enemy buildups along the Cambodian border, where the enemy has every advantage. The record of what happens to us there is not less than ghastly."[10]

Said General Fred Weyand: "When Clark Clifford took over as Secretary of Defense after Tet 1968, he found that the Joint Chiefs of Staff had no concept of victory and no plan to end the war. And that was the case in Saigon as well."[11]

For his part, Westmoreland viewed himself as having been much put upon. "As American commander in Vietnam," he maintained, "I underwent many frustrations, endured much interference, lived with countless irritations, swallowed many disappointments, bore considerable criticism."[12]

WESTMORELAND TOLD PEOPLE that he kept under the glass on his desk this quotation from Napoleon: "Any commander-in-chief who undertakes to carry out a plan which he considers defective is at fault; he must put forward his reasons, insist on the plan being changed, and finally tender his resignation rather than be the instrument of his army's downfall."[13]

During his four years in command in Vietnam Westmoreland had found much he disagreed with in the war policies of his government, and in later years — in speeches and articles and memoirs — he would be even more openly critical of such policies as confining ground forces to South Vietnam, keeping the enemy sanctuaries in Laos and Cambodia off limits to allied forces, and limitations on bombing North Vietnam and refusals to mine or otherwise block North Vietnam's ports. But never, apparently, did he view those limitations, individually or in the aggregate, as sufficiently disabling to suggest resignation in accordance with the Napoleon dictum.

Once, during the period when the memoirs were being drafted, Charles MacDonald was in the car with Westmoreland and Kitsy, and the matter of the Napoleonic quotation came up. "I'm sorry, sir," said MacDonald, "but all these things you've been telling me about interference from Washington, how you were not allowed to go into Laos and Cambodia, and how the air war was run the wrong way, it's hard to reconcile this quotation with all those things you thought were wrong. If you believed in that quotation, then maybe you should have resigned." At that point Kitsy chimed in, "That's what I say, too, Charlie."

But Westmoreland disagreed. While he was unhappy "with many of Washington's decisions and restrictions," he said, he "believed that success would finally be achieved in spite of it, that they would not be, as Napoleon put it, 'instruments of his army's downfall.'"[14]

FOR WESTMORELAND IT was truly a long good-bye, with several preliminary visits to the United States preceding the final departure. On 30 May 1968, visiting President Johnson at the LBJ Ranch, he took the occasion to state his conviction that in the war in Vietnam "time is on our side." And, he said during that same press conference, "The enemy seems to be approaching a point of desperation. His forces are deteriorating in strength and quality. I forecast that these trends will continue." After lunch at the ranch, Westmoreland said in his history notes, the President and his party "departed for a sheep auction, and I departed for Washington."

Aware of Westmoreland's fondness for press conferences and speaking opportunities, Wheeler cabled him some pointed instructions in advance of the U.S. visit, during which Westmoreland would undergo confirmation hearings on his nomination to be Chief of Staff. Wheeler said that he had spoken with the President and Secretary of Defense Clifford about the timing of those hearings. "I believe that everyone is in accord that you should not remain in country [meaning in the United States on that visit] any longer than is necessary. . . . You should not accept any public speaking engagements," he cautioned.

IN A FAREWELL CONFERENCE with senior South Vietnamese officers, Westmoreland issued them each a paper entitled "Final Advice by General Westmoreland, COMUSMACV." The document included nine printed points, with a final tenth point added by Westmoreland in longhand (and which he asked them to similarly add in longhand on their copies): "Maintain the offensive spirit."

Near the time for his departure, Westmoreland briefed members of the South Vietnamese National Assembly. In those remarks, according to a memorandum describing the session, "General Westmoreland stressed that it is important that mobilization be applied impartially, affecting rich and influential citizens as well as those who are not. The sons of ten US generals have been killed in Vietnam, said General Westmoreland, while the sons of many high-ranking Vietnamese civilian and military officers are whiling away their time in Paris. This is difficult for the American public to understand and is an abuse which should be ended." Senator Nguyen Gia Hien thanked General Westmoreland for his remarks and predicted that he would rank with General MacArthur in the annals of American military history.[15]

At a news conference in Saigon held the afternoon before he left Vietnam, Westmoreland told the assembled newsmen that in late morning he had sat at his desk and jotted down "a number of the bench marks" that had occurred during his command. The recitation that followed amounted to nearly seven single-spaced typed pages of detail, including such information as where the trucks used by the North Vietnamese were manufactured, with no apparent sense of perspective or proportion.

When the enemy upgraded his capabilities from porters to trucks, for example, claimed Westmoreland, the change was a result of his having denied them access to porters by pushing enemy units inland toward the Laotian and Cambodian borders. Then he argued that the trucks the enemy began to use to transport heavier tonnages of ammunition for his upgraded weaponry represented "vulnerabilities." Next he described in some detail how the enemy had been provided with the newest and best Soviet and Chinese weaponry, never mentioning that the South Vietnamese had during those same years continued to be armed with castoff U.S. World War II equipment.

Having droned through all of this, he concluded: "Now, I have a summary statement that the TV media may want to record and I will give you a moment to get set up." Then he asked: "Anybody not ready?" Next he read this statement: "At this time our military posture is at its height since our commitment. We are now capable of bringing major military pressure on the enemy. This we are doing and the enemy is beginning to show the effects. The Vietnamese Armed Forces are growing stronger in size and effectiveness. Resolve is still the key to success. Trends are favorable, but it is unrealistic to expect a quick and early defeat of the Hanoi-led enemy."

He then asked (about his summary statement): "Want me to read it again for TV? I will read it again." And he did. Finally Westmoreland told the media, reported the *New York Times*, "that if he had to serve his 52 months in Vietnam over again, he would make few, if any, major changes in the way he has conducted the war."

Afterward Westmoreland's aide distributed statistics on the general's travels: 469,638 miles in Vietnam, for an average of more than 9,000 miles a month and 300 miles a day. In addition he had made eleven trips out of country, racking up another 242,000 miles, for a grand total of 711,638 miles in 52 months.[16] The next day, as a South Vietnamese band played "Auld Lang Syne," Westmoreland climbed into a T-39 jet and flew away.

WESTMORELAND'S DEPARTURE ENDED four years during which, as Secretary of Defense Melvin Laird later observed, "the war had been Americanized." Said Laird, "We had not been giving the South Vietnamese the tools to do the job. We had been doing the job for them." Or, he might have added, trying to.

The attrition strategy had not worked. The enemy had not lost heart, nor given up his intention of unifying Vietnam by force. Under Westmoreland, allied forces had indeed imposed large numbers of casualties on the enemy, a horrifying number really, but that had not diminished the will or magnitude of the opposing force. Instead the communists just kept sending replacements down from the north, year after year, keeping Westmoreland on a kind of treadmill. Meanwhile U.S. losses were also accumulating and, while they were not as large as those the enemy was experiencing, that really did not matter. What did matter was that they became more than the American people were willing to accept as the price for what seemed to be at best a stalemate in a faraway war.

Many years later, when he belatedly decided to speak out on the war, Robert McNamara recalled "the seemingly endless reports and requests made by Gen. William Westmoreland to Washington between 1964 and 1968." His assertions of progress and success of the attrition strategy were, now saw McNamara, "an illusion. At no time during Westmoreland's tenure in Saigon, it now appears, was there the slightest chance of reaching the famed crossover point beyond which the fortunes of the Vietnamese communists would decline, leading them eventually to sue for peace."[17]

General DePuy—the architect of search and destroy—had once said, in what he called a "coldly realistic" assessment of the situation, "We are going to stomp them to death. I don't know any other way."[18] In a much later interview, by then considerably chastened, he explained the outcome: "We were arrogant because we were Americans and we were soldiers or marines and we could do it, but it turned out that it was a faulty concept, given the sanctuaries, given the fact that the Ho Chi Minh Trail was never closed. It was a losing concept of operation."[19]

At least in the views of some observers, responsibility for such failure resided more with LBJ than with his field general. "No capable war President," wrote historian Russell Weigley, "would have allowed an officer of such limited capacities as General William C. Westmoreland to head Military Assistance Command, Vietnam, for so long."[20]

Some senior officers saw it the same way. "If Westmoreland couldn't have done it," said Lieutenant General Samuel T. Williams, "they could have jerked him and put someone in there that could have done it. They had plenty of people who could have done it. Abrams could have done it."[21] Williams had served in Vietnam for five years as first chief of the MAAG and knew a lot about its challenges.

Perhaps, too, fortune played a role in how things played out during these years of greatest American involvement in the war. Westmoreland may have thought so. "The Vietnamese have it that if a gecko outside your house croaks nine times straight, you will have good luck," he said. "Quite a colony of geckos lived around our villa, but no matter how many times I counted, no gecko of mine ever got beyond eight."[22]

21

★

CHIEF OF STAFF

O N   T H E   M O R N I N G of what should have been a day of triumph and celebration, his ascent to the highest post in the Army, Westmoreland received a devastating message. In Vietnam a helicopter carrying Lieutenant Colonel Frederick Van Deusen, commanding officer of an infantry battalion, had been shot down and plunged into the Vam Co Dong River. Colonel Van Deusen and others were missing, cabled General Creighton Abrams, adding, "I thought you should know but not anyone else."[1] Fred Van Deusen was Kitsy's younger brother.

Westmoreland, understandably, initially said nothing of this to Kitsy, for it would have been impossible for her to get through the day. Further details soon came in another cable. Colonel Van Deusen's helicopter had been recovered from the river. Bullet holes were found in the chopper's floor. One body had been recovered from the skids and positively identified as that of Van Deusen. The battalion executive officer would escort the body back to the United States.[2]

Fred Van Deusen had been in command of his battalion for just two and a half weeks. In that brief time he earned the Distinguished Service Cross, two Silver Stars, and the Purple Heart. Brigadier General Elvy Roberts, assistant division commander of the 9th Infantry

Division, wrote to Kitsy that her brother had been "lost in the thick of battle leading his troops magnificently at the location where there was the heaviest combat action."

WESTMORELAND WAS SWORN IN as Army Chief of Staff that morning of 3 July 1968 in the office of Secretary of the Army Stanley Resor, followed by an honors ceremony on the Pentagon Mall.[3] "On this occasion," said Westmoreland, "I cannot help but reflect on the officers and men who have served me so loyally and effectively during the past four years in Vietnam." Five minutes after arriving at his new office, Westmoreland was on the phone with Philip Mallory, Commanding General of Walter Reed General Hospital, setting up an appointment to deal with a persistent intestinal disorder. His assigned doctor, Lieutenant Colonel Boyce, examined him there the next afternoon, the Fourth of July. They conferred twice more by telephone before, on Monday, 8 July, Westmoreland and Kitsy departed for Fayetteville, North Carolina. The following day Fred Van Deusen was laid to rest there after a funeral at St. John's Episcopal Church, the same parish in which Westmoreland and Kitsy had been married twenty-one years earlier, when Fred, still a boy, had been an acolyte. It must have been a surpassingly painful day.

After the funeral Westmoreland went back into the hospital, this time at nearby Fort Bragg. Soon after returning to Washington he was honored at a White House ceremony, where President Johnson awarded him his third Distinguished Service Medal, the citation noting that "by his conduct, by his competence and his compassion, General Westmoreland epitomizes the finest qualities of American fighting men that he commanded so long and commanded so well."

THERE WAS MUCH to be done during Westmoreland's years as the Army's Chief of Staff. The war continued on, although at progressively reduced levels of American involvement. Readiness and morale throughout the Army were at low ebb, largely because the war had deprived other commands of manpower, experienced leadership, materiel, and funds. Before long there would loom the prospect of transitioning to an all-volunteer force, a prospect that filled many with great foreboding. Then too research, development, and acquisition had been badly stunted by the operational costs of the war, with little expectation of improvement any time soon.

Faced with this panoply of challenging tasks, Westmoreland nevertheless concentrated on fostering his own legacy. "He spent a lot of time on that," said an officer who served as his executive officer during a portion of his four years in that post. "He was very interested in the proper telling of the story of the Vietnam War."[4] Major Paul Miles agreed sympathetically. "Vietnam was always at the top of the agenda," he said. "Europe and other major concerns were almost a distraction. He was carrying a heavy burden."[5]

Westmoreland himself told interviewers how he saw his responsibilities. "As you know," he said, "I spoke in every state in the union. I considered myself the military spokesman of the Army, and that I should be exposed to the American public and put forth the Army's point of view." And, most significantly: "I felt that an understanding of the military was the primary mission that fell on my shoulders while I was Chief of Staff."[6]

Lieutenant General Robert Pursley served for several years as Military Assistant to the Secretary of Defense, and he too remembered Westmoreland's extensive speaking schedule. "It was a joke around 3E880 [the Secretary of Defense's Pentagon office] when Westy was in town," said Pursley. "He was always out speechifying."

Wherever he went, no matter the audience, Westmoreland spoke of Vietnam and, more specifically, his tenure in Vietnam. Addressing officers serving in Washington assignments outside the Department of the Army, for example, he told them that in Vietnam "our basic national strategy, which has never been publicly announced for obvious reasons, was to hurt the enemy until he negotiated, and in the process not to widen the war. This has been our basic national strategy." Within the established parameters, he said, "my basic strategy within South Vietnam, my battlefield strategy, was to grind down the enemy" and "concurrently build up the Vietnamese quantitatively and qualitatively and modernize their forces to the point where at some future date we would be able to turn more and more over to them and eventually redeploy some of our forces or phase down the level of commitment."

Not only was that the strategy, Westmoreland told this group of officers, but also "we must make this fact known, that the Army was a winner. We have hurt the enemy to the point that they have now moved to the negotiating table. Our military forces have served their purpose in supporting national policy." Speaking as though the war were over, Westmoreland developed this theme at length, then asked for questions. There were none.[7]

In early autumn, after the Westmorelands had moved from temporary housing into Quarters 1 at Fort Myer, they invited President and Mrs. Johnson to dinner. The President was delayed, so Lady Bird went ahead with her daughter Luci. "At first there were only the four of us," she remembered, "but then General and Mrs. Earle Wheeler joined us. Both Generals were in civilian clothes, looking curiously different and defenseless without those uniforms." The President arrived quite late, maybe nine o'clock, and looking, said Lady Bird, "as tired as I have ever seen him, as worn—the fight temporarily gone out of him." But it was a warm occasion, even sentimental, she said in her diary. "One of the things that Lyndon said in describing the mood of the country concerned the great disservice that has, somehow, been done to the military. Their influence, prestige, place in society, are under attack. The glitter is off the stars."[8]

WESTMORELAND'S VICE CHIEF of Staff was his classmate General Bruce Palmer Jr. Palmer had served under Westmoreland in Vietnam, first as Commanding General, II Field Force, Vietnam, a tactical command, and then, after only a few months in that job, as Deputy Commanding General, U.S. Army, Vietnam, a largely administrative post. Palmer bitterly resented his reassignment, which he attributed to Westmoreland's desire to put someone else into the tactical billet. Now he was going to be backstopping Westmoreland again, and for the entire four-year tour.

A widespread perception in the Army of that day was that, while Westmoreland was out trying to salvage his reputation in the wake of Vietnam, Palmer was running the Army. "Westy went back," recalled Lieutenant General John Cushman, "and of course that's when Bruce Palmer began to run the Army." Another officer recalled that "General DePuy would tell us: 'Send Westmoreland out of town for a week so we can get some work done with General Palmer.'" Colonel S. V. Edgar, who was involved in a major reorganization project during the Westmoreland years, said that "it was widely, I think, admitted that General Palmer was really running the Army and made most of the decisions."[9] That was confirmed, especially in terms of the planned reorganization, by Secretary of the Army Robert Froehlke, who said, "the person really behind that was Bruce Palmer. Bruce ran the Army, and did it very well."[10]

Later, when Palmer was about to retire, then–Major General Donn Starry wrote to him, "I shall always remember those sort of grim years there in the Pentagon when you held the Army together virtually singlehanded."

Palmer himself addressed this issue sympathetically in an interview with Army historian Dr. James Hewes. "Particularly in the early days," he recalled, "the Chief, General Westmoreland, was away, out of town a lot. He had come back from Vietnam in the summer of 1968. We had a new President in 1969, and to President Nixon Westmoreland was a political liability." Thus, "Westmoreland's feelings were pretty badly hurt, particularly in those early days. So he decided he would travel as much as he could."

"We never saw him," Palmer said of Westmoreland during these years. "He made a great point of it—he visited all fifty states." Palmer perceived a political motive at work: "He thought he was going to be another Eisenhower, swept into office. He really thought he was going to become President. He talked to me about it." Palmer knew that was unlikely. "I told him, 'Westy, it's completely different. It [Vietnam] was not a popular war.'"[11]

The extent to which Palmer functioned essentially as the Chief of Staff was later reflected in a gesture spearheaded by the Army Chief of Military History, who arranged for a "Chief of Staff" portrait of Palmer to be painted and hung in the Pentagon corridor where former Chiefs of Staff are commemorated. Done by Herbert Abrams, the same artist who painted portraits of Westmoreland and Abrams, Palmer's portrait hangs between those of his two classmates. The rationale was that Palmer had been Acting Chief of Staff for an extended period, which indeed he had for about three and a half months between Westmoreland's retirement and the confirmation of Abrams, but of course that had not entailed any Presidential nomination, Congressional committee hearings, or Senate confirmation. Clearly the intention of the Army was to honor Palmer for his extraordinary service as Vice Chief of Staff during the years he largely stood in for Westmoreland.

IN THE AUTUMN OF 1968 Westmoreland made his first visit as Chief of Staff to the Army War College. The Commandant, Lieutenant General William McCaffrey, had asked him up to give the usual Chief of Staff's "State of the Army" overview to the class. "I thought the Chief of Staff ought to come talk to his War College students," said McCaffrey, "because that's the future of the Army and they ought to see him and he ought to see them and so on." Things didn't work out as McCaffrey had hoped. "I invited Westy up to talk as the Chief of Staff. But Westy didn't talk as the Chief of Staff," said a disappointed Commandant. "He talked all about Vietnam. He couldn't wrench his mind away from it, and it was an apology for all that had happened."[12]

The following year Westmoreland was the first speaker for the next War College class. Colonel William Greynolds remembered the occasion vividly. The talk began at 11:00 A.M. and continued, despite being scheduled to end at noon, until about 1:15 P.M. "When lunchtime came," said Greynolds, "students from the other services began to drift away, but the Army guys in uniform had to gut it out. Westmoreland went through a seemingly endless series of cards, each dealing with some problem he had identified and the actions he said he had taken to deal with it. He seemed sort of desperate, and the whole thing was rather pathetic."[13]

Former Secretary of the Army Stephen Ailes had anticipated this problem. "The trick in the business of transition is to look ahead, not back, and to become involved in the projects that lie ahead," he said in a letter to General Harold K. Johnson. "In this connection I am hopeful that Westy will concentrate his energies, and certainly his speeches, on his myriad responsibilities as Chief of Staff and on what lies ahead for the Army rather than on any defense or justification of his role in Vietnam—a role that I agree is easily defended."[14]

That was, unfortunately, not to be. "All he wanted to do was write his book," said General Ferdinand Chesarek, then the Assistant Vice Chief of Staff. Chesarek saw there was also going to be another problem for the new Chief of Staff. Secretary of the Army Resor, he said, "was a detail man. Mr. Resor wants to know all of the ramifications. . . . General Westmoreland doesn't function like this. He's not a detail man."[15]

Another place where doing the homework mattered was in the Joint Chiefs of Staff conference room, known as the Tank. Deliberations there were serious business, with virtually every substantive issue overlaid with budget implications and often contentious matters relating to service "roles and missions" as well. Westmoreland, traveling extensively and not that interested in detail, did not show to advantage in that venue. "I was in the Tank a lot," recalled Colonel John VanDerBruegge, then a J-1 staff officer. "I was always embarrassed at the presentations that Westmoreland would make on Army matters. He was inept, and he wouldn't put his glasses on. One time he started briefing from the wrong paper." Lieutenant General Robert Beckel, an Air Force officer working for JCS Chairman Admiral Moorer, recalled being in one meeting with Westmoreland and was not impressed. "He seemed rather stupid," said Beckel. "He didn't seem to grasp things or follow the proceedings very well." General Ralph Haines, the holdover Vice Chief of Staff until Bruce Palmer took that position, formed an even broader view. "Westy certainly never understood the operation of the JCS," he said.

The young majors and lieutenant colonels—"action officers" as they were called—soon encountered their own problems in dealing with Westmoreland. Memoranda for the Chief of Staff were now limited to one page (although attachments were permitted), and most business was conducted on the basis of oral briefings. Remembered Lieutenant General Charles Simmons of that time: "General Westmoreland was intellectually very shallow and made no effort to study, read or learn. He would just not read *anything*. His performance was appalling."

Some briefers were dismayed to find that Westmoreland would occupy himself during one-on-one deskside briefings by signing photographs of himself, one after another, while they made their presentations. Sometimes he would fall asleep while being briefed, leaving the panicked staff officer trying to decide whether to continue as though nothing had happened or wait until the general awoke before resuming.

. . .

"WESTMORELAND CAME HOME," said his close associate Colonel "Hap" Argo, then still in Vietnam as the MACV Historian. "He thought he was regarded as a great hero. Then he got a look at the *Washington Post*. He called me in Vietnam, where it was midnight, and ran on for three hours. Saving his career [reputation] became a major thing with him. He called me every day for a week, and we talked two hours a day. He could not believe what had happened to him. He had stayed at the White House."[16]

Westmoreland brought into the Army Staff a number of senior officers with whom he had served before, including perhaps most importantly General William DePuy. DePuy was serving on the Joint Staff, but before long Westmoreland was able to have him reassigned to fill an important billet as Assistant Vice Chief of Staff. In that role he often coached Westmoreland for important meetings and negotiations, and in a very basic way. Major General Clay Buckingham, then a more junior officer, was a division chief in the Office of the Assistant Chief of Staff for Force Development. He remembered an occasion when DePuy and his staff were briefing Westmoreland on proposed major changes to the Army force structure. "Westmoreland did not seem to be paying attention," he said. "At one point General DePuy, who was sitting next to the Chief, interrupted the briefing and, turning toward Westmoreland, spoke directly to his face from about eight inches away. General DePuy's brief and forceful remarks, spoken in a calm voice but one which could easily be heard by all attendees, went something like this: 'Chief, you *must* pay attention and listen to this. It concerns the future of the Army. It is very important for the Army and you *must* make a decision about our proposal at the end of this briefing.' For a few minutes Westmoreland seemed to pay attention, but before long his eyes wandered and his gaze seemed to be on something far, far away. At the end of the briefing he simply told General DePuy to go ahead with his plans, got up, and walked out. His mind was obviously somewhere else."[17]

ONE OF THE MOST traumatic, and shameful, episodes of the Vietnam War was what came to be known as the My Lai massacre, a bloody day in which elements of the 23rd Infantry (Americal) Division mercilessly gunned down as many as several hundred unarmed South Vietnamese civilians, mostly women and children, including babes in arms.

This had happened (16 March 1968) on Westmoreland's watch as commander, but because leaders in that division covered it up it did not become public knowledge until about a year later. By then, of course, Westmoreland was back in the United States as Chief of Staff.

Lieutenant General William R. Peers was appointed to head an investigation into what happened at My Lai and any subsequent cover-up of the incident. Peers and his team did an admirable job, but encountered many frustrations. Concluded Peers, "[T]he failures of leadership that characterized nearly every aspect of the My Lai incident had their counterpart at the highest level during the attempt to prosecute those responsible."[18]

The Commanding General of the Americal Division at the time of the massacre was Major General Samuel Koster. When those war crimes came to light Koster was, of all unfortunate places, at the Military Academy at West Point as Superintendent. A dozen or so officers had been charged with covering up the war crimes at My Lai, with Koster as the former division commander the most senior of them. He was replaced at West Point and subsequently reduced in rank, stripped of a high decoration he had been awarded by Westmoreland for his service in Vietnam, and officially reprimanded.[19]

All those actions were taken administratively, with no court-martial charges preferred. General Peers was appalled. "I was especially disturbed by General Seaman's dismissal of charges against the senior officers, particularly in General Koster's case," he wrote. "When General Westmoreland informed me of the proposed action against General Koster (the letter of censure), I told him, in effect, that it was a travesty of justice and would establish a precedent that would be difficult for the Army to live down."[20]

Historian Stephen Ambrose wrote that what happened at My Lai "was a failure of leadership on Westy's part, and on down—corps, division, battalion, down to platoon level."[21] Guenter Lewy, a distinguished academician, concluded that "until the My Lai incident, the rules of engagement were not as widely known as they should have been, and the American command can justly be faulted with failing to take all possible measures to enforce the rules."[22] Those were the considered views of favorably disposed and responsible critics.

General Peers wrote in an 18 March 1970 memorandum to Westmoreland: "Directives and regulations, no matter how well prepared and intended, are only pieces of paper unless they are enforced aggressively and firmly throughout the chain of command."[23] He told Westmoreland directly that "I need the directives, but I need the psychology down the chain of command, too. That is more important to me than a written piece of paper." The Peers Report subsequently included the devastating finding that "prior to My Lai 'there had developed . . . a permissive attitude towards the treatment and safeguarding of non-combatants which was exemplified by an almost total disregard for the lives and property of the civilian population . . . on the part of commanders and key staff officers.'"[24]

At one point Westmoreland himself faced war crimes charges, brought against him at the instigation of an Army sergeant who maintained that he had been guilty of dereliction of duty in failing to prevent the massacre at My Lai. The complainant specifically rested his charges on the precedent of a World War II case in which Japanese General Yamashita was convicted of war crimes by an American tribunal and executed.

Subsequently the Army's General Counsel, Robert Jordan, concluded that Westmoreland "did as much as any person could be expected to do to prevent the kind of atrocities which are alleged to have occurred" and that "there is clear evidence here that General Westmoreland knew absolutely nothing about the Son My [My Lai] incident at the time in question." Jordan so advised Secretary of the Army Stanley Resor.[25]

Resor took the matter under advisement, then recorded in a memorandum for record dated 14 October 1970 that "as a court-martial convening authority under Article 22, Uniform Code of Military Justice, 10 U.S.C. 822, I have personally made a preliminary inquiry into the charge against General William C. Westmoreland preferred on September 9, 1970 by Sergeant Esequiel Torres." Then: "I have concluded that General Westmoreland took appropriate measures available to him to prevent the commission of atrocities, that criminal justice proceedings were initiated by appropriate authorities subordinate to General Westmoreland where the evidence supported such atrocity charges, that at the relevant times General Westmoreland had no knowledge of the alleged Son My incident, and that the Yamashita case provided no precedent to support the charge. I have consequently concluded that the charge filed by Sergeant Torres is unsupported by the evidence and that it should be and hereby is dismissed."[26]

"As regards the question of ultimate responsibility," wrote William F. Buckley Jr. at the time, "the public is entitled to be confused. We hanged General Yamashita after the Second World War, and if we applied rigorously the logic of that execution, we would have a case for hanging Gen. Westmoreland. That would be preposterous and cruel."[27]

While all this was going on, an extremely interesting letter was provided to Westmoreland by a friend. It had been written by Major General Russel B. Reynolds, identified as president of the Yamashita court-martial (actually a military tribunal), to Colonel John H. Tucker Jr. in late 1970 and forwarded by Tucker to Westmoreland the following April. Reynolds described how the members of the tribunal had decided, at their final meeting, that they would never speak or write publicly about General Yamashita's trial. Their rationale was that the very voluminous record, over four thousand pages, including testimony from 286 witnesses, would stand on its own. But then, said Reynolds, various commentators began to charge that "the prosecution had failed to produce any evidence that Yamashita ordered the atrocities, or even knew about them." He knew that was not true and quoted from a letter addressing the point that he had written to a friend in 1946: "There was such testimony—direct, responsible, eyewitness, convincing, damning."[28] There were no parallels whatsoever between the Yamashita case and the conduct of Westmoreland in Vietnam.

DESPITE HIS LONG years of service in Vietnam, and his almost daily visits to various units and installations across the country, Westmoreland seemed to understand very little of what the war was like for the combat infantryman there. After *Life* magazine published an article with numerous photographs showing troops just in from the field as ragged and disheveled, he called in Lieutenant Colonel John Galvin, recently returned from having commanded the battalion depicted. Galvin was appalled by the interview. "Westmoreland had no understanding of what the troops were going through out in the jungle," he discovered. "I couldn't believe it." One picture was of a company commander named Utermahlen. He had just come out of the field and was unshaven and unkempt. "Westy didn't understand that," recalled Galvin. "He was fixed on appearance. 'Jack, explain this to me,' he said. But he just didn't get it!"

IN 1969 TOWNSEND Hoopes, a former Under Secretary of the Air Force, published *The Limits of Intervention* in which he described Westmoreland as "a thoroughly decent, moderately intelligent product of the Army system, long on energy and organizational skill, short on political perception, and precluded from serious comparison with the leading generals of World War II and Korea by an unmistakable aura of Boy Scoutism." Turning to his performance in Vietnam, Hoopes said that Westmoreland "relished the challenge of searching out and destroying the NVN regular forces, but was essentially indifferent to the fighting capacities of the ARVN, asking only that it get out of his way; and he gave lowest priority to pacification, indeed seeming not to see the relationship between aggressive search and destroy operations in the populated countryside and the stream of refugees (some 900,000 by October 1967) who were forced from their land by the terror of artillery and air strikes, by burned out villages and ruined crops; and who, pressed into crowded, unsavory camps along the southern coast, were now sullenly anti-American."[29] Westmoreland was furious, assigning staff officers to take the book apart in search of misstatements or errors that could be used to attack and discredit it. He himself resorted to an impassioned ad hominem attack, telling the Army's Chief of Military History that Hoopes's account was "the most fictitious thing I ever saw, based apparently on deep-seated antagonism, complete ignorance, and a lack of integrity."[30]

Even as Westmoreland continued fiercely defending the war of attrition and search and destroy tactics he had devised, one of his chief associates in conceptualizing and implementing that approach admitted its futility. "I guess my biggest surprise," reflected General William DePuy, "and this was a surprise in which I have lots of company, was that the North Vietnamese and the Viet Cong would continue the war despite the punishment they were taking. I guess I should have expected that. I guess I should have studied human nature and the history of Vietnam and of revolutions and should have known it, but I didn't. I really thought that the kind of pressure they were under would cause them to perhaps knock off the war for a while, at a minimum, or even give up and go back north. I understand that, from 1965 to 1969, they lost over 600,000 men. But I was completely wrong on that. That was a surprise."[31]

"What we also didn't anticipate was the massive intervention of the North Vietnamese Army," he further revealed. "Our operational approach was to increase the pressure on the other side (size of force, intensity of operations, casualties) in the belief that it had a breaking point. But the regime in Hanoi did not break; it did not submit to our logic."[32]

To his great credit, DePuy was very candid in discussing these matters. "We fell into that trap," he said, the trap of trading casualties with the enemy in an effort to wear them down. "We thought, and I guess Mr. McNamara thought, and the JCS thought we were beating the hell out of 'em, and they couldn't take it forever. It turned out they controlled the tempo of the war better than we would admit. We beat the devil out of 'em time after time, and they just pulled off and waited and regained their strength until they could afford some more losses. Then they came back again." "So we ended up with no operational plan that had the slightest chance of ending the war favorably."[33]

Sir Robert Thompson, the British counterinsurgency expert, took part in a seminar in London not long after the conclusion of Westmoreland's tenure as COMUSMACV. He noted the near-total neglect of pacification during those years, along with the emphasis on search and destroy operations, and he too commented on how that had in turn created millions of refugees who, instead of being protected in their home villages, had been driven out of them. Thus, he concluded, "the strategy which was adopted, particularly in the past three years, destroyed the American justification for the war, the justification almost for their being in Vietnam." And: "If you want it in one word, it was given when General Westmoreland was asked at a press conference what was the answer to insurgency. He replied, 'Firepower.'" Thus, said Thompson, "When you come right down to it, it was not General Giap who beat President Johnson, it was General Westmoreland."[34]

WHILE HE WAS Chief of Staff Westmoreland had, as did all senior officers, a stable of speechwriters. They understood their primary mission. "I was a speechwriter for General Westmoreland," recalled Major General Frank Schober. "He talked all over the country . . . trying to explain what he had done in Vietnam."

By mid-November 1971, Westmoreland reported to the Army Policy Council, he had spoken in forty-three states. This strenuous schedule necessarily limited the time available for other aspects of his job, a reality he acknowledged. "I had too much to occupy me [to get into details of Army reorganization]," he said in an oral history interview, "and of course I was Chief of Staff during the height of the antiwar movement in this country. And I, frankly, in evaluating the priorities of my time, gave rather high priority to going around the country and giving them the facts of life with respect to the military . . . and offsetting some of the adverse propaganda that had been perpetrated. I went right into the lion's den on this."[35]

During his final two years as Chief of Staff, Westmoreland had as his Executive Officer Colonel (later full general) Volney Warner, a thoughtful and decent man, thoroughly committed to Westmoreland and desirous of helping him. By then, many officers wanted Westmoreland removed from his position. "When I reported to General Westmoreland to begin service as his executive officer," recalled Warner, "he asked me whether I was a GROW charter member." GROW was an informal network of people who advocated "Get Rid of Westmoreland," as General William Knowlton noted in his oral history. "There grew up within the staff—some of the young civilians within the OSD area or within the Army—it became quite faddish to become a member of that group—devoted themselves wholeheartedly to getting rid of General Westmoreland as Chief of Staff of the Army."

According to Ted Sell, writing in the *Washington Post* in March of 1968, a substantial number of relatively junior Army officers also privately criticized Westmoreland for concentrating on "a war of big units maneuvering all over the country," and as early as 1965 "officers who opposed his views wryly [and clandestinely] organized what they called the GROW program."[36] Another version sometimes cited was WON, "Westmoreland Out Now."

Warner sought to persuade Westmoreland to look ahead rather than back. "I feel that most people are no longer interested in the origins of the Vietnam War and the problems in pursuing it," he said in a note to Westmoreland. Likewise, in a speech outline provided to

Westmoreland, Warner suggested that he "emphasize that the Vietnam era is drawing to a close, and that we must now look to the future." Westmoreland would not, or could not, take that good advice, continuing to dwell on Vietnam for the rest of his long life.

"I would often bring to General Westmoreland's attention problems or bad news I thought he needed to know about," Warner recalled. "Often he wouldn't react in any way. I never knew whether this was because he didn't agree, or he just didn't understand."

Warner recognized Westmoreland's shortcomings but also understood and admired him as a person. "I am absolutely convinced," he said, long after Westmoreland had retired, "that Westmoreland didn't understand the war, doesn't understand it now, but everything he did he did because he thought it was the right thing to do." Their relationship was an unusual one, however. Warner believed that Westmoreland had chosen him for the executive officer's position because of his disagreement with Westmoreland's position on Vietnam, "and because he was willing to give me personally the opportunity to argue directly with him on my point of view." That went nowhere, however, Warner acknowledging that he was never successful in changing Westmoreland's mind. In the process, though, he came to have affection and compassion for Westmoreland matched by few others.

WESTMORELAND'S TRAVEL ORDINARILY involved at least some time at military installations, on occasion with unanticipated results. On a trip to Germany he visited elements of a tank battalion in their local training area. Getting ready for the visit, the unit had set up smoke generators, arranged close air support, and even painted the helipad orange (under the snow). Westmoreland arrived and was greeted by various people. The battalion S-3 watched as his driver, a soldier named Chambers, shook hands with the Chief of Staff. Westmoreland asked him if that was the way things were always set up at the range. "Fuck, no!" exclaimed Chambers, who "proceeded to give Westmoreland an impassioned rundown on all the Mickey Mouse things they'd had to do in preparation for his visit, all the while still grasping and vigorously shaking his hand."[37]

Impending visits by the Chief of Staff were not uniformly welcomed. At Fort Hood the 2nd Armored Division was told that Westmoreland would be inspecting their training on a given date. That created something of a dilemma. Recently the division had received a large number of infantrymen, rotating from service in Vietnam, and been told to retrain these men as tankers. The troops involved had only a few months left before leaving the Army, which was about the only thing that interested them. They were in the midst of a strenuous four-week program to make them into tankers and would be on the range firing tank guns when Westmoreland visited. The division's senior officers discussed the situation, realizing that the gunnery would probably not go well and that they would consequently make a poor impression on the Chief of Staff. It was decided that "the Chief of Staff needs to know the trauma we are undergoing resulting from a DA [Department of the Army] decision to convert short-term Vietnam infantrymen into qualified tankers in four weeks." Besides, they reasoned, "the Chief of Staff is an experienced commander with a reputation for fairness and will understand our situation."[38]

On the appointed date Westmoreland arrived. The range was squared away, but when firing began, few rounds hit their targets. "Although the CG had carefully briefed him on the whole situation en route to the range," recalled an officer then serving as division chief of staff, "the Army Chief of Staff is incensed. He calls the firing to a halt, dismounts the [troops], the NCOs, the officers, and gathers everyone around him. He berates everyone for such a rotten example of gunnery, for the waste of ammo, for the poor NCO instruction, for inadequate officer supervision. Then he takes the CG aside, mercifully out of hearing of the troops, but in their full view, and proceeds to tear the division commander apart; he thereafter leaves the range without a single word of appreciation for anyone." In later discussion with the division staff, the division commander made excuses for Westmoreland. "He is exhausted from a killing schedule," he suggested. "He has been under severe attack by the press in recent weeks." That was Westmoreland's last visit to the division during that commander's tenure. And, to the sorrow of the commander's senior associates, who greatly admired him, Major General Wendell Coats, "until then considered to be a rising star, eventually moved on to another major general's position, well out of the mainstream of the Army, from which he retired."[39]

• • •

AT ONE POINT Major John Sewall was the "traveling aide," which meant he made most of the trip arrangements and generally accompanied Westmoreland wherever he went. As was the case with so many others who were close to Westmoreland, Sewall's impression during his year of duty was that Westmoreland's principal preoccupation was with rationalizing and justifying his performance in Vietnam. One day the two men were being driven back from an appointment in Washington, riding together in the back seat of the sedan. Westmoreland was pretty much all business all the time, observed Sewall, so there was very little chitchat when they were traveling together. On this occasion Westmoreland had been gazing out the window for a long time when he suddenly turned to Sewall and said, "out of the blue" as the young officer remembered it, "You know, if we'd used nuclear weapons we could have won that war."[40]

WESTMORELAND HAD LONG WISHED to achieve an Army aviator's rating and wings. Now he took advantage of his position as Chief of Staff to further that quest. He began by setting up "informal" flying lessons at Davison Army Airfield, at nearby Fort Belvoir. His calendar for those years shows numerous entries for "helicopter flying at 0800," especially on Saturdays, and once even at 2130 (9:30 P.M.) after returning from a two-day conference out of town. This routine continued through much of 1970 and then, on Monday, 21 September 1970, he was at Fort Rucker, Alabama, home of the Army Aviation School, for "flight instruction 0730–1500." That same date, according to an Army special order, he was certified as having completed flight training and was designated an Army Aviator, awarded the Army Aviator Badge, and placed on flying status. A week later Westmoreland signed a document waiving flight pay.

Westmoreland proudly wore his wings, of course, even though—as photographs from that period show—fitting them in among his numerous ribbons and other badges pushed his Combat Infantryman's Badge nearly out of sight beneath the epaulet of his blouse.

IN APRIL 1970, prompted by a recommendation from General Peers based on the findings of his My Lai inquiry, Westmoreland directed the Army War College to conduct a review of "certain areas and practices" within the Army which, he said, "were a matter of grave concern to me." Stunned by Peers's report, Westmoreland tasked the War College to examine "the moral and professional climate within the ranks—the state of discipline and ethics."

The result was published later that year as the *Army War College Study on Military Professionalism*. Its conclusions could not have been more troubling. "The findings of this study surprised and, in some cases, shocked many of the Army's senior leaders," said a staff summary. "In general, it discovered that the majority of the Officer Corps perceived a stark dichotomy between the appearance and reality of the adherence of senior officers to the traditional standards of professionalism, which the words duty, honor, and country sum up. Instead, these officers saw a system that rewarded selfishness, incompetence and dishonesty."[41] That devastating finding came from the officer corps itself, not any outside critics or academicians or any other group with hostile motives or lack of inside knowledge.

An oft-quoted passage from the study noted "a scenario that was repeatedly described in seminar sessions and narrative response . . . an ambitious, transitory commander—marginally skilled in the complexities of his duties—engulfed in producing statistical results, fearful of personal failure, too busy to talk with or listen to his subordinates, and determined to submit acceptably optimistic reports which reflect faultless completion of a variety of tasks at the expense of the sweat and frustration of his subordinates."

Colonel Walter Ulmer (later a lieutenant general) and Colonel "Mike" Malone, who wrote the study, went down from Carlisle Barracks to the Pentagon to brief the results to Westmoreland. "Our recommendation was that a copy go to every general officer in the Army," recalled Ulmer. "General Westmoreland decided instead to have it marked 'Close Hold,' which meant that for a number of years very few people were aware of the study's findings."[42]

Ulmer and Malone were present when Westmoreland and his key people discussed how to handle the report. "Westmoreland said that we should use it, but we should keep it close hold because the Army had just been beaten over the head and this was just another reason to be beaten over the head," said Ulmer. "It was then put close hold. There was very limited distribution. We put a couple hundred copies in some bathroom up at the War College and locked the door."[43]

Scarcely two years later, approaching retirement, Westmoreland would claim that these deep-seated systemic problems had been resolved. "Your Army today is a dynamic organization, proud of its traditions and accomplishments, optimistic about the future, and confident of the direction in which it is moving," he wrote in a January 1972 *New York Times* op-ed essay. And, in a 30 June 1972 valedictory letter to President Nixon, he wrote that in his opinion "the officer corps has never been stronger than it is now—certainly not in my 36 years of commissioned service."

In his final summation, *Report of the Chief of Staff,* Westmoreland provided a more realistic assessment, noting that "throughout the four years [of his tenure as Chief of Staff], we also had to deal with a series of social or behavioral problems—absenteeism, dissent, racism, drug abuse, and crime. Some were peculiar to military service, others were akin to problems existing in civil society."

For some reason that last report was not published until five years after Westmoreland's retirement. Despite the lengthy delay, when the document finally appeared Westmoreland dispensed copies to one and all, just as he had done with the economy plan at Fort Campbell, with MacArthur's speech to the Corps, with his final report as Superintendent after leaving West Point, with his *Report on the War in Vietnam,* and with his book *A Soldier Reports.* Sending a copy of this final document to Lieutenant General Andrew Goodpaster, then Superintendent at West Point, Westmoreland wrote: "Enclosed is a report, recently off the press, covering my stewardship in the Army. I thought this may be of interest to you particularly those portions concerned with the Officers Corp."

22

★

SHAPING THE RECORD

WESTMORELAND USED HIS position as Chief of Staff in multi-faceted efforts to shape the history of the war in Vietnam. His vehicles for telling the story his way included a *Report on the War in Vietnam*, a series of monographs he commissioned, a very heavy schedule of speeches, his *Report of the Chief of Staff*, and — after his retirement — his memoirs and two extensive sets of oral history interviews. He had very able help in these endeavors, most notably from Charles MacDonald, the ghostwriter for the memoirs; Colonel Reamer "Hap" Argo, who had been the MACV Historian and was now brought in to continue helping with writing the history in its multiple forms; and Major Paul Miles, a new aide-de-camp who eventually functioned primarily as a research assistant. Miles recalled that when he was summoned to Washington to be vetted by Westmoreland for the job "the entire interview — conversation — was about how historians would look back on the Vietnam War. He was concerned about how the Army had abdicated to civilians the framing of the historical outlook."

During Westmoreland's tenure as Chief of Staff Dr. Robert E. Morris was at the Army's Center of Military History, overseeing the people who were writing the history of the Army in the Vietnam War.

"General Westmoreland came by," he recalled, "and told us 'I had a very limited role in Vietnam.'"[1] Westmoreland made this claim at various other times as well, citing as evidence his not being allowed to take the fight to the enemy in the sanctuaries and his lack of influence over aspects of the war outside South Vietnam such as bombing North Vietnam.

Westmoreland also intervened in plans for a multivolume Vietnam War history being prepared by the Army Center of Military History. Charles MacDonald, then the staff historian in charge of that project, described it as a ten-subject history in which some subjects might take several volumes. "One of them will be a broad overall history," he said in 1970, "which we intended to do last. But General Westmoreland is anxious to get something in print fairly soon, so we have had to transfer that last volume up to the first." At that point, of course, the war itself was far from over, so any overview written then would necessarily increase Westmoreland's prominence.

Westmoreland even sought to persuade the Director of Marine Corps History to revise a draft of the volume on the Marine Corps in Vietnam in 1966 by substituting the term "offensive operations" for "search and destroy operations" wherever it appeared in the text. The Marine historians welcomed this as grist for their mill, adding in a footnote Westmoreland's contention that "search and destroy had been distorted" by his critics and featuring beside it Marine Commandant General Wallace Greene's conviction that search and destroy had been a bad idea, while in the text the term "search and destroy" was let stand.[2]

AS WESTMORELAND BEGAN these energetic efforts to shape the record of his performance in Vietnam, an early vehicle was the *Report on the War in Vietnam*, which he claimed Lyndon Johnson had asked him to prepare. The title itself was revealing, Westmoreland styling it a report on the war even though it covered only the four years of his tenure in command, was written while that war raged on without him, and was in fact only a report on *his* war in Vietnam — and as he would have it remembered.

Westmoreland drew on such sources as Colonel Argo for input, but went outside the Army Staff to General William DePuy to shape the finished product. "He asked me to pull together his first historical account of the Vietnam years," said DePuy. "I did that, with a lot of help, and did it more or less as an additional duty to those which I had down in the Joint Staff."[3] As the document took shape, various officers working for the Secretary of the General Staff, pressed into service as proofreaders, began referring to it as the "Me-My Manual."

Charles W. Hinkle was then Director of Security Review in the Office of the Secretary of Defense and eventually the *Report* came to his office for processing. Hinkle noted two major substantive issues and flagged them for Phil Goulding, the Assistant Secretary of Defense for Public Affairs. There was "no mention of the *large* number of U.S. forces requested of the President by COMUSMACV in early 1968," he noted, "nor are the American killed figures given." The omission of friendly KIA had previously been raised with Westmoreland, who declined to have them included. Thus that point, cautioned Hinkle, "has been avoided as a matter of personal preference by COMUSMACV, although it will be certain to be noted by the press and others." The other point (omission of any mention of the 206,000 troop request), he noted, "was a major issue requiring a Presidential decision. The lack of discussion on this issue is unlikely to go unnoticed." Westmoreland's preferences prevailed and the document was published as he had crafted it.

There were other omissions, deliberately chosen, also determined by Westmoreland. "You will note," he said in a letter to Professor Douglas Kinnard, "no use of the term 'body count' and scant reference to 'search and destroy.'"

Speaking on one occasion with Colonel Argo by phone, Westmoreland said, "The US press has created an image that I over a period of years neglected the ARVN. Want to go through text to point out constant and abiding interest I took in ARVN and programs pursued. Want report to make it loud and clear that have not neglected ARVN." After some further musings, Westmoreland told Argo: "Based on my feel for the situation here, believe the American people think I neglected both ARVN and pacification and went slow in mobilization [of the South Vietnamese armed forces]. Don't worry about controversy in correcting this impression. Step on toes if you have to. Lay it on the line. Must correct this impression, which was caused because I couldn't give the press all the information."

Argo, who had previously been the MACV Historian, knew how

vulnerable Westmoreland was on this point. In another conversation he recalled that when Westmoreland asked for M-16s for the South Vietnamese "this was turned down because of failure of US to go to wartime production. Do you want to use this language?" he asked. "No," Westmoreland responded. "Don't explain why request denied, just say due to reasons beyond my purview they were not immediately available."

In a subsequent telephone conversation with Argo, Westmoreland instructed him that in the *Report* he wanted to "make the point that from '67 on we seized every opportunity to have ARVN take over more and more of the load. This was general policy which should be made known." That claim was grossly inaccurate, of course.[4] Nevertheless Westmoreland urged Argo to note in the *Report* the "tremendous strides that ARVN had made. This will," he stressed, "make a good crescendo to the whole report—demonstrable evidence of success of our efforts. This is the best way to conclude the report. Get the best brains we have working on this to get just the right tone to it."

Despite his personally directed omissions and evasions in the document, Westmoreland was bold enough to assert, in a letter to the Editor-in-Chief of *Reader's Digest*, that "the fact remains that this is the only authentic publication on the war."[5]

During his first year as Chief of Staff Westmoreland devoted much of his time to the *Report*, first in giving such detailed instructions to those who were writing it for him, then in developing an extensive distribution list and personally inscribing large numbers of presentation copies. The manuscript was evidently completed at Admiral Sharp's headquarters in Hawaii, for in late September 1968 Westmoreland cabled Lieutenant General Claire Hutchin, the chief of staff there, to say that, "at my direction, Lieutenant Colonel Edwin C. Keiser is hand-carrying to you the changes I have made to the Report on the War in Vietnam."[6] He thanked Hutchin for his help in having these changes made, adding, "as you know, the report is of great importance to me."

As the document neared publication Westmoreland called Major General Wendell Coats, his Chief of Information, almost daily, quizzing him on the production schedule and then on press reaction and commentary. "Are we still struggling to get the history published?" Westmoreland asked in late December 1968. "I would like to get advance copies to Joe Alsop, Beech, Bob Considine, Ed Sutland (*New Yorker*), Johnny Apple, and Martin of *Newsweek*."

"No news magazines yet?" he asked in another such conversation. "No." Then: "What about Huntley-Brinkley?" Coats reported that in that program's coverage the "twist was that this was an apology by Adm. Sharp and you for not winning the war." And, suggested that prominent news broadcast, the report also served another purpose: "If we pull out, the military are on record as saying they did the best they could under the restrictions imposed."

Ernest Furgurson, then a journalist and Westmoreland's first biographer, wrote of the Westmoreland and Sharp accounts that "the most experienced military observers here are puzzled by them. Never in our history have erstwhile theater commanders published such narratives, essentially justifications of their own stewardship, while a war rumbled on."[7]

It was significant that the *Report* was configured in two separate sections, one by Admiral Sharp and the other by Westmoreland. "Admiral Sharp would not collaborate with General Westmoreland on the 'Report,'" recalled Colonel Fred Schoomaker. "That's why we had to do it in two sections. Westmoreland asked him to collaborate with him on it, but he would not do so." Westmoreland suffered a similar disappointment when he later published his memoirs and asked Admiral Sharp to contribute a blurb. Sharp turned him down.

WESTMORELAND NEXT COMMISSIONED an extensive slate of monographs on various aspects of the Vietnam War, even though, again, the war was still very much in progress, ensuring that any such accounts would be incomplete and possibly out of balance. But they would be about Westmoreland's war, and that seemed to be the point. These documents were produced by authors chosen by Westmoreland and written, or at least begun, while he was still on active duty and in a supervisory position over those who wrote them. Peter Braestrup described these monographs as "highly uneven in quality" and characterized by "a reluctance to explore error, command failure, or confusion."[8] That assessment was accurate, but their even more important shortcoming was that they were truncated treatments of a war still being fought.

Westmoreland omitted any monograph on armored or mechanized forces in the war, even though General Ralph Haines had suggested such a work. Later, after Westmoreland retired, General Donn Starry took on that task and wrote *Mounted Combat in Vietnam*, which turned out to be a best seller through the Government Printing Office. And, since it was not written until after the others in the series, it was the only monograph to treat the war as a whole.

DURING THESE CHIEF of Staff years, apparently looking ahead to writing his memoirs, Westmoreland had his aide-turned-research-assistant, Major Paul Miles, interview various senior officials, both military and civilian. Transcripts of those sessions provided raw material for the book and other purposes. Westmoreland also used his position to task various staff elements to produce studies and analyses that might prove useful to him. In March 1970, for example, he asked the staff to research the origins of the term "war of attrition" as applied to Vietnam. He told them he recalled using the term in 1967.

"Actually," said the response, "you had used the term at least as early as January 1966, during MACV preparations for the February Honolulu Conference."

Besides tasking the staff, Westmoreland wrote to a number of senior officers who had served with him in Vietnam to ask for, as one officer phrased it in his response, "any post-mortem comments I might have on the strategy for winning the war in Vietnam." This canvassing seemed aimed at collecting ammunition for defending Westmoreland's conduct of the war. If so, however, Westmoreland could not have liked the tenor of some of the answers his requests produced. Major General Willard Pearson, for example, was blunt in his response.

"Our inability to precisely locate the enemy in the Vietnamese jungles," he wrote, "together with his operating out of border sanctuaries, gave him two unique capabilities." He could control the level and intensity of combat, and he could control his losses and thereby "insure continuation of war for the period desired." As long as the enemy enjoyed those advantages, said Pearson, we were "basically fighting a defensive war and reacting to enemy initiatives." He also said he "must confess" that, while he was serving in Vietnam as Westmoreland's J-3, he "had an uneasy feeling that in the application of the approved strategy . . . there were a few missing ingredients essential to a more successful prosecution of the war." Among those he cited were "our inability to more effectively employ available combat power" and inability "to win wholehearted support of the Vietnamese peasantry." And finally: "An undue portion of the time is spent in unproductive flailing around, reconnoitering for the enemy, with relatively less time attacking and destroying him."[9]

IN NOVEMBER 1970 Westmoreland tasked General William Rosson, then Commander-in-Chief of U.S. Army, Pacific, to prepare what evolved into a paper entitled *Assessment of Influence Exerted on Military Operations by Other Than Military Considerations*. The thrust was made clear in Westmoreland's instructions to Rosson, directing assessment of those influences having a negative effect on conduct of the war, "to include rules of engagement, problems in security clearance to fire, border restrictions (including the DMZ), the limited incursion into Cambodia and prior proposals, ceasefires, policies on use of air power (including restrictions on the use of B-52's in South Vietnam until June 1965, the 37-day halt in air attacks north of the demilitarized zone beginning in December 1965, the partial limitation of air strikes in March of 196[8], and the complete bombing halt in North Vietnam in October of 1968), the Buddhist uprising in 1966, the prisoner of war issue, limitations in command and control of third country forces, the effect of anti-war sentiment in the United States, and the effect of coverage of the war by the news media." That amounted to a pretty comprehensive catalogue of frustrations. The study was not formally published until the Army Center of Military History issued it in 1993, introduced by a Chief of Military History comment that the manuscript was "a reflection of the era in which it was prepared."[10]

Having done not one but two extended oral history interviews for the U.S. Army Military History Institute at Carlisle Barracks, Westmoreland then took the precaution of having the draft transcripts reflecting his editing, and also the tape recordings of the interview sessions themselves, destroyed. Only the final paper versions of the transcripts survive—no recordings of his voice, of what he said and how he said it. "As requested, all previous drafts of this transcript have been destroyed, as were the tapes," reported the Army's Military History Institute in a letter to Westmoreland.[11] That constituted a significant loss to history.

Some years later, in a November 1986 radio interview, Westmoreland was asked, "How do you want to be remembered in history?" He responded, "I've never really given it any serious thought. . . . You serve in the military, you develop a philosophy that you're going to do the best job you can in every job that you're given, try to do what the government expects of you, and do it honorably. And then you let the chips fall."

23

VOLUNTEER ARMY

MILITARY CONSCRIPTION — the draft—actually continued throughout Westmoreland's tenure as Chief of Staff, but increasingly preparation for transition to an all-volunteer Army became a major issue. "People in DoD [the Department of Defense] thought we weren't doing enough," recalled General Walter Kerwin, the Army's Deputy Chief of Staff for Personnel during most of Westmoreland's years as Chief of Staff. "Westy's answer was to set up a special group, which was called SAMVA [Special Assistant for the Modern Volunteer Army], with General Forsythe in charge."

In both the grand design and many of its specifics Westmoreland was ambivalent about, if not solidly opposed to, an all-volunteer force. Colonel Jack Butler documented this attitude in an insightful analysis of volunteer force issues published near the end of Westmoreland's tenure. Butler called the Modern Volunteer Army initiative "probably the most misinterpreted and misunderstood program ever initiated by Department of the Army," a result, as he viewed it, of rushing the program to the field without any real preparation of those who were expected to implement it, and also of Westmoreland's questionable commitment to a volunteer force. During 1969–1970, when the Army's official position was in favor of an all-volunteer force, wrote Butler, "statements of some high officials suggested less than full support for the concept."[1]

Westmoreland said in interviews that he was for a volunteer force, but expressed reservations in testimony before Congressional panels and in internal discussions. He was also apparently encouraging, and being encouraged by, members of the Congress who shared his view. In May 1969, for example, Westmoreland told a friend that he had breakfasted with Congressmen Rivers and Hebert. "The volunteer Army was discussed," he said, "and both pledged that as long as they are in Congress we'll never have a volunteer Army."

When President Nixon appointed a commission to advise him on the future of the draft, Westmoreland was called to testify before it. Right away he got into trouble by saying he was not interested in leading an Army of "mercenaries." Milton Friedman, a member of the commission, asked, "Would you rather command an army of slaves?" Westmoreland "bristled. 'I don't like to hear our patriotic draftees referred to as slaves,' he said." To that "Friedman snapped back—and pointed out that if they were, then he was a mercenary professor and Westmoreland a mercenary general."[2] Soon the fifteen-member commission unanimously recommended ending the draft.

Regarding the all-volunteer force, Vice Chief of Staff Bruce Palmer stated flatly: "Westy and myself—we wished we weren't doing it."[3] Lieutenant General Eugene Forrester, who as a colonel had been Palmer's executive officer, confirmed that outlook. "I can truly say," he recalled, "that the Army did not want to do it. I think it was very candid on the part of General Westmoreland and General Palmer that they had hoped this cup would pass." Concluded Butler in his study of the issue: "This kind of high level contradiction contributed greatly to unsupporting attitudes" within the Army itself, giving "tacit support to resisting changes advocated to achieve the volunteer concept."[4]

By early autumn of 1970 President Nixon and Secretary of Defense Melvin Laird had had enough of passive resistance by Westmoreland and the Army. Major General John Singlaub recalled being told by Westmoreland that the President had advised him, "We're going to have an all-volunteer Army or we're not going to have an Army at all." According to Martin Anderson, a policy advisor, Laird "ordered" Westmoreland "to cease opposition to the all-volunteer force or risk dismissal."[5] When Laird became Secretary of Defense, Westmoreland had been Chief of Staff for only seven months. Laird was not enthusiastic about his inherited Army chief, but "chose to live with him," wrote Laird's biographer Dale Van Atta. Van Atta added that "Westmoreland later complained in his memoirs that Laird never listened to him as a member of the Joint Chiefs. That wasn't true, Laird said; he listened . . . but he rarely agreed."[6]

Westmoreland decided he wanted to retain his job and so, at the October 1970 annual meeting of the Association of the United States Army, he announced in the keynote address that the Army was committed to "an all-out effort in working toward a zero draft–volunteer force."[7] A few days later Westmoreland emphasized this message to his generals in a *Weekly Summary* article. "After almost 30 years of a draft environment," he said, "the Chief of Staff recognizes that to many commanders the volunteer concept is an emotional issue. Reducing [draftee] inductions to zero may have certain traumatic aspects; however, the time for emotionalism has passed. All commanders should once again examine their approaches to their mission and their personnel, their regulations and procedures, and their attitudes toward traditionalism and against change."

That declaration created its own problems within the Army. Such a "sudden change in direction," said Colonel Butler, "created many false impressions and produced what might be termed 'culture shock' among many professional soldiers," the result of changes "thrust upon the field."[8]

Just as Westmoreland had been wary of the volunteer Army overall, he was uneasy about many of the initiatives proposed by Lieutenant General George Forsythe, his (and the Secretary of the Army's) Special Assistant for the Modern Volunteer Army. Forsythe and a team of bright, able, and aggressive younger officers had studied the problem of recruiting a volunteer force and come up with some provocative findings. What was needed, they told Westmoreland, was a thoroughgoing reform of the Army and the way it operated. "We need to trust our people more. We need to open up a dialogue with them. We need to give them challenging and relevant work, and we need to free them in order to do that work. You can be as tough as you want with these guys, and give them as tough a challenge as you want, as long as it makes sense to them."[9] The implication, of course, was that their proposed reforms were far from how the Army was then being operated.

And, added Forsythe, "the real job was to convince senior Army commanders that Westy was serious about VOLAR," the Volunteer Army initiative.[10]

General Kerwin later concluded that the Volunteer Army office under General Forsythe "did not turn out well. They got a lot of zealots in there, guys who wanted to change the whole Army, period." Kerwin remembered as "just indescribable" some of the things they wanted to do.[11] Forsythe understood quite well the controversial nature of much of what he and his staff were proposing. "I think Westy, at times, really wasn't sure if we were doing the right thing," he recalled.[12]

An emotional issue during these years was the policy on haircut length and styles. The youth culture of that time favored long and often unkempt hair, leading to dilemmas for those in reserve forces (who had to try to survive in two disparate cultures at almost the same time) that reportedly led some to acquire wigs of neat and relatively short hair to wear to military formations. In the Navy Admiral Elmo Zumwalt, recently installed as Chief of Naval Operations, decided to allow sailors to wear beards and mustaches, as they had in an earlier day. The Army was dead set against that. "We knew we could not go with beards," said Brigadier General James Anderson, then a lieutenant colonel assigned to SAMVA (office of the Special Assistant for the Modern Volunteer Army). "The gas masks would leak." Thus, he recalled, they decided on a haircut policy that included a specification that sideburns could not be lower than the bottom of the ear opening, determined to be the maximum length that would not interfere with the fit of a gas mask.

Anderson and Major Peter Dawkins were the "action officers," as the Pentagon styled them, who took the issue to Westmoreland for decision. He approved the recommended policy, but astounded the young officers by telling them it was one of his toughest decisions, perhaps an indication of how wrenching for him was the social and cultural ferment of the day.[13]

The details of the haircut policy, its enforcement, and media interest in the matter were constant sources of contention and confusion. In a message to the field, Westmoreland said: "I have directed a thorough review of the Army's policies on hair styles to bring them more in line with existing customs among the respectable youth of today—not the hippies." General DePuy, as usual getting to the heart of things, said: "This is going to cause more trouble than beer in the barracks."

MOST MODERN VOLUNTEER Army initiatives involved elimination of practices considered harassing by young soldiers—signing out on pass during off-duty hours was discontinued, as was reveille in most cases. Others instituted privileges or benefits thought to appeal to younger soldiers, such as having beer in the barracks. James Binder, the longtime editor of *Army* magazine, characterized these initiatives as "the most thorough assault on things Mickey Mouse ever undertaken by the Army."[14]

All these matters were grist for the mill of *Army Times*, a civilian weekly widely read throughout the service. Its take on the new initiatives was not always what those responsible for the program might have wished, as one staff officer reported to General Forsythe: "The senior NCOs interpret the MVA [Modern Volunteer Army] as permissiveness," he told him, because they were being influenced by such *Army Times* headlines as "Booze in Barracks Says Westy."

Within the Army Staff, including General DePuy's office, there was considerable concern about planned initiatives for the volunteer Army in their total impact. One DePuy staffer wrote about the overall plan: "This is a visionary piece of work which bears little relation to the actual army. It promises more than we can or should produce. . . . It should be quietly ignored—the work being done in the personnel and training areas provide[s] about as much revolutionary change as the Army can stand." Thus his recommended course of action: "Bury [it]."

ALONGSIDE THE WIDELY publicized lifestyle changes were more important initiatives designed to improve Army readiness and the satisfactions of service. Perhaps the most successful ones involved training, both in decentralizing responsibility for planning training and in making it more interesting and challenging for the soldier.

The lead in these matters was taken by a Board for Dynamic Training, headed by Brigadier General (later General) Paul Gorman, himself a dynamic officer of impressive intellect, energy, and vision. The Board's starting point was challenging, as its main findings included a determination that "Army-wide, training is regarded as only marginally adequate." Among the problems, found the Board, were personnel turbulence, inadequate manning levels, inadequate budgets, and underqualified trainers.

Gorman observed that Westmoreland deserved much of the credit for the new training initiatives. Westmoreland directed that training be decentralized, most mandatory subjects eliminated, and the maintenance of training records simplified, along with initiatives to make training more interesting and challenging, more "dynamic," a term that, said Gorman, Westmoreland himself applied to the new board. Adventure training featuring such techniques as river rafting put some fun into the program and was well received by young soldiers.

THE ARMY'S APPROACH to recruiting was also often controversial during the Westmoreland years, especially when a new recruiting slogan was adopted: "Today's Army Wants to Join You." General Forsythe observed that "many of the Army's older soldiers have bristled at that slogan, viewing it as a blatant capitulation to radical fads, a rejection of the traditional conduct of Army affairs." A powerful congressman, Representative F. Edward Hebert, incoming chairman of the House Armed Services Committee, was already deeply concerned about various volunteer Army initiatives. "I'm very fearful of this trend," he stated. "I'm afraid they're trying to make it a country club." He would not have been reassured by the views of the new Secretary of the Army, Robert Froehlke, who told a reporter that the plan was to allow "young men to have a life style in the Army to which they've become accustomed in their homes and communities."[15]

An issue of "Army News Photo Features" had a front-page headline: "Modern Volunteer Army Spirit: 'Soldier, You've Changed. We're Changing.'" On a copy of that publication General William DePuy had penciled: "Unmitigated disaster—we're in the hands of madmen."

AS CHIEF OF STAFF, Westmoreland issued several letters to the officer corps. The first, published in November 1969, dealt with integrity. "I want to make it clear beyond any question that absolute integrity of an officer's word, deed, and signature is a matter that permits no compromise," he wrote. "Competence and integrity are not separable."

There was a second letter, sent in April 1971 to all officers and this time also to all noncommissioned officers, addressing the topic of leadership. "We expect our people, from general to private, to be loyal, honest members of the team," he wrote, "dedicated to giving their best efforts to any assigned mission or task." This document was described in an internal memorandum of General DePuy's office as "a disaster—sent against everybody's advice," because it dealt in such entry-level basics as to be insulting to any but the newest Army leaders. "We must all be constantly concerned and involved with the welfare of our troops," Westmoreland instructed.

A third letter, sent in October 1971 to all officers, addressed the topic of special trust and confidence. "I want the policies and practices of the Army to reflect: More careful selection of commanders who can provide honest, forthright, and productive leadership and who will establish and maintain high standards."

Subsequently Westmoreland had the three letters printed up as a small pamphlet that could be issued to various recipients, and he later referred to them in his memoirs: "I stressed those aspects of leadership that I considered to be imperative."

Given these repeated admonitions to his officers to adhere to high standards of professional behavior, Westmoreland himself was involved in an episode that another senior officer found ethically troubling and disappointing. "I was president of a board to select new brigadier generals," recalled General Donald Bennett. The thirteen members of the board had the daunting task of choosing 65 officers for promotion from a population of 4,600 possible selectees. In a first cut they narrowed the field to about 130, then went through again and chose those to be promoted. "Westmoreland had written in a special efficiency report that [his executive officer] was an exceptional officer who should be selected for brigadier immediately." When the final list was sent in, that officer was not on it. Immediately afterward, said General Bennett, "Westmoreland called me in and said I had disobeyed his orders." Bennett observed that there had been many officers who had such comments in their files, and it had not been possible to select them all for promotion. Then he pointed out that

Westmoreland had two choices: he could accept the board's report, or he could dissolve the board and appoint another that might come up with different results. "Then," said Bennett, "I saluted and excused myself without waiting for a reply."[16]

ON THE EVE of his retirement in June 1972 Westmoreland wrote President Nixon a four-page letter. Still fighting against the volunteer Army, he sounded a cautionary note: "I can give you no assurance that we will achieve our goal of a volunteer force by 1 July 1973." This final attempt to influence administration policy on the draft was futile at best. Secretary of the Army Robert Froehlke "recalled that the letter had 'relatively little effect' and further observed that 'because of the controversy that continued to attend Vietnam and [Westmoreland's] close identification with that unpopular war he was not effective as a spokesman for the Army in general. . . .'"[17]

24

★

VIETNAM DRAWDOWN

BRUCE PALMER RECALLED that "Westmoreland said he was Chief of Staff for four years while the Vietnam War was going on, and the White House never once asked his opinion about the war."[1] While there were exceptions, those occasions were admittedly few and left Westmoreland feeling that the expertise he thought he could offer was neither appreciated nor utilized.

Palmer explained the neglect: "Because of the unpopularity of the war, the administration considered Westmoreland a political liability and treated him accordingly." Nixon may also have been intentionally isolating a challenger. "Both major political parties regarded him as a significant presidential threat because he was so well known nationally and internationally. A proud, sensitive South Carolina gentleman, Westmoreland was deeply hurt by the slights accorded him by administration officials, who rarely consulted him on Vietnam affairs."[2]

THERE WAS ONE AREA of crucial importance in which, although not perhaps specifically asked for advice, Westmoreland was able to wield determinative influence. That had to do with the progressive withdrawal of American forces from Vietnam. The first 25,000 were brought out in July and August 1969, followed by 40,500 more during September through December of that year.

These first increments occasioned a fierce controversy between Westmoreland and his successor in Vietnam, General Creighton Abrams. Westmoreland wanted to bring troops home as individuals, choosing those who had served the longest in Vietnam and were thus closest to the end of their nominal one-year tour. Abrams pushed for unit withdrawals. That would mean bringing out all those in a given unit without regard to how long they had been in Vietnam.

Colonel Donn Starry, then in Vietnam as redeployment planner for Abrams, recalled the point of crisis on this matter. Westmoreland "overrode his [Abrams's] strong recommendation to redeploy units as units instead of as individuals. Individual personnel redeployments destroyed unit integrity, increasing turbulence in units remaining. In the end, it caused leaders to go forth to battle daily with men who did not know them and whom they did not know. The result was tragedy."[3]

"Our fear," recalled Starry, "was that the turbulence rate would be so high that units would become ineffective. And that's what happened: I believe it caused most of the indiscipline in units which plagued us later." He remembered those events vividly. "The confrontation was a direct one between Abrams and Westmoreland. We did ourselves an enormous disservice." The Vice Chief of Staff, General Bruce Palmer, also weighed in, telling Westmoreland, "We went in by brigade; we should come out the same way," but that good counsel was also rejected.[4]

Starry added a personal note. "The night of the final rejection of our proposal to redeploy units instead of individuals, he [Abrams] and I sat long over scotch and cigars. Finally, his eyes watering, he turned to me and said, 'I probably won't live to see the end of this, but the rest of your career will be dedicated to straightening out the mess this is going to create.' How right he was."[5]

General Maxwell Thurman, whose later brilliant performance as head of Recruiting Command literally saved the volunteer Army, also had to deal with the consequences of Westmoreland's stance. "The 'fair' and 'equitable' policy," he confirmed, "was a disaster."[6]

• • •

DAVID HALBERSTAM REPORTED that, while Westmoreland was Chief of Staff, he was "deeply depressed about [the favorable press General Abrams was getting] and wanted to make a public statement saying that the gains which were being made under Abrams had been originated during his tour, and friends had to take him aside and tell him that the last thing a very troubled United States Army could stand at that particular moment was a public split between Westmoreland and Abrams."[7]

Others were also very aware of this jealousy. "Westmoreland lost my respect, pumping me when I got back from Vietnam," said Lieutenant General Sidney Berry. "'Damn it,' he asked, 'how does Abe get such good publicity?'"[8]

At home in his new role Westmoreland was finding it hard to come by not only good publicity but even respectful treatment. One historian reported that the senior officer corps was so thoroughly discredited by the Vietnam War that "in 1972 military audiences booed General Westmoreland from the stage, first at Fort Benning and then later at the Command and General Staff College."[9]

In retirement Westmoreland was also booed at the National War College, an almost unimaginable occurrence. But Dr. Jeffrey Clarke, later the Army's Director of Military History, witnessed it. "There was a tremendous amount of anger, especially among the young colonels," he recalled.[10] Dr. Alan Gropman, Westmoreland's escort officer for the occasion, recalled: "The students, all of them, booed loudly." He also remembers what happened next: "Westmoreland went white, his hands gripping the sides of the podium. I was sitting in the front row, and I saw him put a nitroglycerine pill under his tongue. Then things quieted down, and he continued, with both the class and the speaker acting as if nothing untoward had happened."[11] A contrary report was provided by Major General Lee Surut, who stated that he had "no recollection of students at NWC booing General Westmoreland. Since I was Commandant at the time, I would have remembered any such breach of conduct. The occurrence is highly unlikely."[12]

While these accounts differ, there is little doubt that the alleged attitudes existed. Brigadier General Allen Grum, who served as an officer for thirty-four years, observed that he "could not remember a Chief of Staff so disdained as Westmoreland." At the Army War College, he remembered, the students in his class created a fake prize, the William C. Westmoreland Award, and presented it to "the dumbest, best-looking officer in the class."[13]

Westmoreland remained defiant: "Nobody has suffered more anguish as to the plight of the Vietnam veterans than I have. Nobody has taken more guff than I have, and I am not apologizing for a thing—nothing, and I welcome being the point man!"[14]

A critical part of that stance was maintaining that whatever good things happened in Vietnam after his departure were simply the result of building on what he had accomplished and that his successor had made no significant changes in the conduct of the war. At most, Westmoreland asserted, any changes made were simply adaptations to different enemy tactics.

A lot of this spinning came out in Westmoreland's confrontation with Marine Corps historians over their history of the war. The Marines published some of the comments Westmoreland had made on a draft of one volume, as for example his take on the "One War" approach to the war under Abrams, Bunker, and Colby. "It was not Abrams that did it," insisted Westmoreland, "it was the changed situation which he adapted to."[15] Indeed, Westmoreland refused to give Abrams credit for *any* changes he might have made: "Any changes that Abrams made in strategy and tactics that was presumably mine—and it was my watch and I assume full responsibility for them—or any policies as practiced by me, there was any changes made in my view were the function of the changed situation after the defeat of the Tet offensive."[16]

As time went on Westmoreland broadened the scope of these comparisons and contrasts, as in his oral history: "I came back in the summer of '68," he recalled, "and discipline collapsed after I left Vietnam, not while I was there, it collapsed later on. And the drug thing started after I left, too. And a lot of the [problems with lack of] discipline had crept in, which was a spin-off of the anti-war syndrome. It started after I left Vietnam. When I left during the '67–'68 time frame, I don't think America has ever put a better force in the field than we had then, a more professional force."[17]

In October 1971 the *Washington Post* published a series of articles, "Army in Anguish." This material caused great distress to both civilian and uniformed military officials. Ironically, Westmoreland may have been the precipitating factor. At one point he had said to Ben Bradlee, managing editor of the *Post*, "I'm not sure I can hold this Army together." Bradlee told his reporters George Wilson and Haynes Johnson that that sounded like something worth looking into, and the hard-hitting series resulted. The traumas of incremental withdrawal from Vietnam, racial divisiveness, widespread drug problems, and erosion of discipline were laid out in telling detail.

Westmoreland soon cabled Abrams, in Vietnam, with a long description of the articles and their fallout. Generals Palmer and Kerwin had been called to testify before the House's Mahon Committee for a full day, he reported. "The Committee expressed deep concern for, what appear to them, to be a deteriorating discipline and lack of leadership in the services, particularly the Army. The questioning was intense and, at times, hostile."

Westmoreland devoted the bulk of his long message to the appearance of soldiers in Vietnam. One of the *Post* articles, he noted, included a photograph of a soldier reading his mail. Westmoreland had many problems with how that soldier looked, describing them in detail — his haircut, the appearance of the uniform, his wearing "what appear to be peace and religious medallions." In short, said Westmoreland, "This soldier is a sorry sight." Then came the rebuke: "I assume that Army policies regarding personal appearance and wear of the uniform have been received in Vietnam and are known to the chain of command. Enforcement by the chain of command is critically important. There must be no misunderstanding on this point." And finally: "With competent and alert professional officers in the chain of command, a communication such as this should not be necessary. Nevertheless, it has become necessary and I expect you to take those measures you deem required to correct the situation and reestablish the Army's traditional standards."

WESTMORELAND'S CLASH WITH the prevailing youth culture had begun at home. He had problems with his children during these years, as did many other senior officers who through long service in Vietnam had only sporadic contact with their families. A family friend, very sympathetic to the situation, recalled that "the children were out of step with Westy all through the war. It created an atmosphere of coolness between them and their father which took some time to resolve.

One daughter became involved with an anti-war group, and that was very difficult."[18]

Rip, the only son, sandwiched between his sisters, was candid on these matters in notes he wrote as a high school freshman. "My sister and my father get into a lot of arguments," he revealed. That was his older sister, Stevie, then twenty years of age, who went to school in California. "She is strongly opposed to the war," he said. "She brings home students and friends with long hair whom my father doesn't like at all." The elder Westmoreland made comparisons to his own youth, said Rip, and they were predictably unfavorable to the way his children were conducting themselves. Added Rip, "I respect my father's ideals," even though he was getting flak on the length of his own hair. "I guess he doesn't like long hair because it is a symbol of being against the system, and my father is all for the system." In concluding this account Rip demonstrated considerable insight and compassion, despite his youth. "He likes his job a great deal," he said of his father, "although I don't think he enjoyed staying in Vietnam for five years. That really changed him."[19]

Westmoreland's sister, Margaret, was also a sympathetic observer of his family situation. "The children resented their father's long absence during the Vietnam War," she recalled, "but later they were reconciled." Said Rip on one occasion, "I got to know my father during lunches at the Brook Club," a New York establishment to which Westmoreland belonged in his retirement years. It had taken that long for reconciliation to take place.

25

★

DEPARTURE

As RETIREMENT IMPENDED, Westmoreland was aggressive in taking what final steps he could to shape the historical record of his service. In published materials and in a letter to President Nixon he claimed that the Army was in good shape and that the officer corps had never been stronger. Those were empty boasts. Just two years earlier, the internal study conducted by the Army War College at Westmoreland's request had documented deep-seated and widespread failings of professionalism among the Army's senior leadership, not the kind of thing susceptible to quick fixes. And the trend line of Westmoreland's four years as Chief of Staff had been, especially in terms of resources, relentlessly downward. As documented in the Department of the Army's *Historical Summary* for Fiscal Year 1970, for example: "Whether the subject is funds, personnel strength, training, combat, casualties, construction, research, development, procurement, or production, the tendency was toward reduction. Curtailment, consolidation, withdrawal, retrenchment, adjustment, constraint — these are the watchwords that set the tone of Army operation in 1970 and charted the directions for the coming year."[1]

The noted military sociologist Charles Moskos called the years 1970–1973 "the worst times in modern Army history."[2] General James

Woolnough, Commanding General of Continental Army Command, had observed in late 1970 that "the military services, particularly the Army, are actually fighting a delaying action at this time, giving ground as grudgingly as possible and hoping for reinforcements in the form of a changing public attitude before it is too late."[3]

May 1972, observed the Army Center of Military History, "was the month in which the Army 'bottomed out' in its personnel situation," a crisis artificially induced when, well into the fiscal year, Congress reduced the authorized end strength by 50,000, producing near chaos in Army personnel.

Not all of this was Westmoreland's fault, of course, but his denials of the prevailing realities were disingenuous and further undermined whatever credibility he might otherwise have retained.

Some problems were just too difficult to be solved, given the context of the times and the available resources, a situation that even Westmoreland, despite his upbeat public pronouncements, was finally forced to acknowledge. "It may be that a combination of personnel turbulence, the readiness reporting system, and cumulative mandatory training requirements all taken together represent an impossible situation at the lower unit levels," said Westmoreland very late in the game.[4]

Speaking at a Commanders Conference at Fort Monroe in May 1972, Westmoreland said that "our Army is understrength. Our Army has enormous, conflicting, often overly extensive commitments." After a recital of widespread problems with training and in other areas, Westmoreland told these senior commanders that "excellence should be our by-word. At this time we are falling short of such a standard." With no sense of irony he added, "The key to our success will be honesty . . . honesty in facing the facts of life . . . honesty in developing realistic goals and priorities . . . and honesty in conducting our day-to-day business." He was then just six weeks from retirement.[5]

Westmoreland did not want to retire. He asked his executive officer to contact Brigadier General Michael Dunn, then Army aide to the President, "to reaffirm that the only Federal job he is interested in is a fifth year as Chief of Staff. He is perfectly willing to delay his civilian plans if the Administration desires in order to make a major contribution as the Army moves through one more year of transition. If the President is really serious about achieving a volunteer Army, General Westmoreland is prepared to work toward that end, but he does not want the Administration to feel obliged to find another Government position for him."[6] Of course nothing came of that.

WHEN HERBERT ABRAMS painted Westmoreland's portrait as Chief of Staff, Westmoreland chose to pose in jungle fatigues, with his badges nailed into the plain wooden frame of the painting.[7] He stands out in the long corridor lined with portraits of successive Chiefs of Staff, the only incumbent depicted in field uniform rather than the dress uniform of the day. What message Westmoreland intended to convey by this departure from custom is somewhat obscure, but it makes him a very sad figure, permanently out of uniform, as it were, for the duties he could never bring himself to turn his attention to, obviously mired in another time and place that continued to obsess him.

Lieutenant General Phillip Davidson, his chief intelligence officer in Vietnam, later wrote about Westmoreland's legacy as Chief of Staff. "The consensus of the army," he concluded, "is that Westmoreland tried hard as chief, but that the times and its problems overwhelmed him."[8] General Walter Kerwin was succinct in describing Westmoreland's years as Chief of Staff: "An unhappy time."[9] Said General Bruce Palmer Jr., Westmoreland's classmate, associate in Vietnam, and now for four years Vice Chief of Staff: "With great nostalgia we bade him farewell. It was the end of an era for the U.S. Army—a drama-packed and unhappy era."[10]

MANY TRIBUTES MARKED Westmoreland's final days on active duty. There was a dinner hosted by the Joint Chiefs of Staff, who gave Westmoreland a plaque. He was invited to lunch by the Marine Corps Commandant, and on another day lunched with long-retired Major General Louis Craig, his division commander from World War II. He went to Fort Campbell, where he awarded the Presidential Unit Citation to the 3rd Brigade of the 101st Airborne Division, an element of the division he once commanded. He also visited Fort Bragg, site of his first airborne duty after World War II, for a parade, reception, and dinner, and presented a uniform he had worn while in Vietnam to a museum at the post. Admiral and Mrs. Moorer hosted a black-tie dinner at their quarters. Secretary of the Army Robert Froehlke gave a dinner at which the Army Band and Chorus presented a musical "This

Is Your Life," recalling highlights of Westmoreland's career. Among these engagements he fitted in calls on a dozen or so senators and congressmen at their offices on Capitol Hill.

Westmoreland was presented the Distinguished Service Medal, his fourth award of that decoration, by President Nixon. On his last day in uniform he said, "I am reminded of the responsibility and trust which have been reposed in me, of old friendships which have warmed and sustained me, and the comradeship in arms which I have shared with American soldiers for more than 36 years."

It rained that day, with the planned parade at Fort Myer replaced by an indoor ceremony. Westmoreland wore his white uniform. His family was with him, all except his daughter Stevie. Mamie Eisenhower attended, but the Chairman of the Joint Chiefs of Staff, Admiral Thomas Moorer, did not, citing a "conflict in our schedules." Westmoreland and Kitsy were taken by horse-drawn carriage to the following reception, then it was home to the South Carolina of his youth.

26

★

IN RETIREMENT

CHARLESTON BECAME THE Westmorelands' final home. Things there got off to a happy start, Westmoreland writing to an old Army friend in January 1973: "Retirement I have found quite satisfactory and have experienced no trauma in the least." That euphoria was short-lived, however, as a letter to that very same friend in October admitted: "We speak from experience and the knowledge that making the transition from that of a military life to a civilian environment is not easy and is fraught with frustrations. We have not overcome ours yet."

A part of the problem may have been that the Westmorelands were not accepted into Charleston's old-line social circles quite as readily as they might have anticipated. Westmoreland's sister, Margaret, remembered, "[A]fter they moved to Charleston, Kitsy complained that nobody was paying any attention to them. I told her that she had to start making friends one at a time."[1] They decided to build a house at 107 Tradd Street, in Charleston's historic district, on a lot so narrow that the house had to be oriented sidesaddle, facing not the street but the house next door. In December 1972 the Westmorelands obtained the services of their old friend Leslie Boney as the architect of their new home. Soon Boney wrote to them of his plans for "a garden between house and slave quarters."

Westmoreland wrote to a friend that "we plan to build a modern house in the midst of antiquity." Completing the house turned out to be an ordeal, as the site proved marshy and much money had to be spent on expensive underpinnings. In late November 1973 Westmoreland wrote to a friend that "for the last year and a half, Mrs. Westmoreland and I have been struggling to build our first home and we have encountered every obstacle and frustration imaginable."

Orr Kelly, the veteran military affairs correspondent for the *Washington Star*, wrote at the time of Westmoreland's retirement that, "in other times and under other circumstances, the career of Gen. William C. Westmoreland was such that he might have found himself propelled into politics—into the presidency itself."[2] Of course that was out of the question. Westmoreland needed something else to do. Only four months after retiring, he reached out to Richard Nixon. "If at any time I can be of assistance in lending, for what they are worth, my views or insights," he wrote to the President, "I stand ready to be of service, either as an individual or as part of an advisory group."

Westmoreland also made some fairly strenuous efforts to obtain a full-time appointed position. Submitting a Personal Data Statement to a White House staffer, Westmoreland offered to put the securities he had inherited from his father into blind trusts, "as I did while serving as Chief of Staff of the Army." He also wrote to Ambassador Bunker, asking his help in being appointed to the Panama Canal Commission Board.[3] Somehow investigative reporter Seymour Hersh learned about these overtures, writing that Westmoreland "waited embarrassingly for an offer to join the Nixon administration," but that "he heard nothing and retired to South Carolina to work on his golf game."[4]

Then, in the autumn of 1972, only a few months after his retirement, South Carolina's Governor John West offered him a part-time position as head of a state task force on economic growth, charged with developing increased economic opportunities in the state. This involved Westmoreland in extensive travel throughout the state, and may have been viewed by him as an opportunity to become better known in advance of a political opportunity.

Meanwhile, the White House did ask his advice. Westmoreland had complained of not being consulted while Chief of Staff about matters in Vietnam. Now, only months into retirement, he was called to the White House for a séance with Richard Nixon on the Vietnam cease-fire agreement then being negotiated. After being briefed on the proposed agreement, wrote historian Stephen Ambrose, Westmoreland urged the President "to delay action on the new agreement and to hold out for better terms." He "emphasized that it was 'vital' that North Vietnamese troops be compelled to withdraw from South Vietnam. As to the National Council of Reconciliation, Westmoreland thought it was 'impractical, almost absurd, nothing more than a façade.'"[5] Admiral Thomas Moorer commented acidly: "This is typical Westy. He always had a solution after the fact."[6]

Henry Kissinger was stung by Westmoreland's stance. "This was amazing," he wrote in a retrospective analysis of the Vietnam War, "since a stand-still cease-fire had been part of our position since October 1970 and had been endorsed then by the Joint Chiefs of Staff, of whom Westmoreland was one."[7] Westmoreland took issue, arguing in a letter to Kissinger that, since he was now retired, he was "not a party to the development of the concept of a cease-fire as it was being applied at that time."

That was disingenuous at best. In October 1970, when President Nixon made an important speech to the nation, Westmoreland had been Acting Chairman of the Joint Chiefs of Staff. Thus Secretary of Defense Laird asked Westmoreland to review a copy of the speech the President planned to give. "I informed him that the speech caused me no problems and that I thought the plan was well conceived and the wording appropriate," said Westmoreland. Laird then took Westmoreland with him to the White House, where the President was meeting with the cabinet and then with the Congressional leadership. At one point Nixon asked Westmoreland to comment. "I stated in effect as follows," Westmoreland said in a memorandum for record dated the following day: "In my opinion, a ceasefire in place entails little if any risk." Then, he continued, "before the speech I called each member of the JCS and covered the major points of the speech. All agreed that the proposal was consistent with positions taken by the JCS."[8]

AS HIS FIRST Christmas in retirement approached, with the concomitant need to do some shopping, Westmoreland wrote to Brigadier General James L. Collins Jr., the Army's Chief of Military History, thanking him for having sent twelve copies of a photograph of Westmoreland's Chief of Staff portrait. Please send me thirty more, said Westmoreland. "I would like to give copies to some of my close friends during the holiday season."[9]

Once established in Charleston, Westmoreland told a correspondent that hanging on the wall of his study was his most prized possession, his commission as a 2nd Lieutenant in the United States Army. To another he confided that what he really missed was "flying that helicopter."

Another thing he probably missed was all the people who had taken care of him when he was a senior officer on active duty. An Army major was in an airport when he was approached by someone in civilian clothes. "Do you know who I am?" the man asked. "Yes, sir," responded the officer, "you're General Westmoreland." Westmoreland then explained that he had recently retired from the Army and was on a trip. "But I don't know how to get my luggage. They took it when I checked in for my flight, but now I'm here and I don't know where my bags are." The major explained about baggage claim.

On a visit to West Point, Westmoreland was escorted by Brigadier General Walter Ulmer, the Commandant of Cadets. Ulmer introduced him to the King of the Beasts, the senior cadet in charge of "Beast Barracks," where new plebes got their initial training. That young man had just been selected to be First Captain. He stood about 5'9" or 5'10", causing Westmoreland to exclaim to Ulmer, "He's not very *tall*, is he?" Ulmer explained that he had other redeeming features.

DURING HIS RETIREMENT Westmoreland struggled with an unfortunate situation involving veterans of the 9th Infantry Division. That division was his old World War II outfit and had later served under him in Vietnam, with Major General Julian Ewell as its commander. Under Ewell's leadership the Vietnam veterans formed what they called the Octofoil Association. Ewell wrote to invite Westmoreland to serve as honorary president of the new organization. Westmoreland "accepted with pleasure," but also suggested to Ewell that they merge with the existing Ninth Division Association. That group had not opened its membership to those who served in it at later times, only

World War II veterans. This situation gave Westmoreland an opportunity to take a stand on principle.

A respected sergeant major, still in uniform, had attended a 1970 division reunion in New York City and been booed for suggesting that Vietnam veterans be admitted to membership. Wrote a retired lieutenant colonel to Westmoreland: "I have been told that the opposition is essentially racist on the assumption that Vietnam veterans would include a high proportion of black people." He said he understood Westmoreland had withdrawn his support from the organization because of the position it took on the issue. "Would you please reconsider your position and actively support this project?" he asked. Westmoreland declined to do so.

In 1977 it was reported to Westmoreland that at a meeting in Chicago the Ninth Division Association "defeated a motion to allow the division's Vietnam veterans to join their ranks. That was not unexpected," he was told. Soon thereafter Westmoreland was invited to be the main speaker at the Association's 1978 meeting. "I can not in good conscience continue to affiliate myself with an organization that rejects the younger comrades-in-arms of our division who saw service in Vietnam," he responded. "For the Vietnam veterans to be 'shafted' by his fellow division veterans is beyond my comprehension, and I will not be a part of an organization that conducts itself in such a bigoted and irresponsible manner." Over a decade later, with the issue still unresolved, Westmoreland wrote that "my position is clear and will never change."

As late as 1992, responding to a correspondent who had written about an upcoming division reunion, Westmoreland again laid it on the line: "My position on the attitude of the Association is I hope well known and I will not compromise." Apparently, though, he did just that. There is in his papers the text of an address prepared for delivery at the reunion that very year. It is identified as "50 Years Ago," 9th Infantry Division, 1992.

WESTMORELAND CONTINUED MAKING other speeches at every opportunity, many of them exculpatory. Lieutenant General William McCaffrey was sympathetic to his plight. "One of the saddest things I've seen of my contemporaries," he said, "is what happened to Westy. It's been a monkey on his back ever since. I don't think he's ever had the fun I've had, a lot of others have had, of retiring and saying, 'Well, screw them all but six, and save them for pallbearers.' He's trying to rationalize what happened. He did the best he could."[10] Westmoreland was proud of what he was doing, saying in an oral history, "I haven't dug a foxhole, like McNamara has, and hid out. I made myself accessible." Eventually, he said, he had spoken in every state in the Union, plus Guam and Puerto Rico. "When I retired," he said, "I made a commitment to myself that I would accept any invitation to talk about Vietnam, and I did." That determination never wavered, Westmoreland telling an interviewer more than thirty years after his retirement, "The Vietnam War is my number one priority."[11]

It is clear that Westmoreland also considered his speechmaking a money-making enterprise. Queried by a former fellow soldier about a possible 60th Infantry Regiment staff reunion in Lincoln, Nebraska, he responded: "I am now an author/lecturer and one thought would be for the University to ask me to lecture during the time of the reunion. That would assure my presence since they would pay my expenses. If the University is interested they will have to go through my agent and I will provide the details on that on a timely basis."

Invited to speak at the 1988 annual meeting of the Association of the United States Army, Westmoreland gave what had become a stock lecture, one in which he said that "Rogers Hammerstein characterized the Vietnam veteran, and the American soldier, when he said, 'Give me some men who are stout hearted men,'" then went on to quote those lyrics at considerable length.

General John Galvin suggested that, "if Topeka had a garage sale going on, Westmoreland would speak." Given some of the engagements he actually undertook, that did not seem too far off the mark. Among them were the Hampton County Watermelon Festival, the Junior National Team Handball Champions Recognition Ceremony, the South Carolina Subsection of the Society of American Foresters, the Lees-McRae Junior College Gymnasium Dedication, the Annual Installation of Officers of the Oak Cliff Chamber of Commerce in Dallas, and a meeting of the National Soccer Coaches Association in New York. Besides such esoterica, of course, were rafts of military, veteran, and civic organization events all across the country.

While traveling for such speaking engagements Westmoreland also made an effort to maintain physical fitness, or so he apparently convinced reporter Elisabeth Bugg. Writing in the *Richmond News Leader,* she reported that Westmoreland, then nearly sixty-five years old, "was out of bed by 7 o'clock to jog in place for 40 minutes in his hotel room here." No mention was made of how this may have been received by the people on the floor below.

Among Westmoreland's other pursuits was an effort to have "tackle football" adopted as an Olympic sport, advocacy of a five-year moratorium on new federal spending (for which he was paid a monthly stipend by backers of that idea), and a suggestion that nuclear weapons be used to affect hurricanes (a proposal that horrified a scientist at the Los Alamos National Laboratory to whom he suggested it).

Westmoreland's preoccupation with the war and its outcome continued as the long struggle played out. "The important thing," he wrote to General Alexander Haig in early 1973, during the brief period when Haig was Army Vice Chief of Staff, "is to get our POW's back and, if we can accomplish this, we should pull out and let them fight it out among themselves until they are exhausted and conclude that there is a better way to settle their differences."[12] When we did just that Westmoreland, apparently appalled by the consequences, would dramatically shift his position on abandoning the South Vietnamese.

General William DePuy maintained a correspondence with Westmoreland when both were in retirement, by turns consoling and counseling him. "I believe the public in general now accepts the fact that the whole country was involved," DePuy wrote in response to a letter from Westmoreland complaining about how *Reader's Digest* had characterized the war. "They don't much blame anybody, and certainly don't blame the military for getting in or getting out. Oddly enough I think the military man came out amazingly well—a soldierly kind of image—doing what they were told to do—doing it reasonably well—and operating under complex limitations."

Then DePuy offered a suggestion. "You have borne the brunt," he told Westmoreland. "You have borne it well—you have been a gentleman about it. There is a great reservoir of *respect* for that. History will get it straight some day. In the meanwhile I hope you will be somewhat philosophic and maintain the marvelously effective and impressive stance of a distinguished gentleman soldier."

• • •

IN DECEMBER 1974 Westmoreland and Kitsy accepted an invitation to visit the Rose Festival in Pasadena. Westmoreland sent a letter to Doubleday asking them to pay for the plane tickets as a promotional trip for his book. While they were still in California, early in January, Westmoreland suffered a serious heart attack. He was in Palm Springs at the time and was hospitalized at the Eisenhower Medical Center in nearby Palm Desert, where he remained for twenty days. Then it was on to Walter Reed Army Medical Center in Washington for another ten days. After some six weeks of recuperation at home Westmoreland wrote to the Veterans Administration asking that his medical records be reviewed to determine whether his retirement disability could be modified "in view of the possibility that there were indications of a possible heart malfunction at the time of my departure from active service." Ironically, that letter was written on 30 April 1975, the day Saigon fell to the communists and the Vietnam War at last came to an end.

AFTER THE U.S. CONGRESS decided to abandon the South Vietnamese, thereby assuring their conquest by the North, Westmoreland was bitter—as were, of course, many of those who had made such great sacrifices for the cause. "Our erstwhile honorable country betrayed and deserted the Republic of Vietnam after it had enticed it to our bosom," Westmoreland wrote in *Military Review*. "It was a shabby performance by America, a blemish on our history and a possible blight on our future." In sum, "The handling of the Vietnam affair was a shameful national blunder."[13]

In later years Westmoreland also criticized Lyndon Johnson's conduct of the war, something he had not done while the former president was alive. "Johnson . . . hoped the war would go away . . . ," wrote Westmoreland in an essay, "Vietnam Blunders," "but his key decisions were destined to drag the war out indefinitely." Then Westmoreland offered some political advice: "We should choose our leaders carefully, broad-gauged statesmen, not slaves to the public-opinion polls."[14]

Nearly two decades after the end of the war that had consumed him then and later, Westmoreland offered a considered but surprising judgment: "In the scope of history, Vietnam is not going to be a big deal. It won't float to the top as a major endeavor."[15]

27

★

MEMOIRS

WESTMORELAND ARRANGED FOR Charles MacDonald, the professional military historian working for the Army with whom he had already had many dealings, to take a year's leave of absence to help him write his memoirs. The task became for MacDonald something of an ordeal, with Westmoreland's requests for assistance of various kinds spilling over well beyond the dedicated year and burdening MacDonald when he was back at work for the Army. Westmoreland's gratitude, at least as expressed in the book's acknowledgments, was minimal, confined to a single sentence mentioning that MacDonald "took leave of absence from his usual duties to assist."

During the process Westmoreland sent MacDonald a remarkable document, a typed summary by Kitsy of her views on the Vietnam War, along with Westmoreland's note saying he agreed with some of them. Kitsy stated her approval of amnesty for those who refused to serve and her compassion for what her own offspring went through: "Our poor children. There seemed to be no middle crowd or ground. They were either *violently* for or violently opposed." She added that Vietnam brought out the very best and the very worst, dividing the country in two camps, "and neither was right, neither was completely wrong." Finally, she concluded, "somehow, no matter if one served in

VN with honor, became a student and accepted deferment, [or] ran away . . . we will all carry the scar of Viet Nam, with or without honor."[1]

MAJOR PAUL MILES played an important part in the development of Westmoreland's memoirs. While aide-de-camp to Westmoreland during the Chief of Staff years, Miles interviewed various senior officers, primarily on Vietnam-related matters, and even visited a prospective publisher on Westmoreland's behalf. "In March 1971, Mr. Thomas Congdon of Harper & Row wrote that he was joining Doubleday as a Senior Editor and hoped that I would discuss with him your plans for writing," Miles said in a later memorandum to Westmoreland. "I visited Doubleday in the Spring of 1971 and met Congdon and Sam Vaughn." In a separate note in which Miles identified himself as, rather than aide-de-camp or research assistant, now historian, he reminded Westmoreland that "Congdon is the young editor recommended by Halberstam."[2] As things turned out, Doubleday did publish the memoirs, with Stewart Richardson as the editor.

Westmoreland dictated comments, some brief, others running to several pages, on a number of people he had associated with, fellow soldiers like Maxwell Taylor, Hank Emerson, George Forsythe ("He had the appearance and many of the mannerisms of Danny Kaye"), and Bruce Palmer. These were passed to MacDonald as grist for the mill. Periodically MacDonald went down to Charleston. "We would work all day long," he said. "I'd have his records, and he would read every word in that record. He wouldn't skip a word. He was *loving* every minute of it, particularly when he came to *his* words. He absolutely adored it. And I would have to sit there sometimes, *my god*, trying to stay awake." They'd take a break for dinner at six o'clock, then go back to work. MacDonald remembered the abrupt transition, "as though a mask, a shade, went down over his face, and he'd become a different person. I mean, there was never any banter while we were working. The minute he decided we went back to work, we went back to work."[3]

During the book's development MacDonald also of course got to know Kitsy, whom he termed "a delightful person" with "an earthy quality that Westy can't possible have, and you wonder how if he can't have it he can even admire it in somebody else." MacDonald also met Kitsy's father, "an old horse artilleryman" and "a great old boy." He remembered being offered a drink by Colonel Van Deusen, then ninety-three years old, who told him, "[Y]ou can have sherry or a martini. Those are the only two things I remember how to make."

Westmoreland wrote to many of his former aides and other close associates, asking that they send him "antidotes, reminiscences, and events" that could be used in the book. The results of such appeals were mixed at best, some respondents attempting to pattern their input on those short snippets often featured in *Reader's Digest*. Most were pretty lame, especially those submitted by former aides-de-camp.

In mid-November 1973 MacDonald sent Westmoreland a draft of Chapter 2. Westmoreland responded that he found it "somewhat heavy going" and suggested that a good approach to revising it might be to "give a very brief summary of the history and then move into the character of the country and its people in a lighter vein throwing into the account as much color as we can find. As an example, the Cao Dies are a peculiar religious sect and have a beautiful and ornate temple in Tay Ninh where they have a pope and are organized dissimilarly to the Catholic church."[4]

Paul Miles noted that Westmoreland "thought he was an excellent editor, and he liked to add dependent clauses that made things longer and more ponderous." That tendency, and his fondness for rather lame stories he thought were amusing, posed continuing challenges for MacDonald. Of course Westmoreland had the controlling role.

THOSE WHO WRITE memoirs can be expected to put the best face on their actions and motives, and Westmoreland's version of the genre is no exception. Those seriously interested in the history of his era, however, need to be aware of the distortions by omission in his volume. Dr. John Carland, longtime staff historian at the Army Center of Military History, then later at the Department of State, cited an important example in Westmoreland's discussion of the Vietnam War battles of the Ia Drang Valley in the autumn of 1965. Two separate but related encounters constituted the larger battle. In the first, at Landing Zone X-Ray, the 1st Battalion, 7th Cavalry, fought a desperate and prolonged battle with NVA elements, sustaining heavy casualties but inflicting even heavier ones on the enemy, who eventually withdrew. In the second, immediately following, the 2nd Battalion, 7th Cavalry, failing to follow good security practices, was ambushed by NVA elements while moving on foot to Landing Zone Albany, suffering in the process very heavy one-sided casualties.

Immediately after these events Westmoreland visited the units involved and was briefed on the fighting at X-Ray, but told nothing of what had happened at Albany. Talking with some of the wounded in the hospital, he began to realize he had not been given the full story. Westmoreland contacted Lieutenant General Stanley Larsen, commanding general of I Field Force, Vietnam, and asked for an explanation. Larsen personally investigated the matter and was told by the division commander, assistant division commander, and brigade commander of the forces involved that they had said nothing to Westmoreland about what happened at LZ Albany because they had not known about it at the time. Larsen later spoke of this matter to Joe Galloway, coauthor with General Moore of two books about the battle and its aftermath, and even gave Galloway a signed affidavit detailing those events. Of what he had been told by the division's senior officers, Larsen stated: "They were lying and I left and flew back to my headquarters and called Westy and told him so, and told him I was prepared to bring court-martial charges against each of them. There was a long silence on the phone and then Westy told me: 'No, Swede, let it slide.'"[5]

Yet in writing of these events, Dr. Carland noted, "Westmoreland in his text conflated the two battles but implied that he was discussing only one." Thus, "if a reader knew nothing about Albany before reading . . . he would still know nothing afterward."[6] The aspect omitted, of course, was the one in which Westmoreland's troops had performed poorly and suffered accordingly. In this he mimicked the commanders on the scene of the original battle. There is no doubt that Westmoreland was aware of all this, for he dictated extensive coverage of it in his history notes, none of which made it into the memoirs. In fact, coaching MacDonald on what to say in the memoirs, Westmoreland told him: "Never a defeat of a US unit of battalion size or larger. Some hard knocks, yes; no defeat."[7]

BEFORE CHARLES MACDONALD became a historian he was a soldier, commanding an infantry company during World War II and writing of those experiences in *Company Commander*, which became a classic of small unit leadership. He later rendered a judgment on Westmoreland derived from their long discussions while working on the book. "I think Westy was a little hard when it came to being inured to casualties," said MacDonald. "I don't think he was emotionally affected by heavy losses, as I think some commanders might be. And perhaps this is a virtue. Maybe a senior commander *has* to be this way." But: "From a *human* angle, you sort of question it."[8]

MacDonald was also realistic when it came to reviews of the book. One particularly critical treatment was by Kevin Buckley, formerly the *Newsweek* bureau chief in Saigon. Writing in the *New York Review of Books*, Buckley said Westmoreland believed "a general is a very special person, and people should be made aware of this" and that "some officers who worked with him in Vietnam say that he reached the limit of his talents [during World War II]." Also, he observed, "from the beginning Westmoreland probably expected to write a memoir of victory similar to *Crusade in Europe* and the books of other successful American generals of the past" and "the defeat in Vietnam has not deterred him from this." Said MacDonald, in a talk at the Army War College, "Kevin Buckley has got[ten] to Westmoreland in many ways. In many ways, *very* correct."[9]

Those comments were rendered by MacDonald in the course of a lecture on a topic of great interest to the War College students, a comparison of Westmoreland and Abrams. When he told Westmoreland about that talk, said MacDonald, "he almost blew his stack. He said, 'What the hell? Has the War College got no better thought than to think up some subject like that?!' He simply didn't *want* it." Admitted MacDonald, "I regret to say that I simply didn't have the guts to tell him it was my idea."

In the discussion following MacDonald's talk the student officers expressed a great deal of hostility toward Westmoreland. One recalled that he had been at the Army Command & General Staff College when Westmoreland, as Chief of Staff, came to lecture. "He gave us a speech in the morning, telling us why we were in Vietnam. And I think that some 99 percent of that class had been to Vietnam once, 70 percent twice, and 40 percent three times. And he was really trying to convince us why we were there. That turned me off. And that afternoon he gave the same speech to the University of Kansas student body—you know, just *totally* the *wrong* thing."

Another student officer asked MacDonald whether Westmoreland read a lot. "No," answered MacDonald. "He just simply doesn't have any interests. I would venture to guess that the man has not read a book from cover to cover in a *hell* of a long time." That was in decided contrast to Westmoreland's self-portrayal, especially in talks given after his retirement, when he would sometimes describe himself as "a student of the history of war." General John Galvin, a four-star officer who had served with Westmoreland repeatedly, confirmed MacDonald's assessment: "Westmoreland had an astonishing lack of interest in a wide range of things."

Several years after the end of American involvement in Vietnam Westmoreland received a letter from Lieutenant General James M. Gavin, long since retired but the proponent of the much-debated "enclave" strategy for Vietnam. Gavin had read Westmoreland's memoirs and had a question about the assigned mission when Westmoreland had been in Vietnam. Citing the relevant page number, Gavin wrote, "You state that the mission . . . in Vietnam was to assist the Government of Vietnam and its armed forces to defeat externally directed and supported communist subversion and aggression and attain an independent South Vietnam functioning in a secure environment." Question: "Did this remain your mission throughout your tour, or was it modified in any way?" Westmoreland apparently did not know, for he passed the letter to MacDonald, sidelining the question and writing in the margin: "Charlie Would you research this?"

WESTMORELAND HAD PLANNED to call his memoirs *The War Nobody Won*. That had to be scuttled when, during the book's preparation, somebody did win that very war. Dozens of other titles were then considered before settling on the final choice: *A Soldier Reports*. Said MacDonald, "It was a second-choice title resorted to when the original selection became untenable after the North Vietnamese occupied Saigon."[10]

They had missed the manuscript deadline by quite a margin, submitting the finished product in December 1974, some nine months after the contracted date. Then Doubleday decided to hold up publication until the following autumn, apparently to see how the situation evolved in Vietnam and whether the manuscript might need to be revised accordingly. Publication finally took place in early 1976, in the midst of a season of soul-searching and continued controversy in the wake of South Vietnam's abandonment by the United States and subsequent 30 April 1975 conquest by the North Vietnamese communists. Those delays turned out to be prudent, since they provided time to change what had become a wildly inapt title.[11]

When the book finally appeared Westmoreland was not happy with the presentation. The jacket bore only the title and his last name, separated by four stars (inappropriately gold rather than silver). "*Look at this cover,*" Westmoreland exclaimed, suggesting it had been "done by a B/O [maybe meaning "back office"] drafts man during his lunch break." And, he said, "I goofed on the book by giving it a bland unsexy name."[12] The publisher had also persuaded Westmoreland that his author photo, which comprised the entire back of the jacket, should be taken in civilian clothes.

When the paperback edition appeared, Westmoreland was much happier: "*Dell has done better.*" The front cover depicted Westmoreland in what he himself called "a favorite stance," hands on hips, standing on the hood of a jeep while addressing a crowd of soldiers gathered beneath him. Westmoreland stands out clearly, but the soldiers are masked by a sort of sepia tint that puts them very much in the background.

REACTION TO THE book was, as might have been expected, decidedly mixed. Westmoreland devoted just two chapters to his life before Vietnam, with the remainder of the 425-page text dealing entirely with the war and its aftermath. Peter Braestrup, the highly respected journalist who had exposed shoddy Tet Offensive reporting in his book *Big Story*, said that Westmoreland's book was "notable for showing how LBJ manipulated the U.S. commander in Vietnam into becoming, in effect, a spokesman for the administration in the domestic political arena."

Norman Hannah wrote that "on finishing the book, one has the feeling that even today, General Westmoreland is perplexed as to what happened [in Vietnam] and why." A like view was expressed sympathetically by Frank Getlein, writing in the *Washington Star*. "The poor man," he concluded. "That's really about all you can say of Westmoreland either before or after you read his story of a sincerely dedicated career of service to his country and the utter destruction of that career in Vietnam." And, he added: "No American in a position of authority in either Saigon or Washington had any remote notion of what the war was all about, least of all poor Westmoreland."[13]

George Wilson, long the military affairs correspondent for the *Washington Post*, made a telling point. "My own biggest disappointment . . . ," he wrote, "is that the general writes so unfeelingly; saying so little about the pain both his troops and the Vietnamese people endured day after day as we waged modern warfare in that small country."

Westmoreland was also brought up short by Larry Laurion, a West Point classmate, who wrote: "I'm concerned that an uninformed reader, who doesn't know you as well as I do, might conclude that you personally claim credit for everything that panned out well in Vietnam and that those knuckleheads in Washington were to blame for everything that didn't turn out right."

Hanson Baldwin provided what was probably the most positive assessment, recording his belief that "the verdict of history will be that Gen. Westmoreland did his duty as a proud leader of a proud outfit—the U.S. Army—and that he was vastly more sinned against than sinning."

S.L.A. Marshall, another widely known writer on military affairs and an earlier critic of Westmoreland's performance in Vietnam, now commiserated with him on the circumstances of his service, agreeing that, as Westmoreland had argued, "he commanded with one hand tied behind him. When the Joint Chiefs and Defense Secretary were not preempting his command prerogatives," said Marshall, "he was getting flak, interference, directives or naysayers from the Man in the White House and his coterie of crisis managers." Thus, he concluded, Westmoreland "remains the goat symbol of the country's most mournful misadventure abroad ever."[14]

28

★

CAMPAIGNER

IN 1974 WESTMORELAND, scarcely two years into retirement and already working on a memoir and building a house, nevertheless decided to run for governor of his native state of South Carolina. The campaign was an interesting and in some ways historic one, for the ultimate outcome was election of a Republican governor for the first time since Reconstruction.

A number of influential people tried to tell Westmoreland in a friendly way that he was not cut out for politics. "Several of us suggested," recalled former Secretary of Defense Melvin Laird, "that he enjoy his retirement for a while, but he was bound and determined to get into politics." A very old friend wrote with similar advice: "You have done your part. Relax and enjoy 107 Tradd with Kitsy and the children."

An apparent stranger also sought to dissuade Westmoreland from a campaign. "I am writing to you personally in hopes of stopping you from running for Governor of South Carolina," began a letter from Charleston. "I feel that the elite class that you associate with has and is giving you some bad advice. I read the text of one of your speeches given last year in the Georgetown area and from a business point of view it sounded like you had copied it out of a third grade reader. The middle and lower class people of this area just aren't too impressed with retired generals and admirals."

But others encouraged Westmoreland to get into the campaign, including then–Vice President Gerald Ford, California's Governor Ronald Reagan, Senators Strom Thurmond and Charles Percy, and Secretary of Commerce Fred Dent.

"Kitsy pushed for it," said a family friend, William Morrison. "She wanted him to be recognized in every way possible. She thought he was presidential material." There is contrary evidence in a contemporary letter Westmoreland wrote to his World War II commander, Major General Louis Craig. "Kitsy is not overjoyed about the situation," he said, "but, as usual, she will go with me."

A group calling themselves the Draft Westmoreland Committee launched a petition campaign urging Westmoreland to enter the race. Lee Atwater, then a relative unknown but later one of the political superstars of the Republican party, managed that endeavor. Eventually some 20,000 people signed the petition, enough to provide considerable optimism about the outcome of a Westmoreland candidacy.

Westmoreland couldn't seem to make up his mind, though, first about whether to run at all, and then if he did run which party's nomination he would seek. It appeared that he hoped for a brokered nomination, one bestowed by the party leaders, sparing him a primary run, but that was out of reach. Although for many years South Carolina Republicans had handled things that way, this time they decided on their first-ever primary, and of course the Democrats would also be holding a primary. Lieutenant Governor Earle Morris had talked with Westmoreland about these matters. "I told him the day of the silver-platter nomination is over," he said.

In early February 1974 *The State*, South Carolina's leading newspaper, published the results of a statewide survey assessing the strengths of announced and potential candidates. Westmoreland was still pondering whether to run, but his name was prominent in the results. "Westmoreland had more than a three-to-one margin of support over other GOP possibilities included in the survey," the newspaper revealed. Westmoreland drew 43.0 percent of the votes. Next was Frederick Dent at 12.1 percent, then James Edwards with 6.4 percent. Commenting on these results, the report noted that "Westmoreland

LEFT: Named "Man of the Year" by *Time*, Westmoreland was riding high. But the laudatory article also suggested ominous prospects: "As the price of the war begins to crimp Great Society programs and boost taxes, Americans may find it harder than ever to accept the long war predicted by the Administration." © 1966 *Time Inc.*

RIGHT: Especially after the family moved to the Philippines Westmoreland was, he wrote to his mother, able to "get over about every two weeks for a short visit." *South Caroliniana Library*

During LBJ's 1967 "Progress Offensive" Westmoreland, here briefing in the White House Cabinet Room in late April, made several visits to the United States, speaking optimistically in public forums while in private asking for more U.S. troops. *LBJ Library*

Invited during the April 1967 U.S. visit to address the South Carolina General Assembly, Westmoreland escorted his mother to the event. The legislators subsequently adopted a resolution to "prevail upon" their native son to become a candidate for the Presidency of the United States. *The State*

RIGHT: Yet another appearance during Westmoreland's very busy April 1967 visit to the United States was an address to a Joint Session of Congress, which he later described as "the most memorable moment in my military career" and "my finest hour which gave me the greatest personal satisfaction." *South Caroliniana Library*

BELOW: In December 1967 President Lyndon Johnson made a brief visit to South Vietnam, addressing the troops at Cam Ranh Bay and awarding Westmoreland the Distinguished Service Medal. Back in Washington, LBJ told his senior associates: "I like Westmoreland.... Westmoreland has played on the team to help me." *LBJ Library*

In the wake of the enemy's 1968 Tet Offensive, Westmoreland and JCS Chairman General Earle Wheeler presented LBJ with a request for 206,000 more American troops in Vietnam. That ploy backfired and soon led to Westmoreland's reassignment and a changed approach to conduct of the war. *LBJ Library*

In July 1968 Westmoreland was sworn in as Army Chief of Staff, then spent much of the next four years on the road. "I spoke in every state in the union," he said. "I considered myself the military spokesman of the Army, and that I should be exposed to the American public. . . ." That, he concluded, "was the primary mission that fell on my shoulders while I was Chief of Staff." *South Caroliniana Library*

"General Westmoreland's Compliments, Private . . . And After the Coffee Would You Kindly Join Us at the Ten-Thirty Reveille Gathering?"

Anticipating the end of conscription, the Army's efforts to make service more attractive to volunteers earned Westmoreland considerable comment editorially and from such cartoonists as the famous Pat Oliphant. © Universal Uclick

Chief of Staff Westmoreland agonizing at an Army football game. Earlier, when he was Superintendent, he had been told by President Eisenhower to "buck up the football team" but, despite changing coaches, he had gone 0–3 against Navy. *South Caroliniana Library*

In early 1969 the "My Lai massacre" became publicly known. The March 1968 slaughter of innocent Vietnamese civilians by U.S. troops had taken place when Westmoreland was in command, and now — unfortunately — it was up to him as Chief of Staff to sort the matter out. Secretary of the Army Stanley Resor, Lieutenant General William Peers (running the investigation), and Westmoreland address reporters. *U.S. Army Center of Military History*

On his retirement in June 1972 Westmoreland, accompanied by Kitsy, was awarded another Distinguished Service Medal by President Richard Nixon. Westmoreland had sought a fifth year as Army Chief of Staff, but that was not to be. *South Caroliniana Library*

LEFT: Westmoreland's portrait stands out in the Pentagon gallery of those who have held the Chief of Staff post, the only person portrayed not in the dress uniform of the day but in jungle fatigues, perhaps appropriate in view of his continued obsession with the Vietnam experience. *U.S. Army Center of Military History*

RIGHT: At a West Point reunion Westmoreland talked with his classmate General Bruce Palmer Jr., who served with him in Vietnam and then was his Vice Chief of Staff for the entire four years, playing an indispensable role during difficult times for the Army. *South Caroliniana Library*

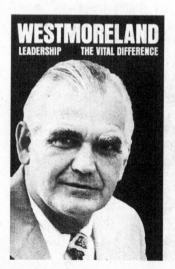

LEFT: Westmoreland ran for governor of South Carolina, telling voters he was "the *only* candidate with the proven leadership and administrative ability to carry South Carolina to greatness," but he proved an ineffective campaigner while overspending the budget. *South Caroliniana Library*

LEFT: Westmoreland and Kitsy watch the election returns on television. He was defeated by State Senator James Edwards, who went on to become the first Republican governor of the state since Reconstruction. *The State*

BELOW: Bob Hope and Westmoreland became friends during Hope's Christmas trips to entertain the troops in Vietnam. Later, Hope held benefits that helped retire Westmoreland's campaign debt. Westmoreland's golfing skills may be gauged by a letter he wrote to a friend: "The dozen golf balls arrived. In fact, only today I tried them out and lost 6 of them." *South Caroliniana Library*

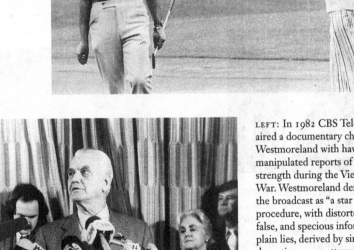

LEFT: In 1982 CBS Television aired a documentary charging Westmoreland with having manipulated reports of enemy strength during the Vietnam War. Westmoreland denounced the broadcast as "a star chamber procedure, with distorted, false, and specious information, plain lies, derived by sinister deception — an attempt to execute me on the guillotine of public opinion." *South Caroliniana Library*

RIGHT: Against the advice of high-powered attorneys Westmoreland sued CBS for libel. Some veterans, styling themselves "Westy's Warriors," raised money for his legal expenses, but after a lengthy trial Westmoreland withdrew his suit just before the case would have gone to the jury, claiming vindication. *Tony Bliss Jr. Collection*

LEFT: Westmoreland was very proud of his support for Vietnam veterans and their regard for him, and in 1988 he was honored by the Association of the United States Army with a special award for that devotion. The citation read: "He made their cause his own. He has helped bind up a nation's wounds. There is no more noble work." *South Caroliniana Library*

BELOW LEFT: "Westmoreland's life since Vietnam has been miserable," observed a former aide, but Westmoreland remained combative: "I have no reluctance to talk about Vietnam. There is nobody that has given it more thought than I have. Nobody has suffered more anguish as to the plight of the Vietnam veterans than I have. Nobody has taken more guff than I have, and I am not apologizing for a damn thing — nothing, and I welcome being the point man!" *Roger Pettengill, Academy Photo, West Point, New York*

RIGHT: Westmoreland died in July 2005 and was buried at West Point, at a gravesite he had selected when he was Superintendent there. Members of his former Scout Troop 1 in Spartanburg raised funds to help pay for the special monument, a contribution recognized on its base. *Author photograph*

drew strong backing from many white voters, but he had little success with blacks."

Finally, after months of temporizing, Westmoreland wrote to Maxwell Taylor to report what he called his "bold decision to throw my hat into the political ring for the Governorship of South Carolina on the Republican side." There were then just four months to go until primary day. The correspondence files show that Westmoreland also wrote to at least three other people that they were "the first to know" of his candidacy.

Some idea of the campaign's likely tone was provided by a publisher's comment on the decision. "Retired General William C. Westmoreland has condescended to permit the people of South Carolina to elect him Governor on the Republican ticket," wrote Fred Sheheen in the *Allendale County Citizen*. But a political writer, Robert Liming, reported in *The State*, on the very day of the announcement, that "most observers predict that Westmoreland will easily capture the party's formal nomination."

The next day, his first on the campaign trail, someone handed Westmoreland a three-by-five card with some advice: "If you don't make it for governor—keep running for *President*—you will be elected. You need campaign experience." Earlier in the year Westmoreland himself had testified to the paucity of his political involvement. "Up to now," he confided, "I've never even voted."

His campaign literature said that "William Westmoreland has vast managerial, administrative, executive and educational experience" and that he had "inbred understanding of people and has committed himself totally to public service," adding that he held decorations from sixteen foreign countries and had received the Silver Buffalo Award from the Boy Scouts of America. He was going to need more than that.

Near the end of the month someone who had apparently known Westmoreland for a long time (she referred to him as "Childs") wrote to a campaign worker with a suggestion for the candidate: "He should not let himself be 'pop' interviewed by the press which catches him completely by surprise and with no answers." That highlighted one of Westmoreland's most basic problems—he did not know the issues, did not know the constituency, and did not know the arcane workings of South Carolina politics. He was, despite being a "native son," an outsider, and he had not done his homework. Later Westmoreland ruefully admitted as much, saying, "[T]here were certain animosities within the South Carolina Republican Party that I was unaware of to which I became a victim."

THE CAMPAIGN WAS very slow getting started. On 21 March 1974 Hal Byrd, a member of the Westmoreland for Governor Steering Committee, wrote to the candidate to say "I hope that you will soon find the right person to be your campaign manager and state finance chairman." Westmoreland had announced his candidacy without having lined up people for those key positions, and they remained unfilled two weeks after the announcement. The following week, on 26 March (his sixtieth birthday), having announced on 12 March that he was running for governor, Westmoreland "officially" announced his candidacy and then — said a press account — "gave generally vague answers to the questions of a handful of reporters."

Getting organized seemed to be a problem even when it came to basic logistics. On 1 April 1974, for example, almost three weeks after entering the race and with only three and a half months to go until the primary, Westmoreland was forced to respond to a supporter in apologetic terms: "You asked for a bumper sticker and two pins. I regret to say that my campaign has not developed to the point where we have those at this time."

Westmoreland eventually recruited Dick Edwards, a television broadcaster, as his campaign manager. Their first joint appearance, a news conference in Columbia six weeks after the candidacy was announced, was a disaster. An account entitled "An Unauspicious Start" told the story. Westmoreland introduced Edwards, "then retired to a seat in the rear, declining to answer questions about his campaign." And: "Westmoreland wouldn't respond and the best Edwards could do was promise the general would answer the questions later." Commented an influential newspaper editorially: "The general had better find his voice soon. Buck-passing may have worked in the Army, but avoidance of the issues doesn't go over too well with the voters. They want to know where a candidate stands."

In the Republican primary Westmoreland was pitted against State Senator James B. Edwards, a popular Charleston native. "Jim Edwards was a very loyal, hardworking man," said Westmoreland's friend William Morrison. "Westy made a lot of mistakes in the things he said." Charlie Montgomery helped Westmoreland out on weekends. "I drove him in his Olds 98," he recalled. "We went to little towns. Two-thirds of the meetings were in colored churches, in their meeting halls. He would give his speech to a whole lot of blacks."

This was all pretty agonizing for Westmoreland. "I've had difficulty breaking the personal barrier on three items," he wrote to a friend, "—shaking hands with everybody, asking people for favors, and talking about myself."

Political correspondent Lee Bandy commented on Westmoreland's campaign trail demeanor. "Westy was not very comfortable as a candidate," he observed. "He was kind of reserved and kind of stiff, as you'd expect from a general. He wasn't used to taking orders, or to having people push him around." But "Kitsy was a real asset. She enjoyed the campaign. She was more of a politician than he was." Bandy was very savvy in the ways of his state and its political mores, and also quick to size Westmoreland up: "Westy thought all he had to do to get the nomination was to declare his candidacy."[1]

Efforts to gather support seemed at times pathetic and almost desperate. Writing to someone in Carrollton, Georgia, Westmoreland said, "If you are disposed to throw the [Georgia!] Junior Forest Rangers behind me, I am sure it would be most helpful support and I would deeply appreciate it."

A California firm hired by the Westmoreland campaign submitted its plan on 2 May 1974, just ten weeks before the primary, proposing a campaign budget of $300,000 for the primary alone. Campaign records include suggestions from Westmoreland about people (most of them outside South Carolina) who might make contributions and then (in some cases) the amounts contributed. The very first contribution came from Kitsy's mother, a generous $500. Among the other amounts recorded were $50 from Henry Cabot Lodge, $50 from Maxwell Taylor, $100 from Hugh Bullock, $250 from Nelson Rockefeller, and $100 from Bob Hope. Another supporter is shown as contributing a case of Old Crow (valued at $60). Westmoreland himself is down for $1,000. By mid-March the campaign had, reported its treasurer, raised a total of $2,120. Eventually they reduced the goal from $300,000 to $120,000, collected $126,800, and overspent their way into debt.

Meanwhile Westmoreland was working with Senator John Stennis to have a special bill passed in the U.S. Congress that would "protect" him from a provision of the South Carolina constitution relating to loss of his military retired pay should he succeed in being elected governor.

The catalogue of ineptitudes in the campaign is long. A very late start. A bunch of retired military officers trying to help run a political campaign. No in-state financial support of any consequence. No coherent message. The bulk of Westmoreland's campaign speeches in the files are not finished manuscripts but sketchy notes handwritten on five-by-eight cards. He seemed to have gotten very little staff support in shaping or articulating his campaign theme or stands on the issues. And, perhaps in part a result of competing claims on his time and attention (the memoirs, the house), Westmoreland gave almost no leadership to the campaign, a fatal deficiency in what he was attempting to showcase as his primary asset.

WITH TWO WEEKS to go Westmoreland wrote to an old friend: "I am reasonably confident that I can surmount the primary." Yet, in those crucial closing days of the campaign, he lacked the funds for radio and television advertising.

Then, during the final days of the primary campaign, Westmoreland completely lost his voice. "You would have shared my pride in Kitsy in responding to the situation," he wrote to a friend. "She made three stump speeches in my behalf and stole the show. She was poised, warm with a nice sense of humor." Afterward some people said it was too bad Kitsy wasn't the candidate.

The Republican party, historically almost nonexistent in the Democratically solid South, still had no viable slate of candidates at local levels. Voters could cast their ballots in either the Republican primary or the Democratic, but they had to choose. If they voted Republican, they would as a consequence have no influence on the choice of many local officials. The dramatic outcome of this circumstance was illustrated in Union County, which reported 7,842 Democratic votes versus just 43 Republican in the primary. Two statewide surveys published during the primary campaign in *The State* had depicted Westmoreland as a two-to-one favorite, convincing his supporters that he was a sure winner and they could safely skip the Republican primary and support their local candidates in the Democratic primary. As a result, Westmoreland's fate depended upon a highly unrepresentative tiny sample of state voters.

The primary was held on 16 July 1974. When all the returns were in, James Edwards had tallied 58 percent of the Republican vote, Westmoreland 42 percent. Even so, Westmoreland had received the majority of the votes in five of the six congressional districts. Only in Charleston, where Edwards had a rock-solid home base (not seriously challenged by the arriviste Westmoreland), were the results overwhelmingly lopsided against Westmoreland. There Edwards piled up an 8,500-vote advantage, greater than his statewide margin of victory.

The voter turnout for both parties was extremely low. There were at that time over 958,000 registered South Carolina voters. Just 320,000 voted in the Democratic primary, another 34,950 voting as Republicans.

A friend wrote to Westmoreland: "I am still at a loss to determine what happened to the 20,000 people who signed a petition asking you to run for Governor. I really don't know where they were on election day, but they certainly were not at the polls."

John Courson, who had worked on the campaign, analyzed the results, noting three principal causes of Westmoreland's defeat: Intraparty fighting which, he said, "political novice Westmoreland was unaware of." James Edwards's "strong, personal following among GOP regulars." And the fact that Westmoreland's supporters, believing he would win easily, "voted in the Democratic primary where most local races were decided." Courson also mentioned how poorly the Westmoreland campaign had handled its finances, noting that "an 'expensive and ineffective West Coast consulting firm' was hired and a large and expensive staff of politically inexperienced people was assembled."[2]

Shortly after the votes were counted Westmoreland headed for Washington, where he checked himself in at the Army's Walter Reed Hospital for treatment of his laryngitis.

During the campaign Westmoreland, apparently viewing his membership in the Council on Foreign Relations as a political liability in South Carolina, tendered his resignation. He later concluded that the ploy had not worked, writing to a friend after the primary defeat that one of the things working against him was "a suspicion of my political orientation because I was a member of the Council on Foreign Relations. The hard core of the conservative wing of the South Carolina Republican Party, dominated by John Birchers, consider the CFR an organization whose objectives are contrary to the best interests of the country." With the campaign behind him, Westmoreland tried to regain his membership. "I shall bring to the attention of the secretary of the Membership Committee your request to be considered for reentry into the Council membership," he was told rather formally by the Council's president. They never took him back.

The commentary of political veterans in the state, and of the press, was fairly homogeneous in laying most of the blame on Westmoreland's inexperience and ineptitude as a candidate. It began with the basics. "He was very, very deficient in his knowledge of South Carolina politics and the issues," observed Raymond Moore, a former University of South Carolina professor. "I must tell you quite frankly I was never comfortable in a capacity so strange to my experience," Westmoreland wrote to a friend in the election's aftermath. "I have proved to myself that I have no appeal as a politician."

IN THE NOVEMBER general election, thanks to some unprecedented circumstances, Republican Jim Edwards was elected governor. The Democratic primary had featured a tight race between Charles "Pug" Ravenel and Congressman William Jennings Bryan Dorn, with Ravenel coming out on top by fewer than a thousand votes out of more than 300,000 cast. In a runoff held two weeks later, Ravenel topped Dorn, this time by a larger margin, to win the nomination.

Then a bizarre twist occurred. Ravenel had spent a number of years on Wall Street, where he amassed a fortune, before returning to South Carolina. Now a lawsuit was filed charging that he was not eligible to become governor because he failed to meet the constitutional requirement of having been a resident of the state for five consecutive years before the election. That challenge was upheld in the state Supreme Court, and the U.S. Supreme Court declined to overrule. Ravenel was disqualified with just over a month remaining until the general election.

The Democrats, understandably in some disarray, caucused and decided to resurrect their runner-up, Congressman Dorn, who of course had to restart his campaign more or less from scratch, and with not very satisfactory results. That allowed Jim Edwards, the Republican, who had all along been running a vigorous campaign, to prevail in November by a margin of some 17,000 votes of over half a million cast. It was an extraordinary accomplishment in an overwhelmingly Democratic state.

Writing to Governor John Rhodes of Ohio, Westmoreland gave his successful opponent a lukewarm endorsement: "Jim Edwards, a dentist, ran a very skillful campaign and he should make an excellent governor but will need a great deal of support." Surely Westmoreland was not unaware that the dentist was also a highly respected and experienced state senator.

WHEN WESTMORELAND'S CAMPAIGN was over there remained bills to pay. He received a number of dunning letters, including one from a woman who said she had been hired by the campaign for telephone work but never been paid. "I regret the delay in paying you," Westmoreland responded. "I have found politics most confusing and raising funds and paying bills has been most frustrating."

Westmoreland's very close friend from childhood days, Conrad Cleveland, was serving as campaign treasurer, and it was not easy duty. Negotiating a partial payment to one creditor, Cleveland wrote: "We regret that the disastrous end of the campaign so bankrupted us as to make no further settlement possible."

Senator Barry Goldwater came to Columbia for a fund-raising dinner to help retire the campaign debt. Half the proceeds would go to the successful Republican candidate, James Edwards, said a press account, the other half to Westmoreland, "whose loss to Edwards in the July primary stunned outsiders." At least a year and a half after the campaign Cleveland rendered a report on the current status of efforts to raise funds to pay off the remaining debts. "Looks like we are beginning to see a glimmer of light at the end of the tunnel!" he said with obvious relief. That proved to be premature.

The California political consulting firm had been both expensive and worthless. Cleveland, saying that he and Westmoreland agreed that "their guidance of the campaign was disastrous," described a computerized letter sent by the firm to 50,000 people at a cost of $10,825:

"From 50,000 letters, 52 return envelopes were received, 45 of these were empty or negative, and *seven* contained contributions to a total of $67.00!"

The campaign lagged in paying the consultants, who finally took the matter to court, rejecting proposals to settle for payment of less than the full amount owed. As the case was about to go to trial, the campaign's attorney wrote to Westmoreland and several associates to describe the situation. "We have a problem, from the standpoint of jury appeal," he said, in that our own campaign treasurer will testify that the amount "was justly owed and that the primary reason for the settlement was lack of funds." Thus "I greatly fear" that we will not be able to avoid payment of the amounts claimed "when it is developed that *all* other debts of the campaign were paid, *including* money advanced by General Westmoreland."[3] Confronted with those realities, Westmoreland and the campaign decided to pay up, to the relief of their attorney. "I think this relieves you Gentlemen of lost time and perhaps some embarrassment from having this case publicized in the newspapers," he said.[4] It was January 1977, two and a half years after the primary.

WITH WESTMORELAND ONCE more in need of something to do, John West, the outgoing South Carolina Governor who had earlier given Westmoreland the economic development position, now wrote to President Nixon: "I would like to suggest that you give consideration to the utilization of the talents of our mutual friend, General William C. Westmoreland. As you know, 'Westy' had an unfortunate political experience, not of his own making." Nothing came of that letter, and not surprisingly, since by then Nixon was consumed by some extremely serious problems of his own. In fact, the letter itself seems somewhat bizarre, dated 13 August 1974, four days after Nixon resigned the presidency.

Even before that Westmoreland's prospects with Nixon had not been good, as Ernest Furgurson wrote in March 1974: Westmoreland "could never convince me that [being elected governor] would be the pinnacle of his ambitions. There was a time . . . when the thought of becoming President passed through his mind." Given that, concluded Furgurson, when Westmoreland retired as Chief of Staff Nixon was not going to risk giving him a major appointment. "He offered him the Immigration Service. Westmoreland refused." He would have accepted the Veterans Administration, "but that was not offered."[5]

Congressman Dorn, Chairman of the House Committee on Veterans Affairs and also an unsuccessful gubernatorial candidate, now also tried to be helpful in Westmoreland's job search, telegraphing the President (still Nixon, but barely): "Should the position of Administrator of the Veterans Administration become vacant, I recommend for your earnest consideration General William C. Westmoreland." Westmoreland wrote in longhand on a letter from Dorn the reply he wanted typed up: "You and I have shared a most unusual gubernatorial campaign and I must say the results are perplexing."[6]

"Mrs. Westmoreland is one happy woman that he lost," said Charles MacDonald. "What he would like is some government appointment in which he could work part-time. He doesn't do anything now," this said in May 1976. "He has no real job."

The aftertaste of the campaign was bitter for Westmoreland. "It was one of the more tiring and frustrating experiences of my life, and I have had many in that category," he wrote six weeks after the defeat. As so often in the past, though, Westmoreland tailored his message, or outright reversed it, for different recipients. "I found the campaign enjoyable, interesting and educational," he wrote to someone else at almost that same time. "I hold no animosities." On that latter point he told another person that "the world is filled with some peculiar people with extreme and radical views and, as you pointed out, the South Carolina Republican Party is no exception."

Asked a few years later about his political future, Westmoreland responded that "I would never get elected for office. I'm too controversial, too forthright, and I can't act. It would be totally repugnant to me."

29

★

PLAINTIFF

On 23 JANUARY 1982 the CBS Television Network aired a documentary entitled "The Uncounted Enemy: A Vietnam Deception." Said Mike Wallace in the introduction: "Tonight we're going to present evidence of what we have come to believe was a conscious effort—indeed, a conspiracy at the highest levels of American military intelligence—to suppress and alter critical intelligence on the enemy in the year leading up to the Tet Offensive."

Westmoreland cooperated in preparation of the program, traveling to New York for interviews with Wallace. George Crile, the program's producer, told Wallace before the session that he had read Westmoreland a letter describing the proposed areas to be discussed and that Westmoreland had made no complaints. "He puzzles me—," said Crile, "seems not to be all that bright."[1] He also told Wallace about "Westy's rather vigorous request for an honorarium." Westmoreland had told him, he said, "that he makes his living giving speeches and thinks it only fair of us to pay him something. Suggests $3,000."[2]

In a note to Wallace before the interview, Crile was optimistic about how the program was coming along. "Now all you have to do is break General Westmoreland and we have the whole thing aced," he said.[3]

On the program Westmoreland did not come across well. He looked and acted nervous, even evasive, stammering and licking his lips and clearly uncomfortable. Observed William F. Buckley Jr., "I am prepared to concede that Gen. Westmoreland must have strengths. But he has none at all in front of a camera."[4] Peter Braestrup also commented tellingly on Westmoreland's paltry skills: "His memory for details, even when the facts favor him, is now rusty. He occasionally shoots from the hip. He is easily exasperated by questions he considers unfair or malicious."[5] The result was that Westmoreland often looked guilty, whether or not he was in fact.

At one point Wallace and the general discussed losses inflicted on the enemy during the 1968 Tet Offensive. Wallace noted that MACV's first order of battle report after the offensive credited the enemy with 204,126 men. That was down from some 224,000 claimed just before Tet, and so reflected a reduction of 20,000. "How many troops did he [the enemy] lose, General?" asked Wallace. Westmoreland responded that, out of about 84,000 troops committed, in "the early days of the Tet offensive, he lost 35,000." Wallace confirmed that Westmoreland meant 35,000 killed, then asked how many wounded. "Well, I—we have no way of knowing that, but usu—usually the ratio is about three-to-one, three wounded for one that is killed."[6]

Wallace, having set his prey up for the kill, then pounced. "If you take General Westmoreland at his word," he observed, "here is the logical problem you run into. It begins with MACV's official estimate of total combined enemy strength in the South just before Tet—224,000. Five weeks later, on March seventh, Westmoreland reported 50,000 of those enemy had been killed. Now, according to his own standard ratio, for every one killed three were wounded. So, even disregarding the enemy soldiers who defected or were captured, the bottom line figure just doesn't make sense. If so many Viet Cong had been taken out of action, the question had to be asked: Whom were we fighting?"[7]

Wallace recalled a crucial part of that interview when he asked about Westmoreland's decision to drop the Self-Defense Forces from the order of battle. By that point in the interview, said Wallace, Westmoreland was "acutely irritated with the whole tenor of our discussion, so instead of answering my question, the general decided that the time had come to put me in my place with a verbal reprimand." Thus,

Westmoreland: "This is a non-issue, Mike. I made the decision. It was my responsibility. I don't regret making it. I stand by it. And the facts prove that I was right. Now, let's stop it!"[8]

Wallace responded by rephrasing his question: "Isn't it a possibility that the real reason for suddenly deciding in the late summer of 1967 to remove an entire category of the enemy from the order of battle—a category that had been in the order of battle since 1961—was based on political considerations?"[9]

Westmoreland: "No, decidedly not. That—that—." Wallace pressed: "Didn't you make this clear in your August twentieth cable?" Westmoreland: "No, no. Yeah. No."[10]

Wallace: "I have a copy of your August twentieth cable—." Westmoreland: "Well, sure. Okay, okay."[11]

Wallace then quoted from what he said was Westmoreland's cable, but it was not. It was instead a cable General Creighton Abrams, then Westmoreland's deputy in Vietnam, had sent in Westmoreland's absence, stating the "command position." It really did not matter, except that Wallace had misstated the sender, since the message was the same: "We have been projecting an image of success over the recent months. The self-defense militia must be removed or the newsmen will immediately seize on the point that the enemy force has increased. . . . No explanation could then prevent the press from drawing an erroneous and gloomy conclusion."[12]

During a break in the taping Westmoreland exclaimed angrily that he had been "rattlesnaked" by Wallace, meaning apparently that he had been quizzed on matters for which he was unprepared. He claimed that he had not had access to his files and records in preparation for the interview.[13]

In fact, however, in subsidiary materials held with the Westmoreland biographical file at the U.S. Army Center of Military History, there is a typescript account, "General Westmoreland's Use of His Personal Records in Connection with His Litigation with CBS," dated 13 October 1982. Noting that Westmoreland had upon his retirement temporarily given the Center "a large collection of official and personal files," the document notes that he used these materials extensively in the preparation of his memoir and, "most recently, to refresh his memory during the filming of the recent CBS special on the order of battle controversy during the Vietnam war and to answer or correct erroneous statements made by the network."

Also, states this document: "[F]rom time to time since General Westmoreland first learned of CBS's intention to produce a program dealing with the order of battle question and intelligence collection during the Vietnam war, he has asked to examine portions of his official and personal papers to refresh his memory in preparation of interviews with members of the CBS staff or to correct mistaken information following these interviews and the airing of the program itself."[14] Vince Demma, the staffer who pulled the documents requested by Westmoreland, remembered that "he was honing in on the August 1967 order of battle conference. That was what interested him."[15]

Wallace's "lengthy, harsh interview with Westmoreland"[16] took place at the Plaza Hotel in New York on 15 May 1981. On apparently even simple matters Westmoreland came across as defensive, unfocused, and inarticulate. A question about his 1967 trips from Vietnam to address the Congress and other audiences produced this response: "I was ordered to come to Washington." "And—and I—I wasn't happy about it, but I was ordered back. And I said if this is the President—what—is—if this is what the President wants me to do, well, I'll—I'll do my best."[17]

Apparently it was something of an ordeal for both parties, Wallace later stating, "[T]hat was hard work." What is puzzling is why—given his obvious unhappiness with how the interview was progressing—Westmoreland did not just walk away from it rather than endure a three-hour interrogation that portrayed him in such a negative way.

The ninety-minute program aired on 23 January 1982, at nine-thirty on a Saturday evening, drawing what was reportedly the smallest audience of the week for a prime-time broadcast, ranked seventy-ninth out of seventy-nine. Nevertheless several million people viewed the documentary in whole or in part.

Westmoreland was devastated by the program's allegations that reporting on the enemy order of battle had been intentionally falsified. Within days he assembled a group of supporters (prominent people, but not all of them in a position to know the truth or falsity of the program's substance) to join him in a press conference held at the Army and Navy Club in Washington. Joining Westmoreland were Ambassador Ellsworth Bunker, Lieutenant General Daniel Graham, George Carver of CIA, Lieutenant General Phillip B. Davidson Jr., and Colonel Charles Morris. Barry Zorthian was there in the audience. Westmoreland had sought to have Major General Joseph McChristian attend, but McChristian declined, instead sending Westmoreland a telegram in which he said: "I have defended the integrity of Army intelligence when I was in Vietnam. I cannot speak to that after I left. Those who served later should defend it if they can, and if they cannot you should find out why."[18]

Westmoreland led off, accusing CBS of "a vicious, scurrilous, and premeditated attack" on his character and charging that the interview with Mike Wallace was "a star chamber procedure, with distorted, false, and specious information, plain lies, derived by sinister deception — an attempt to execute me on the guillotine of public opinion."[19] Westmoreland also maintained that "he had 'done no research and had brought no documents' to refresh his fourteen-year-old memory of events."[20]

He also sought to obscure what he had done to the order of battle, stating that "to include the disputed categories in the Order of Battle would have been 'to introduce a substantial jump in enemy strength when in fact there was no increase in combat strength.'" But of course the issue was not introducing such categories, which had been included all along, but Westmoreland's taking them out, thereby reducing the total. "I refused to include those,"[21] said Westmoreland, when what he had done instead was to take them out.

There was also, as it turned out, a little manipulation of evidence at the press conference. Westmoreland introduced Graham to offer a detailed rebuttal of the CBS broadcast. Graham used a videotape machine to show Colonel Gains Hawkins appearing to acknowledge that the statistics in question referred only to the political order of battle, not to armed Viet Cong. But, reported Robert Kaiser for the *Washington Post*, "[T]he clip from the documentary that Graham showed was edited to cut out Hawkins' final words, when he said that the political order of battle included 'the Vietcong's political bureaucracy and the guerrilla strength.' The guerrillas were armed." Asked about this use of editing to distort Hawkins's remarks, said Kaiser, "Westmoreland made no reply."[22]

Naturally the broadcast and its immediate aftermath occasioned a cascade of media commentary, much of it notably well informed in commenting on the complicated and somewhat arcane details of the matter. This was particularly the case with the enemy categories known as Self-Defense Forces and Secret Self-Defense Forces. "They had been included as part of the enemy force in the order of battle since MACV was formed in February 1962," observed one press account, "but MACV was now claiming they were merely old men, women and children who planted punji stakes."[23]

That characterization tripped Westmoreland up during a discussion of body count in the course of his later deposition during a libel suit he brought against CBS and others. After he had directed removal of several categories of enemy (Self-Defense Forces and so on) from the order of battle, Westmoreland was asked whether he nevertheless continued to include them in body count when they were killed. His first stab at an answer was that, after an action, "an estimate based on an inspection of the battlefield was put in," and an effort was made "to relate those casualties to a very specific category or a specific unit. But," he said, "that was very difficult to do."[24]

That assertion proved puzzling to the CBS attorney questioning Westmoreland: "One of the problems that I have, General, is that on the one hand you stated that the self-defense militia were almost all women, youngsters and old men, and on the other hand you say that it is hard to distinguish them from the rest of the enemy order of battle. I can't understand why those two things would both be the case."[25] Westmoreland could only backtrack: "Well, I, uh — I think that I overstated the first point. I really hadn't concentrated on the practicality of it, and I think I overstated that."[26]

STRONG CRITICISM OF the CBS documentary soon came from an unlikely source, *TV Guide* magazine. Don Kowet and Sally Bedell's article, "Anatomy of a Smear: How CBS Broke the Rules and 'Got' Gen. Westmoreland," obviously drawing on at least one inside source at CBS, alleged numerous very serious violations of network policy and journalistic ethics by Crile and his production team. These included failing to interview important witnesses who might have supported Westmoreland, such as Lieutenant General Phil Davidson and the CIA's George Carver. Some of those who were interviewed and gave evidence favorable to Westmoreland were excluded from the broadcast, in particular Walt Rostow. Others who appeared on the program were misidentified, including a DIA officer said to have been the chief of the MACV delegation to the SNIE conference.

While the *TV Guide* writers were sympathetic to Westmoreland and how he had been treated by CBS, they were also somewhat dismayed by the man they encountered when they interviewed him in Charleston in February 1982. "Westmoreland seemed a mere shell of the jut-jawed general whose confident face was featured on newscasts throughout the late 1960s," they reported. "His eyes teared easily. He slurred some words. He seemed to suffer spasms of forgetfulness."[27]

Kowet later wrote a book about the case in which he revealed a particularly unflattering opinion of Westmoreland, despite having so harshly criticized the way CBS produced its documentary. "Westmoreland," he wrote, "was a pariah whom most Americans viewed as unsympathetic, the stigma of Vietnam clinging to him like swamp stench."[28]

A more compassionate view was expressed by a former congressman, Otis Pike, writing in *Newsday:* "CBS made Gen. Westmoreland appear evil, and he was not evil. He may not have been either the best or the brightest, but he was doing the best he could for his country and he deserved better than CBS gave him."[29] Many others agreed with that assessment, including a group of veterans who organized a campaign to help pay Westmoreland's legal expenses.

STUNG BY THE CRITICISM, CBS began an internal investigation into the program's fairness and accuracy. Burton Benjamin, a respected senior executive at the network, directed the inquiry. At one point, interrogating Crile, he asked, "Do you think Westmoreland was somewhat inept?" Crile agreed: "Yes. He seems stupid."[30]

Among the serious shortcomings Benjamin discovered in analyzing Crile's work were repeated violations of CBS News standards—matching a question to an answer given to a different question, associating an answer with a referent other than the one intended, conflating portions of answers to produce a position not stated by the interviewee, coaching interviewees friendly to the program's thesis, reinterviewing such witnesses to get a stronger position, failing to interview important persons who would probably have supported Westmoreland's position, unwarranted use of the word "conspiracy," and a serious imbalance in reflecting opposing sides of the issues. Benjamin's summary judgment was that the documentary was "seriously flawed."[31]

Even so, he concluded, "as for the substance of the broadcast, that enemy strength in the Vietnam War had been intentionally undercounted, nine former military and intelligence officers, on their own volition, had made that allegation, and none had recanted. I was convinced that if opposing views had been given more time, which they were entitled to, the thrust of the program would have remained the same, and it would have been a stronger broadcast."

Also there was a critical incontrovertible fact, "a damaging disclosure" admitted by Westmoreland himself in the Mike Wallace interview, noted Benjamin. "Westmoreland dropped a whole category of the enemy—the self-defense militia, a force of 70,000—from the order of battle, thus skewing the enemy-strength total."[32]

What was dropped by Westmoreland in fact amounted to much more than that. In a sworn affidavit Major General McChristian recalled the estimates put the "political order of battle at 88,000 and the irregulars at 198,000," for a total of some 286,000.[33] But later, when the matter went to trial, Westmoreland said under oath that "certainly there was no ceiling involved in the actions that took place. I could have cared less whether a few more of this and a few more of that."[34]

Having carefully examined the program and how it had been developed, and documenting a multitude of unethical and unprofessional practices involved, Benjamin nevertheless concluded that the essential story had been on the mark. "I think the piece itself is accurate," he wrote, "that it faithfully represented what went on back in 1967 and 1968."[35]

Later Benjamin wrote his own book about the documentary, at one point recalling the testimony of Brigadier General George Godding and noting that "under vigorous cross-examination, he conceded that the enemy-strength figure he was carrying to the CIA meeting at Langley could not be exceeded without permission from Saigon. In spite of all the semantics," concluded Benjamin, "it certainly appeared that there was a ceiling or cap dictated by MACV, as the Vietnam program had asserted."[36]

General Bruce Palmer Jr. was later asked whether Brigadier General Phillip B. Davidson Jr., the MACV J-2, had been involved in putting a ceiling on the enemy order of battle. "Sure he was," said Palmer. "Westmoreland put it on."[37]

ON 13 SEPTEMBER 1982 Westmoreland sued CBS, Mike Wallace, George Crile, Van Gordon Sauter (later dropped from the suit), and Sam Adams for libel, asking $120 million in damages. David Halberstam suggested that Westmoreland "sued CBS not just because he felt the documentary had done him damage but also because he wanted a second chance at history, a larger vindication."[38] That might help to explain why Westmoreland ignored or overrode all those who sought to dissuade him from bringing the lawsuit, and indeed why he had agreed to be interviewed for such a program in the first place. He was still trying to shape the historical record on Vietnam.

Several experienced and prominent attorneys who wished Westmoreland well had advised him not to file such a suit, explaining that the bar of proof for a public figure was just too high and that the odds against success were thus very long. These included Clark Clifford, Edward Bennett Williams, and Stanley Resor. Said General Walter Kerwin, "Bruce Palmer and I got together and tried to talk him out of it. He wouldn't listen." Phillip Davidson confirmed: "Westmoreland doesn't take advice. Barry Goldwater (who had sued for libel) told him it was not worth it. Clark Clifford told him not to do it. Edward Bennett Williams told him not to."[39] Also, according to one source, "Kitsy Westmoreland . . . had been 'violently opposed' to filing the suit."[40]

Recalling the press conference a few days after the documentary was broadcast, Davidson said he thought then they had "pretty much refuted the charges which Sam Adams and CBS had leveled against Westmoreland. But of course Westy wasn't satisfied. Here was a chance . . . for Westy to get back in the front pages, or at least the second pages, of the United States newspapers and magazines. And so, if he could find anybody to represent him, he was going to do it [sue CBS]."[41]

Given the outlook of highly competent attorneys on the unlikelihood of Westmoreland's prevailing in such a suit, it was not easy for him to find representation. He finally wound up with Dan Burt, head of an outfit known as the Capital Legal Defense Corporation, a not-for-profit entity known primarily for involvement in public policy matters.

Later Dean Rusk, former Secretary of State, wrote to Westmoreland to tell him "on a very personal and private basis" that he did not think Westmoreland's lawyers "were up to the task they had and were no match for the high stepping group assembled by CBS." Westmoreland's response said in effect that that was the best he could do: "They were the only people in the legal profession who actively came to the rescue."[42]

CBS meanwhile hired as outside counsel the New York firm of Cravath, Swaine & Moore. David Boies became their lead attorney. Larry Worrall, whose firm provided CBS's libel insurance and who had seen a lot of attorneys in action, said, "David Boies was probably the best lawyer that I've ever come across."

THE PRETRIAL PROCESS of taking depositions from potential witnesses was lengthy and arduous. The deposition of Crile "was taken over fifteen days and ran more than nineteen hundred pages. General Westmoreland's was equally long."[43] George Godding from MACV J-2 was also deposed, with the questions put by David Boies. The transcript of that session ran to 348 pages, much of it the result of Godding's efforts to avoid having to acknowledge that he had been given an enemy strength figure approved by General Westmoreland and told to present and defend that figure at the SNIE conference in Washington in August 1967. That figure was 297,790, and it did not include the categories that Westmoreland had unilaterally decreed were to be newly excluded from the overall order of battle.

Godding and Davidson had gone in to see Westmoreland before Godding left for the conference, ascertaining that Westmoreland approved that number and wanted it defended. Godding subsequently did everything he could to avoid having to admit that fact. Eventually Boies wore Godding down, the crucial point coming even before the trial in the deposition. "You went in and had a conversation with him [General Westmoreland] and General Davidson, and, based on that conversation, General Westmoreland instructed you to take that 297,000 figure to Washington and present and defend it; correct?" Godding finally admitted, "That's correct."[44]

A key issue at every stage of the proceedings was exactly that, whether Westmoreland had put such a ceiling on the number of enemy that MACV would accept in the order of battle and in preparation of the Special National Intelligence Estimate. In a pretrial telephone interview with Crile, Godding had, according to Crile's notes, said that, while he was representing MACV at the Washington order of battle conference, "The initial concern was over the total number—guidance from Westmoreland—all negotiations in Washington were held to the MACV official figure for May OB [Order of Battle report]. This was the gospel, the party line." Even more definitively: "Godding confirms that before going to Washington he met with General Westmoreland and Gen. Davidson and Westmoreland instructed him to keep the MACV position to the existing Order of Battle totals. . . . Stay with what we had. 'There was a command figure. . . .'"[45]

Of course Westmoreland himself had reduced the order of battle dramatically by decreeing that whole categories of enemy forces should be deleted. Whether, by specifying that his representatives must adhere to the "command position," he had stipulated a de facto ceiling remained to be adjudicated, or at least determined by a jury.

THE TRIAL BEGAN on 9 October 1984 in the U.S. District Court for the Southern District of New York, in Manhattan, Judge Pierre Leval presiding. It was destined to continue for eighteen weeks, with over three hundred hours of testimony, before a surprise ending brought it to an abrupt halt.

In some respects the trial was over even before it began. Dan Burt, Westmoreland's attorney, made a major decision at a session with Judge Leval the day before the trial began. There he dropped a large part of his case, deciding he "would not challenge CBS on a key assertion of the documentary: that Westmoreland had conspired to deceive the press, the Congress, and public about enemy troop strength in Vietnam." Instead, he would confine his efforts to proving "that CBS had defamed Westmoreland on a much narrower issue: that the general had deceived his superiors, including the Joint Chiefs of Staff and the President."

As the trial progressed Westmoreland was of course an essential witness, just as he had been in the CBS documentary. The credibility of his testimony was undercut at several points by discrepancies between what he said on the witness stand and what he had said in earlier depositions, and in some cases even three-way inconsistencies between his interview testimony as shown in the CBS broadcast, his pretrial depositions, and his testimony in court. For example, on the witness stand Westmoreland said he was aware that LBJ was "convinced that we were making substantial progress in Vietnam and he wanted hard facts so that the progress would be recognized." But in the earlier deposition he had stated he "never had the impression that the President wanted hard evidence on progress in the war." One juror observed that on the witness stand "Westmoreland doesn't seem to be that quick or bright. He has a tunnel-type vision—he sees what he sees on his own terms."[46]

Time after time Westmoreland was caught in contradictions between what he had said on earlier occasions and what he now stated under oath. "In the final analysis," he said at one point in his testimony, referring to his 1967 troop request, "I got basically my minimum essential force." Boies took exception to that, suggesting that in fact Westmoreland's "minimum essential force was scaled down sharply, cut almost in half." He was reading from Westmoreland's book *A Soldier Reports*. Then, continuing his questioning, he asked: "You were extremely disappointed with that, were you not, sir?" Westmoreland responded, "I wouldn't say extremely, no." Boies returned to the book. There Westmoreland had stated, "I was extremely disappointed" not to get the troops he had requested. "Well," now commented Westmoreland, "I was disappointed at the time." And so it went.

ON 13 DECEMBER 1984 Judge Leval issued some important instructions. "The issue in the libel suit," he explained, "is whether the publisher recklessly or knowingly published false material." And "the libel law does not require the publisher to grant his accused equal time or fair reply. It requires only that the publisher not slander by known falsehoods (or reckless ones)."

Later, in a crippling ruling, Leval would frame the issue as whether Westmoreland had *attempted* to deceive, not whether he had succeeded in any such effort.

Setting the bar for a successful libel prosecution was a case known as *New York Times vs. Sullivan* in which the Supreme Court had decided in 1964 that "free speech was so important that even false and defamatory statements made about the official conduct of public figures deserved the protection of law. To recover damages, an official had to prove that the statement was made with 'actual malice,' or with 'reckless disregard' as to whether the statement was true or false." There was a reason so many attorneys had counseled Westmoreland against suing in the first place.

AT ONE POINT Westmoreland sought to share the blame by getting the Army involved in dealing with CBS. "I would hope," he wrote to Secretary of the Army John Marsh in late March 1982, "that in some way the Department of the Army would stand up and expose the hoax and sham of the so-called CBS documentary which attacks the integrity of the U.S. military."[47] That characterization fooled nobody. It was not the "U.S. military," but Westmoreland personally and specifically, who was being charged with data manipulation and misrepresentation. Now the Army stayed out of the matter.

WESTMORELAND DEEPLY RESENTED an interview with CBS given by Major General Joseph McChristian, MACV J-2 during 1965–1967, saying, "I was puzzled by his conduct and telephoned him as an old friend and former colleague and expressed my surprise that he would go on a national tv program and criticize me without the courtesy of informing me. I told him in essence that it was an unusual performance and alluded to the confidence that had traditionally prevailed between West Point graduates. I added that I was disappointed to have him add to the burden that I had carried in major degree for those of us who fought an unpopular war."[48]

In another criticism made during a 15 May 1981 interview Westmoreland sought to demean McChristian and undermine his credibility by assigning to him ignoble motives for his testimony. "If he has a vendetta because he didn't get promoted," said Westmoreland, "well, I'm sorry. But that seems to be the case."[49]

Westmoreland had specifically requested McChristian as his J-2 at MACV, then frequently complimented him on the job he did. Later, as Army Chief of Staff, he brought McChristian in to be the Assistant Chief of Staff for Intelligence, the top Army post in that field. McChristian, having in the meantime commanded an armored division for two years, then went on to serve for three years as the top Army intelligence officer while Westmoreland was Chief of Staff. None of that background supported the vengeful motives Westmoreland later claimed underlay McChristian's order of battle testimony. For his part, McChristian's principal subordinate for order of battle intelligence in Vietnam, Colonel Gains Hawkins, eloquently described McChristian as "your white knight serene, impeccable and untouchable."[50]

During direct examination of General McChristian two important disputed matters were brought out. First it was stated that "General Westmoreland has testified at this trial that, in substance, he asked you at that meeting about the components of the category known as irregulars. Did General Westmoreland ask you about the components of the category known as irregulars at that meeting?" McChristian answered, "I don't recall General Westmoreland asking that question, but if he had, I certainly would have pointed out what happened at the Honolulu Conference, that all of the categories that we have were approved by that conference, and I would have had to point out other things to him, and I think it would have stuck in my mind. I don't recall any discussion on that."[51]

Then: "General Westmoreland has also testified at this trial . . . that at that meeting he said to you, 'Joe, with respect to the self-defense and secret self-defense, we are not fighting these people. They are basically civilians. They don't belong in any representation, numerical representation, of the military capability of the enemy.' Did General Westmoreland say that to you at that meeting in words or substance?" McChristian: "No, sir. Here again, had he have said that, I definitely would have to come back and tell him about what happened at that conference and explain that—as far as I am concerned, that could not have been covered."[52]

In his memoirs, which appeared long before the CBS documentary, Westmoreland had expressed a view he now sought to distance himself from: "in insurgency warfare a civilian political cadre, even if unarmed, also constitutes the enemy."[53] But of course he had been right at that time and wrong before and after.

General McChristian underscored that when he brought the message to Westmoreland describing the better intelligence on irregular forces and the resulting increased order of battle numbers. Westmoreland told him that sending such numbers forward would "create a political bombshell."[54] During the trial Westmoreland was asked why he thought that would have been the outcome. "Because people in Washington were not sophisticated enough to understand and evaluate this thing, and neither was the media," said Westmoreland.[55] In his pretrial deposition Westmoreland had also stated that he was "sensitive to the fact that," had he submitted such a report, "there could be a political problem involving my—that could be used to embarrass my commander in chief."[56]

Asked by David Boies whether he believed it was improper for General Westmoreland to hold his cable, McChristian had no doubts: "I think that for a military man to withhold a report based upon political implications would be improper."[57] Later he would say that he thought Westmoreland, "in being loyal to the President, was disloyal to his country." Was he also being self-serving? "Yes," concluded McChristian.[58]

McChristian's testimony was as dramatic, consequential, and determinative of the outcome as had been his interview comments on the original CBS program, demolishing the idea that, among other things contended by Westmoreland, the enemy was in deep trouble, was running out of men, and had launched the 1968 Tet Offensive because he knew he was losing the war.

Burt's case on behalf of Westmoreland was in a shambles. Meanwhile, the witnesses on whom Westmoreland depended most did not create a favorable impression, at least as viewed by one juror who later wrote a book about the experience. "We've had a lot of witnesses," recalled Patricia Roth. "Some I liked, with others I was uncertain. But with the ones that Westmoreland seemed to feel most credible, I definitely didn't like them and didn't believe them."[59]

ON 18 FEBRUARY 1985, only days before the case would have gone to the jury, Westmoreland suddenly and unexpectedly agreed to settle. In exchange for withdrawing his suit against CBS and the others, he got a statement by CBS that read in pertinent part: "CBS respects General Westmoreland's long and faithful service to his country and never intended to assert, and does not believe, that General Westmoreland was unpatriotic or disloyal in performing his duties as he saw them." There was another provision of the agreement on discontinuance, less often quoted: "General Westmoreland respects the long and distinguished journalistic tradition of CBS and the rights of journalists to examine the complex issues of Vietnam and to present perspectives contrary to his own."[60] Each side, it was also agreed, would pay its own legal expenses.

That was, claimed Westmoreland, a victory for him. "I finally decided to withdraw my complaint in return for a joint statement signed by the corporate general counsel, producer, and a commentator, which I considered a clearing of my honor and integrity as a responsible commander in a controversial and much misunderstood war."

As he spun it to a correspondent, "CBS gave me a public statement that contradicted the theme of the broadcast and I dropped my charge against them, encouraged by the fact that my lawyer had run out of money." Later Westmoreland said the settlement was influenced by the assessments of his legal team's "jury watcher," who reported an anticipated unfavorable outcome. On yet another occasion he wrote a letter to the editor of the *American Spectator* in which he stated that "my attorney did not betray me" and that he had "weighed with Mr. Burt my chances with the jury and decided to accept a settlement after the judge imposed an inordinately high burden of proof on my attorney before a jury dealing with matters unfamiliar to them."[61]

Thus, concluded Westmoreland, "the effort to defame, dishonor and destroy me and those under my command had been exposed and defeated. I, therefore, withdrew from the battlefield, all flags flying."[62]

Others saw the statement and its import quite differently. The *New York Post* headline was "Westy Raises the White Flag." The *Washington Post* noted that CBS's only apparent concession was agreement "not to try to force Westmoreland to pay its court costs and legal fees." The *New York Times* succinctly stated the prevailing reaction, observing that "the general surrendered to the evidence that whether or not his superiors in Washington were in fact deceived, he and some of his aides in Vietnam in 1967 manipulated the estimates of enemy strength, apparently for political effect." At the trial, continued the *Times*, "General Westmoreland had trouble proving any falsehood. At the end, he stood in imminent danger of having a jury confirm the essential truth of the CBS report. For, in court, as on the original program, the general could not get past the testimony of high-ranking former subordinates who confirmed his having colored some intelligence information."[63] Said one of the jurors, speaking to members of the press on the way out of the courthouse for the last time: "The evidence in favor of CBS was overwhelming."[64]

The *Baltimore Sun* also commented editorially, saying: "Now he has lost it, no matter what face he chooses to put on it, and he deserved to lose."[65]

THE ISSUES RAISED by CBS did not go away, despite Westmoreland's claims of victory. A decade later, in an A&E Biography segment about Westmoreland, Mike Wallace was featured, stating his retrospective view of the broadcast. "I believe that the documentary was accurate," he opined. "I do believe he cooked the books in pursuit of what he believed was the right thing to do."[66]

George Crile had his own last say on the matter in a speech he gave several years after the trial, suggesting that some "people who watched General Westmoreland trapped, I think trapped in his own lies, at some point came to resent the fact that his patriotism was being taken from him."[67]

Once, appearing on William F. Buckley Jr.'s program *Firing Line*, Westmoreland asserted that America did not lose the Vietnam War. "Tell that to the boat people!" Buckley exclaimed. Now, in the wake of the aborted libel suit, Westmoreland told a National Press Club audience that he did "not believe that either side in *Westmoreland vs. CBS* obtained what it had been seeking."[68]

In a memoir published the same year in which Westmoreland died, Mike Wallace stated that "Westmoreland's big mistake was not withdrawing when he did, but his decision to initiate the costly libel suit in the first place. In that respect, his legal action stands as a parable of the U.S. involvement in Vietnam."[69]

30

★

DUSK

W ESTMORELAND'S LIFE SINCE Vietnam has been miserable,"
said his former aide-de-camp, Lieutenant General Dave Palmer.
Problems of racial disharmony, drug abuse, and indiscipline had
plagued the Army during his tenure as Chief of Staff. Defeat in the
Republican primary of the gubernatorial election had been a major
disappointment, later described by Westmoreland as his "most de-
grading experience." Then the libel trial against CBS Television had
been a disaster.

Captain Richard Wright, a longtime Charleston resident, got to
know Westmoreland during his later years and developed considerable sympathy for him. "I always considered the general a tragic figure," he said. "Although he carried himself as erect as a West Point
plebe and had a somewhat arrogant demeanor, one could, I think,
sense that his pride had been mortally wounded." Historian Michael
Beschloss summed up the amalgam of sadness and futility, describing
Westmoreland as a general "who, chained forever to the Vietnam ca-
tastrophe, spent the rest of his life trying to explain himself to people
who would not listen."[1]

In one of his last public declarations, Westmoreland sought to counter the perception that regrets and recriminations about the war had blighted his later years. "In my mind," he said, "Vietnam is ancient history and I've kind of divorced myself from it over the years. It was only one chapter in my history."

Conflicted views of Westmoreland's legacy were reflected in an agonized decision at West Point, where in 2002 a bicentennial history of the Military Academy was published. Westmoreland was on the cover with a number of other graduates, recalled Colonel Lance Betros, editor of that work and now head of West Point's Department of History, "then (for about six months) off, then (after a lot of discussion) back on. Rationale: He was the top U.S. commander in a major war."

At some point the National Rifle Association offered to present Westmoreland a scrimshawed powder horn decorated with the highlights of his life and career, and to that end queried Westmoreland about what might be included. He suggested that they "picture the dramatic spread of activities such as horseman, paratrooper, three wars, teacher, my wife Kitsy, father of three, talking to a Joint Session of Congress, TIME man of year."

WESTMORELAND CONTINUED AN active speaking program until Alzheimer's disease, which afflicted him in the usual progressive way during about the last decade of his life, curtailed his activities.

One of his finest days came in November 1982 when he led a veterans march down Constitution Avenue in Washington, D.C., to dedicate the Vietnam Veterans Memorial. Westmoreland called it "one of the most emotional and proudest experiences of my life."

Westmoreland was always very proud of his work with and for Vietnam veterans. This devotion was honored in 1988 by a special award from the Association of the United States Army, which described how, on behalf of veterans, Westmoreland "set about the self-imposed task of restoring their honor and bringing recognition and dignity to their service. For over fifteen years, he has given his time, energy and loyalty to this unselfish cause." The citation, also noting Westmoreland's "massive" correspondence with veterans, concluded with these admiring words: "His is a special case. He made their cause his own. He has helped to bind up a nation's wounds. There is no more noble work."

Westmoreland took advantage of the award ceremony to compliment Kitsy. Since marrying him, he noted, she had lived in three foreign countries, seven states, and thirty-three houses, and had during four years of near-total separation taken complete responsibility for raising their three children and much else. And he cited the hundreds of hours (and probably should have said thousands) she had contributed as a nurse's aide in military hospitals, including a special trip on a medevac aircraft to Japan, "where she gave a rose to every veteran from Vietnam that we had hospitalized there."

General "Dutch" Kerwin observed of Kitsy's outlook that, in her opinion, "Westy carried the whole cross. And in a sense he did." Westmoreland had also addressed that perspective, saying that when he retired "my wife wanted to go home." He spoke of how she had lost a younger brother, "killed in Vietnam on the day I was sworn in as Chief of Staff." But Westmoreland felt he couldn't go home. "I decided to speak about the war, whenever and wherever I could. I was spat on, students and faculty were rude, one group set off the fire alarm while I was trying to speak, and so on." But he persevered, despite the entreaties of friends who advised leaving off the crusade and devoting himself to family in these later years.

WESTMORELAND ALSO RETURNED to his opposition to the volunteer force, saying in the 1980s, "I firmly believe that, in the national interest, we must return to the draft," and at about the same time predicting "5 years from now selective service—the draft will probably have to be resumed." To a correspondent he complimented for "wise thoughts" in expressing the view that "the draft will be back," Westmoreland wrote: "I agree with you. It is only a matter of time."

In a new "Afterword" in the paperback version of his memoirs Westmoreland was hard over on this issue. "Time has also demonstrated," he wrote, "that the political maneuver by President Nixon of setting aside the draft was not in the national interest. The all-volunteer force has not produced the military posture required by the leader of the free world. Reappraisal of the ill-advised concept is essential."[2]

But when an all-volunteer force performed brilliantly in the first Gulf War, Westmoreland sought some of the credit. Appearing before the House Armed Services Committee's Defense Policy Panel on 26 April 1991, he described his view "that the battle in the Gulf was a valid test of the Volunteer Army and it passed the test." And "in that context," he added, "I submit for the record a copy of my public address on October 13, 1970 wherein I committed the Army to the achievement of a modern volunteer force." That was the speech at the AUSA annual meeting that saved his job.

ANOTHER POLICY ISSUE that caused Westmoreland great distress was the admission of women to West Point, begun in the summer of 1976. Rick Atkinson quoted Westmoreland's views on the matter: "Retired West Point superintendent General William Westmoreland extended some best wishes for their success: 'Maybe you could find one woman in 10,000 who could lead in combat,' he said. 'But she would be a freak, and we are not running the Military Academy for freaks.'"[3]

Lieutenant General Dave Palmer, Westmoreland's former aide-de-camp, recalled another blow to Westmoreland in the summer of 1989. "The day that we announced that Kristin Baker was to be First Captain at West Point, the first female to hold that position, was when the annual conference for retired Army four-star generals was in progress," said Palmer. "They were stunned. Westy was livid that we had done this. They almost had to restrain him." Palmer was West Point's Superintendent when that appointment was made.

Another participant in those annual four-star conferences recalled them as "delightfully patronizing" affairs at which the old boys were briefed on current issues and asked for their opinions. But most, knowing full well that old soldiers should just fade away, only "ask questions and praise the noble work being done by those still serving." But then there was Westmoreland, who instead offered advice, and lots of it. Thus "it was as if Westy assumed seniority among the retired that was neither challenged nor granted by the others, just politely ignored. That did not deter him from attempting to anoint himself at the following year's gathering, but with the very same results."

IN THE SPRING OF 1992, soon after the first Gulf War, General Norman Schwarzkopf visited West Point, where he was feted for his role as commander of allied forces in that highly successful campaign. Schwarzkopf also spoke to the cadets, telling them among other things of his view that in Vietnam the Army had lost its integrity. Inflated reports of body count were what he had in mind, a charge he repeated in his book *It Doesn't Take a Hero.* These comments infuriated West-moreland, who sought through intermediaries to pressure Schwarz-kopf into retracting or softening the indictment. Nothing came of those efforts, and the criticism stood.

BY THE MID-1990s, still making appearances at veterans events and the like, Westmoreland was relying primarily on a very short stock speech, maybe three minutes in length, consisting of a mildly humor-ous anecdote about addressing Korean cadets through an interpreter and a recital of the lyrics of "Stout-Hearted Men," which he knew by heart but attributed to the composer "Rogers Hammerstein." The veterans still crowded around him, still asked for his autograph, which he now rendered simply as "Westmoreland," written rather scrawly.

And he still wore his uniform, still made pronouncements on the world scene. When the issue of establishing diplomatic relations with communist Vietnam was being debated he put out a brief statement. "I do not believe that the incumbent political leadership in Hanoi merits recognition of that country at this time," he said.

As the United States was considering lifting its economic embargo of Vietnam, Westmoreland adopted the stance that "when Vietnam throws off its communist yolk," as he wrote to a friend, "then, and only then, should we do business with them." Then he completely reversed his stand, signing a statement in support of lifting the embargo. "It is well past time to put this divisive and difficult era behind us," he main-tained. President Clinton wrote to express "deep gratitude" to West-moreland for supporting him on the matter.

WESTMORELAND DONATED his papers to the library at West Point, then took them back and gave them instead to the University of South Carolina. Apparently he was not satisfied that the people at West Point had been moving fast enough to make them available to researchers, something he was eager to have done.

In 1996 Westmoreland was named a Distinguished Graduate by his alma mater, the U.S. Military Academy, along with General Alex-ander Haig and astronaut Frank Borman. Westmoreland's citation noted that he had "rendered a lifetime of extraordinary service to the United States Army and his fellow soldiers."

• • •

WESTMORELAND DEPARTED THIS life the evening of 18 July 2005 at the Episcopal retirement home in Charleston where he and Kitsy then resided. He was ninety-one. The *New York Times* obituary referred to him as the officer "who failed to lead United States forces to victory in Vietnam from 1964 to 1968 and then made himself the most prominent advocate for recognition of their sacrifices, spending the rest of his life paying tribute to his soldiers."

A funeral service was held on Thursday, 21 July 2005, in Charleston, in the historic and beautiful St. Michael's Episcopal Church, with a bishop and two other clergymen officiating. The casket was draped in a cream-colored pall with a maroon cross and was carried into the church by a multiservice detail of military pallbearers. Kitsy was of course joined by their three children. Bishop C. FitzSimons Allison delivered a eulogy for "the general," describing him as "a trustworthy gentleman" and reciting the Scout Laws, which he pointed out began and ended with "Trustworthy" and "Reverent." Westmoreland, he said, "was a throwback to a time when the word 'gentleman' meant something."

On Saturday, 23 July 2005, Westmoreland was buried at West Point at a gravesite he had chosen when he was Superintendent there. It was an idyllic summer day, with a deep blue sky dotted with puffy white clouds, and hot.[4] The provost marshal had anticipated throngs comparable to those on a football weekend. Thus military policemen spaced at intervals ringed the cemetery grounds, prepared to deal with traffic and crowds that never came.

EPILOGUE

Taken altogether, the life story of William Childs Westmoreland turned out to be infinitely sad. It had begun with such promise — Eagle Scout at fifteen, journeyer to Europe and the World Scout Jamboree, president of the high school senior class, Citadel cadet, First Captain at West Point, battalion commander at age twenty-eight, with the Presidential Unit Citation earned in combat in North Africa, full colonel at thirty, then a brigadier at thirty-eight while leading the airborne regimental combat team in Korea, major general — youngest in the Army — at forty-two, serving at the right hand of the famous Maxwell Taylor, then sent by him to command the storied 101st Airborne Division, on to West Point as the dashing Superintendent with the young and beautiful Kitsy at his side, and the three attractive children she had borne him, familiar then with so many greats of an earlier day — MacArthur, Eisenhower, Omar Bradley — then corps command, again with his beloved airborne, and the third star. The future seemed to hold limitless possibilities, perhaps — said some — even the presidency.

Then came Vietnam, where the great arc of triumph and achievement peaked and fell, never to be recovered, although doggedly rationalized, explained, rued, regretted, for the rest of his days. The lifelong aspiration—Army Chief of Staff, the crowning achievement for a professional soldier, was attained, but only as a bitter and unsatisfying post in the shadow of Vietnam. Then retirement, a book, endless rounds of speeches, all exculpatory but ultimately unconvincing. A plunge into politics, against the advice of wise friends and advisors, leaving a baffled neophyte defeated almost before he began. Then a lawsuit, charging a powerful and wealthy network with libel, again bucking the advice of those who wished him well and knew what they were talking about, and again abject failure, a last-minute withdrawal rather than face a jury's verdict, followed by lame and painful efforts to portray that outcome a victory.

The young Westmoreland, from early days prideful and image-conscious, had developed into an adult of incredible industry, driving himself to achieve, forever in a rush—"This is the way I operate," he once said. "Don't talk long to any one person, but talk to as many people as I can."—unbounded ambition, no apparent sense of personal limitations, doing it by the book, even though he hadn't read the book or studied at any of the Army's great schools.[1] Along the way he shed what sense of humor he might once have had, seldom smiled, held himself in a rigid posture that often seemed a pose, took himself very seriously, expected others to regard him thus as well.

His ultimate failure would have earned him more compassion, it seems certain, had he not personally been so fundamentally to blame for the endless self-promotion that elevated him to positions and responsibilities beyond his capacity. "It's the aggressive guy who gets his share—plus," he insisted. "That principle applies to most anything."[2]

"*Great* division commander," concluded another famous four-star. "He had this great appearance, and this charisma, to lead the 101st." Those were the best days, the balance between tasks and abilities still viable, the conceptual requirements minimal, much physical work, and a managerial span he could handle. Despite a certain silliness—most often displayed in a lifelong penchant for giving "cutesy" names to programs and operations—he was taken seriously, and at that stage not only by himself. He looked and acted like a great general, and the troops were convinced of it. Small wonder that he remembered those days as the most satisfying of his military career.

In later years Westmoreland, widely regarded as a general who lost his war, also lost his only run for political office, lost his libel suit, and lost his reputation. It was a sad ending for a man who for most of his life and career had led a charmed existence.

In his final days it could never be said he'd been broken, for he still maintained the mien and demeanor of the glory days, much of the "look of eagles" his splendid countenance had always afforded him, the obvious expectation of being admired and courted by others. But, at least with family and, back in his native South Carolina, certain very old friends and some new admirers, elements of his youthful charm seem to have reemerged. Said his daughter Stevie, describing those last years, "every day we're seeing less and less General and more and more Daddy."

ACKNOWLEDGMENTS

Work on this biography has consumed a number of years, during which time very many people have encouraged and assisted me. First, of course, tribute must be paid to the indispensable community of archivists, librarians, researchers, and historians upon whom so much depends. Then those who served with Westmoreland and talked with me about that experience—some 175 people in all—provided invaluable and often unique and previously undisclosed elements of the account. Also of great importance were the encouragement and assistance of friends who believed in the historical importance of the story and were sympathetic to the uncongenial task of telling it.

General Westmoreland's personal papers are in the South Caroliniana Library at the University of South Carolina, a charming venue situated on the Horseshoe of this historic campus. There I benefited greatly from the assistance of Herbert Hartsook, Director when I began my work there; his successor, Dr. Allen Stokes; and Beth Bilderback, Graham Duncan, Sam Fore, Henry Fulmer, Craig Keeney, Nicholas Meriwether, and Elizabeth West of the staff. Brian Cuthrell was particularly helpful and knowledgeable about the collection.

Other valuable Westmoreland materials, including copies of his remarkable "History Notes," are held by the U.S. Army Center of Military History at Fort McNair in Washington. There the Chief of Military History, Brigadier General John Sloan Brown, and his successor, Dr. Jeffrey Clarke, provided valuable assistance. I also thank Patricia Ames, Dr. Richard Davis, Vincent Demma, Dena Everett, Joseph Frechette, Lenore Garder, Dr. Joel Meyerson, Mason Schaefer, Jamaal Thomas, and Dr. Erik Villard. Frank Shirer, Chief of the Historical Resources Branch and custodian of much valuable Westmoreland material, was especially knowledgeable and helpful. Also thanks to Renee Klish, curator of the Army Art Collection.

Yet another essential repository is the U.S. Army Military History Institute at Carlisle Barracks, Pennsylvania. Among its holdings are two separate and very useful Westmoreland oral history interviews running to several hundred pages, as well as similar interviews with other senior officers who served with Westmoreland, plus the papers of many such senior officers. The Director, Dr. Conrad Crane, and also Dr. Richard Sommers, David Keough, and Randy Rakers, all valued longtime friends, were extremely helpful. Thanks also to Rich Baker, David Birdwell, Tom Hendrix, and Jessica Sheets. Randy Hackenburg was very helpful with photos.

Some Westmoreland material can also be found at West Point in the U.S. Military Academy Archives and Special Collections, headed by Suzanne Christoff. My valued friend Alan Aimone was also most helpful in the USMA Library. Elsewhere on post Carolyn Knicht helped me with access to Westmoreland's cadet records.

Also at West Point tribute must be paid to the superb work done by Lieutenant Colonel Julian M. Olejniczak, longtime Editor in Chief of the invaluable *Register of Graduates* and of *Assembly*, the West Point alumni magazine, and to Sylvia Graham, Managing Editor of the *Register*. Both publications are of crucial importance to anyone writing about a graduate of the Military Academy. Syl Graham also generously provided the author photo for this book.

The Lyndon Baines Johnson Presidential Library in Austin, Texas, has copies of some of the material held elsewhere, as well as valuable oral histories, a superb photo collection, and—a unique and poignant artifact—the model of Khe Sanh kept in the White House Situation Room for contemplation by LBJ during the 1968 siege of that remote outpost in Vietnam. Archivist John Wilson, Margaret Harman in the audiovisual archives, and Renée Gravois, Mike McDonald, and Tina Houston were very helpful to me.

I had only limited access to the Westmoreland materials at the National Personnel Records Center in St. Louis, but thank William Seibert and Brittany Helfin for the information provided me.

At the South Carolina Military Museum in Columbia there is an extensive collection of Westmoreland-related artifacts, including his personal copy of the Boy Scout manual, *Handbook for Boys*. Ewell "Buddy" Sturgis manages this fine facility, and on my visit Michael Lott kindly showed me through the Westmoreland exhibit.

Debra Hutchins assisted me at the Spartanburg County Public Library, while Brigadier General Ed. Y. Hall (South Carolina State Guard) showed me the Westmoreland materials at American Legion Post 28 in Spartanburg. Walter Oates provided a tour of Troop 1's Scout Hut, virtually the same as it was in Westmoreland's day, and showed us other Westmoreland Scout artifacts on display nearby.

Professor Marc Selverstone generously invited me to spend several days at the University of Virginia's Miller Center of Public Affairs, where I was able to harvest much valuable material from their collection of edited transcripts of Lyndon Johnson's recorded telephone conversations. Sean Gallagher, Sheila Blackford, Ken Hughes, and Shane Vanderberg were very helpful to me during my visit.

Once again Albert McJoynt has assisted me by preparing excellent maps, for which I am most grateful.

My West Point classmates, some of whom served as aides-de-camp to General Westmoreland or in other assignments close to him, have been an enormous source of encouragement and often of substantive help as well. I thank them all, and particularly Lieutenant General Charles Bagnal; Brigadier General John Bahnsen; Brigadier General Zeb B. Bradford Jr.; Lieutenant General Frederic Brown; Colonel Victor T. Bullock; Colonel William Cody; Lieutenant Colonel Maury Cralle Jr.; Colonel William Crews; Colonel Rudolph DeFrance; Colonel Gene Dewey; Major General James Ellis; Colonel Donald Ernst (USAF); General John Foss; Major General Eugene Fox; Lieutenant General Thomas Griffin; Colonel William Haponski; Colonel Jerry Huff; Lieutenant Colonel Kenneth Lang (USAF); Lieutenant Colonel George Martin; Lieutenant General Dave Palmer; General H.

Norman Schwarzkopf; General John Shaud (USAF); Major General Perry Smith (USAF); Colonel Thomas Winter; Lieutenant General John Woodmansee; and Major General Stephen R. Woods Jr.

Colonel Paul Miles, a longtime teacher of history at West Point and now at Princeton University and for many years a close associate and advisor to Westmoreland, was of very great assistance. His understanding of and insights into the personality and motivations of General Westmoreland are unsurpassed. Colonel Reamer "Hap" Argo Jr., now deceased, also had long and close associations with Westmoreland and, also a historian, was an acute and accurate observer. General Volney Warner was also very close to Westmoreland as his executive officer during the last two Chief of Staff years and continued as his sympathetic friend and advisor in later years. His articulation of the most positive view of Westmoreland has been very much appreciated. These three stand out, among the many aides and executive officers who served Westmoreland over the years, in having established and maintained such close personal relationships with him.

Lee Bandy and John Monk at *The State* in Columbia, South Carolina, were very friendly, helpful, and knowledgeable, and Lee Hemphill assisted with photos there.

At the Library of Congress, where I made my way through wartime issues of the *Saigon Post* and searched for an obscure and long-defunct journal in which a stunning Westmoreland photo had appeared, I received valuable assistance from Arlene Balkansky and Thomas Jabine. Jerome Brooks, Liz Faison, and Travis Wesley also helped.

Two earlier Westmoreland biographers, Ernest Furgurson and Samuel Zaffiri, have been very receptive to this project, Zaffiri even lending me the tape recordings of his interviews of Westmoreland. I thank them both for their cordiality.

Others who helped with advice, photographs, documents, and insights include Brigadier General Creighton Abrams; Colonel Thomas Adcock; Thomas Ahern; Brigadier General David Armstrong; Patsy Bagnal; Douglas Becker; Kimberly Becker of Becker Design; Colonel Lance Betros; Colonel Harold Birch; Tony Bliss Jr.; Mara Borack of Academy Photo; Colonel Fred Borch III; the late Colonel Paul Braim; Brigadier General Charles Brower IV; Brigadier General John Sloan Brown; Major General Clay Buckingham; B. G. Burkett; Dr. John

Carland; Colonel Jack Chase; Dr. Roger Cirillo; Allen Clark; James Codla; Lieutenant General Robert Coffin; Major General Neal Creighton; Kelvin Crow; Dr. Edwin Deagle; Maude DeFrance; John Del Vecchio; William E. DePuy Jr.; Brigadier General Robert Dilworth; Lieutenant General David Doyle; Lou Dunham, computer whiz; Brigadier General Karl Eikenberry; Colonel David Farnham; Brigadier General Alan Farrell (VMI); Dr. John Feagin; Colonel Zane Finkelstein; the late Lieutenant Colonel Philip Gage, Scribe of the West Point Class of 1936; General John Galvin; Colonel Martin Ganderson; Dale Garvey; Eric Gillespie; Dr. Russell Glenn; Major General George Godding; Dr. Alan Gropman; Dr. Stephen Grove; Elise Haldane; Brigadier General Ed. Hall (South Carolina State Guard); Andrew Hamilton; Corporal Christopher Hamilton; Colonel Morris Herbert; Dr. Paul Herbert; Colonel John Hesterly; Colonel Jon Hoffman (USMCR); Donald Hogan; Christopher Holmes; Dr. James Hooper; Warren Howe; Richard Howland; Colonel Seth Hudgins; Bobby Jackson; Diane Jacob at the VMI Archives; Lena Kaljot of the Marine Corps History Division; Brigadier General Douglas Kinnard; General Frederick Kroesen; Charles A. Krohn; the late Colonel James Leach; Harold Lyon; Colonel Thomas McKenna; Brigadier General H. R. McMaster; Major General Guy S. Meloy III; Major General John Murray; Brigadier General Harold Nelson; Lieutenant General Max Noah; Walter Oates, the longtime adult leader of Spartanburg's Troop 1, Boy Scouts of America, in which Westmoreland became an Eagle Scout; General Glenn Otis; Mark Perry; H. J. "Jack" Phillips; Rufus Phillips III; Colonel Homer Pickens; Dr. Walter Poole; Dr. John Prados; Robert Previdi; Merle Pribbenow; Dr. Robert Pringle; Lieutenant General Robert Pursley (USAF); Charles Ravenel; Lieutenant Colonel David Rice; Colonel Gordon B. Rogers Jr.; Dr. Herbert Schandler; Max and Loren Schluder at the Carolina Café in Columbia; Major General Frank Schober; Paul Schonberger; Jason Schull; Detective Sarit Scott of the Montgomery County, Maryland, Police Department; the late Colonel Donald Shaw; Brigadier General James Shelton; Stephen Sherman; Dr. Jack Shulimson of the Marine Corps History Center; Judith Sorley Simpson, my sister; Major General John Singlaub; General Donn Starry; Colonel Michael Starry; Kirstin Steele at The Citadel's Daniel Library; Thomas Stites; Colonel William M. Stokes III; Cookie Strong; John Taylor; Dr. Robert

Turner; Dr. John Votaw; Jerry Warner; Robert Watz; Colonel William Weber; Dr. Andrew Wiest; Richard P. W. Williams; Charles Wilson; Dr. Edward Woods; SMA William Wooldridge; and Barry Zorthian.

Bruce Nichols, my editor at Houghton Mifflin Harcourt, has been a delight to work with and has made this a better book. Sincere thanks also to Christina Morgan, Summer Smith, Laurence Cooper, Luise Erdmann, and Gordon Brumm for their valuable help. Over now many years my agent, Peter Ginsberg, has also become a valued friend. I am most grateful for his belief in the importance of the work. This book is dedicated with profound gratitude to my wife, Virginia Mezey Sorley, who, in addition to her multitude of other fine attributes, is a superb professional research librarian who has devoted many years to faithful public service.

GLOSSARY OF ACRONYMS AND ABBREVIATIONS

ABF   attack by fire

AFB   Air Force Base

AID   Agency for International Development

APC   armored personnel carrier

ARVN   Army of the Republic of Vietnam

AUSA   Association of the United States Army

CAS   close air support

CHICOM   Chinese communists

CIA   Central Intelligence Agency

CIDG   Civilian Irregular Defense Group

CIIC   Combined Information and Intelligence Conference

CINCPAC   Commander in Chief, Pacific

CJCS   Chairman, Joint Chiefs of Staff

CMH   Center of Military History

COMUSMACV   Commander, U.S. Military Assistance Command, Vietnam

CONARC   Continental Army Command

COSVN   Central Office for South Vietnam (enemy)

CRS   Congressional Research Service

CTZ   Corps Tactical Zone

CY   Calendar Year

DEPCOMUSMACV   Deputy COMUSMACV

DEPCORDS   Deputy to the COMUSMACV for Civil Operations and Revolutionary Development Support

DIA Defense Intelligence Agency

DMZ Demilitarized Zone

DOD Department of Defense

DPM Draft Presidential Memorandum

DRV Democratic Republic of Vietnam (North Vietnam)

DSDUF Declassified and Sanitized Documents from Unprocessed Files, LBJ Library

FFORCEV Field Force, Vietnam

FSB Fire Support Base

FWMAF Free World Military Assistance Forces

GVN Government of (South) Vietnam

HES Hamlet Evaluation System

H&I harassment and interdiction (artillery fire)

JCS Joint Chiefs of Staff

JGS Joint General Staff (Republic of Vietnam)

JUSPAO Joint U.S. Public Affairs Office

KIA killed in action

KP kitchen police

LBJ Lyndon Baines Johnson

LZ landing zone

MAAG Military Assistance Advisory Group

MACSOG Military Assistance Command Studies and Observations Group

MACV Military Assistance Command, Vietnam

MACOV Mechanized and Armor Combat Operations in Vietnam

MAF Marine Amphibious Force

MAW Marine Air Wing

MFR memorandum for record

MHI Military History Institute

MIA missing in action

NSAM National Security Action Memorandum

NVA North Vietnamese Army

NVN North Vietnam/North Vietnamese

OACSI Office of the Assistant Chief of Staff for Intelligence

OCMH Office of the Chief of Military History

OSD Office of the Secretary of Defense

PA public address

PAVN People's Army of Vietnam (North Vietnam)

PBS Public Broadcasting System

PF Popular Forces

PRG Provisional Revolutionary Government (Viet Cong)

PSDF People's Self-Defense Force

RF Regional Forces

ROK Republic of Korea

R&R rest and recuperation/rest and recreation

RVN Republic of Vietnam (South Vietnam)

RVNAF Republic of Vietnam Armed Forces

SACSA Special Assistant for Counterinsurgency and Special Activities (to the Joint Chiefs of Staff)

SAMVA Special Assistant for the Modern Volunteer Army

SASC Senate Armed Services Committee

SEA Southeast Asia

SEACOORD Southeast Asia Coordinating Committee for U.S. Missions

SEATO Southeast Asia Treaty Organization

SECDEF Secretary of Defense

SOG Studies & Observations Group

SVN South Vietnam

TSN Tan Son Nhut (Air Base) (Saigon)

USAID U.S. Agency for International Development

USARPAC U.S. Army, Pacific

USARV U.S. Army, Vietnam

USGPO U.S. Government Printing Office

USIS United States Information Service

USMA United States Military Academy

USMACV U.S. Military Assistance Command, Vietnam

VC Viet Cong

VCI Viet Cong Infrastructure

VN Vietnam/Vietnamese

VNIT Vietnam Interview Transcript

NOTES

A NOTE ON THE NOTES

The documentary record of the Vietnam War is very rich. Published historical sources, official records and accounts, oral history interviews, after action reports, and the collected papers of participants in various archives and other repositories run to millions of pages.

Given this abundance of material, it would have been possible to footnote virtually every line, but that seemed unnecessarily distracting and pedantic. All materials drawn from the work of other scholars have of course been credited in the notes in the usual manner.

At the U.S. Army Center of Military History the Westmoreland Papers were pending reorganization at the time I made use of them. Thus box numbers I have cited may have changed by the time future researchers consult those materials, but all the documents cited should be easy to locate chronologically in the papers.

In most cases where Westmoreland or another person quoted used incorrect grammar or pronunciation, that has been rendered as stated and without inserting the [sic] notation, as have Westmoreland's misspellings in quotations from materials he has written.

KEY TO NOTE ABBREVIATIONS

These abbreviations are used throughout the notes to indicate sources. In the Selected Bibliography a full citation may be found for each item. When items appear only in the notes and not in the bibliography, a full citation is given in the relevant note.

CMH   U.S. Army Center of Military History
CRS   Congressional Research Service

HKJ	Harold K. Johnson
LBJ	Lyndon Baines Johnson Library
MCHC	Marine Corps Historical Center
MFR	Memorandum for Record
MHI	U.S. Army Military History Institute
NARA	National Archives and Records Administration
SCL	South Caroliniana Library, University of South Carolina
USAC&GSC	U.S. Army Command & General Staff College
USAWC	U.S. Army War College
USMA	United States Military Academy Library and Archives
WP	Westmoreland Papers

PROLOGUE

1 Westmoreland Oral History (Cameron/Funderburk), MHI.
2 SMA Leon L. Van Autreve interview, 12 February 1988.

1. ORIGINS

1 Although Westmoreland's father had no military service, he somehow acquired the informal title of "Colonel," apparently a result of his service with various Citadel boards and associations. On 1 April 1950 Governor Strom Thurmond made it somewhat official, appointing J. R. Westmoreland a Colonel in the Unorganized Militia of the State of South Carolina with date of rank from 4 April 1924.

2 When Westmoreland had been at the Military Academy for perhaps five months, his father wrote to him about a recent conversation when he met Senator Byrnes on the street. "He is much pleased that you are doing so well. He says that your position in your class is a surprise to him. He said that he had had great apprehensions as to you making the grade based on his knowledge of the Academy."

3 Westmoreland Remarks, Memorial Day Ceremonies, Gettysburg Battlefield, 30 May 1988, Box 49, WPSCL.

4 Westmoreland's father also sought to motivate his son academically by paying him for good grades. "I have decided to put you on a different basis when you get to West Point," he wrote in the spring of 1932, when Westmoreland was still at The Citadel. "I will pay you your monthly bonus every time you register in the first half of the class." In another letter he specified the amount as $6.00 a month.

5 This exchange of correspondence with Col. Robert C. Brown is in Box 17, WPSCL.

6 Letter, Col. Garrett S. Hall to Westmoreland, 11 February 1991, Box 17, WPSCL.

7 One of 1936's three-star generals was Benjamin O. Davis Jr., mentioned previously, the only black in the class, who retired at that rank in 1970. In 1999

he was promoted on the retired list to four-star general. Clinton D. Vincent, also 1936, became an Army Air Corps brigadier general at age twenty-nine and thus, according to the West Point *Register of Graduates*, "the youngest West Point graduate promoted to general in the 20th century."

2. EARLY SERVICE

1   "Reminiscent Thoughts About the Class," USMA 1936 50th Reunion, Box 49, WPSCL.

2   As quoted in Gwen Moseley, "The General's Lady," *Sunday Telegraph* (Canberra, Australia) (6 February 1972), Box 67, WPSCL.

3   In telling this story many times over the years, to interviewers in particular, Kitsy sometimes reversed it and claimed she had told Westmoreland she would wait for him until *he* grew up. A childhood friend remembered Kitsy fondly from those days, with her thick dark braids, huge hazel eyes, and "foot-long eyelashes." And, she added, "I can still see you in those wonderful old dresses of your mother in your quarters in New Post, with your hair piled high upon your head, gazing into the mirror in the guest room and saying, 'Patty, someday I'm going to be somebody. Someday I am going to be a great lady.' And indeed you are, Kitsy, my dear — in every sense — and it has surprised no one, least of all me." Letter, Patricia Metcalf to Kitsy, 20 February 1968, Box 38, WPSCL.

4   Westmoreland's sister, Margaret, visited him while he was stationed in Hawaii, and remembered Millie Hatch and her brother's affection for her: "He wanted to marry her, but she threw him over and married somebody else."

5   Camden McConnell the younger, son of the officer at Fort Sill, posted this anecdote involving Westmoreland and his father on a West Point eulogy website honoring Westmoreland.

6   Correspondence in Box 1, WPSCL.

7   The 8th Field Artillery was apparently seriously short of officers at this time, since orders published in July 1940 appointed Westmoreland, only a first lieutenant, the 2nd Battalion's "Battalion Adjutant, S-2 and S-3," or in other words to most of the principal staff positions of the unit.

8   In a January 1969 speech to new Air Force brigadier generals at Maxwell Air Force Base in Alabama, Westmoreland mentioned the cooks and bakers school, saying he had "flunked the course." He misremembered. There is a certificate of proficiency from this course in his papers at the South Caroliniana Library. It should also be noted that, while Westmoreland did not attend any of the usual run of formal service schools, as a junior officer he did receive instruction in "Troop Schools" run by the unit in a variety of such subjects as gunnery, animal transport and stable management, signal communication, tactics, and training methods.

3. WORLD WAR II

1   Westmoreland told this story in "Riding to Battle," *Army* (April 1993), pp. 43–44. Even at that remove there was no indication on his part that he real-

ized the potential seriousness of his disobedience of orders, that by slyly (as he thought) arranging for his battalion to remain behind to "police the ship" so they could also feast on the special meal he risked their destruction had the anticipated German bombing of the port actually come to pass. There is also one minor anomaly: Westmoreland remembered it as a New Year's feast, whereas other sources seem to indicate it was a Christmas Eve repast. Elsewhere Westmoreland says of Casablanca, "I got there just before Christmas," so in the article he has apparently misremembered it as New Year's. Only the year before publishing this article, at the 9th Infantry Division Association's annual meeting, Westmoreland had told a significantly different version of the event: "We were all set for a turkey dinner when we were ordered to go ashore, where we 'feasted' on 'C' rations and [were] bombed that evening in our bivouac area." Speech, "50 Years Ago," 9th Infantry Division [Association], 1992, WPSCL.

2   This anecdote is found in notes dictated by Westmoreland for use in preparation of his memoirs. Box 41, WPSCL.

3   Taylor's recollections of the combat interaction were equally positive. He wrote in his memoirs of "Lieutenant Colonel W. C. Westmoreland, whose sure-handed manner of command led to the entry of his name in a little black book I carried to record the names of exceptional young officers for future reference." *Swords and Plowshares*, p. 50.

4   Westmoreland was equally impressed with Taylor: "This was the beginning of a long association with General Taylor during which I found him to be, himself, very much of a workman-like individual. Extremely alert, always thinking ahead, questioning every statement, weighing alternative courses of action in dealing with a problem, and sometimes imperious in his manner, he was a most unusual man." Notes, Box 41, WPSCL.

5   Mittelman, *Hold Fast*, p. 13.

6   In Westmoreland's papers at the South Caroliniana Library reference is made to a 53-page pamphlet, published on 6 September 1943, entitled "Operating Practice and Procedures" for the 34th Field Artillery Battalion.

7   Westmoreland was apparently serious about this. In 1965 a Vietnam War battle lasting ten days at the Duc Co Special Forces camp in the Central Highlands involved Major (later General) Norman Schwarzkopf. Toward the end of the action Westmoreland arrived to have a look. Right behind him came a helicopter full of reporters. All converged on Schwarzkopf. Westmoreland asked the press if he could have a moment alone with his commander on the scene, then drew Schwarzkopf aside. "Schwarzkopf expected a quiet, soldier-to-soldier pep talk from the general," reported columnist Jack Anderson, "but Westmoreland instead simply asked him if the mail had gotten through during the fighting. Schwarzkopf was deflated." *Stormin' Norman*, p. 31.

8   Westmoreland wrote to his mother: "Please send me some insignia of rank. I have one (1) eagle and cannot get any more. They are not available in the country we have been operating in. I begged one from a friend the day I was promoted [and] have been unable to obtain a supply since."

9   Address, 50th Anniversary Commemoration, Battle of the Bulge, Chicago, 20 December 1994, Box 50, WPSCL.

10 These were matters that obviously troubled Westmoreland greatly, for he made these remarks nearly twenty-five years later in a talk given to new Air Force brigadiers at Maxwell Air Force Base in Alabama. Box 49, WPSCL.

11 Years later, while Westmoreland was commanding U.S. forces in Vietnam, Father Connors accepted on his behalf the "Honor et Veritas" award presented by the Catholic War Veterans of America. On that occasion Connors remembered Westmoreland in World War II as "expert in all phases of his job, well-disciplined, courageous, a forceful leader—but always with compassion and understanding—a reverent man, a truly dedicated, a self-sacrificing American." He also recalled a comment made to him by General Craig not long after he became the 9th Infantry Division's commander and had observed Westmoreland for only a short time: "I believe that your friend, Colonel Westmoreland, is going to go all the way."

12 As quoted in Mittelman, *Hold Fast*, p. 17.

13 Ibid., p. 44.

14 At this time he commented on the new job to his father, saying, "It's an unusual assignment for an artillery officer but considering myself an army officer first & a artilleryman second, I'm quite enthusiastic and consider it a very good break." Also: "Confidentially *please* Gen. Craig ask[ed] for me as his corps chief of staff but Gen. Patton would not allow it because he thought I was too young. He's probably correct but I don't feel that way any longer."

4. AIRBORNE DUTY

1   Westmoreland Oral History (Ganderson), MHI.

2   Memorandum, Lt. Col. Thomas J. H. Trapnell to Commanding General, 82nd Airborne Division, 17 January 1949. Box 1, WPSCL.

3   Col. Thomas P. McKenna, e-mail to Sorley, 1 May 2005. Looking back on the episode later McKenna, who retired as a full colonel, acknowledged that that was "certainly a rather cheeky thing for a PFC to say to a LTC [lieutenant colonel]."

4   After leaving the division General Byers wrote to Westmoreland that "the first letter has to be written to you because, from the date of my reporting to the division, your work has been outstanding. Your varied experience brings to the task of chief of staff a background from which you have drawn examples with unusual intelligence. The energy, tact, drive and patience with which you have undertaken tasks have produced splendid results but"—and this next would come as a surprise to many people—"topping all of these from my standpoint, is the sense of humor which makes it pleasant to work with you." Letter, Maj. Gen. Clovis E. Byers to Westmoreland, 21 July 1949, Box 1, WPSCL.

5   In November 1949 Brig. Gen. W. B. Palmer was sent to jump school preparatory to (or perhaps immediately after) taking command of the 82nd Airborne Division. He had an accident with a bad knee while at the school and

was hospitalized. It appeared he would have to command without jump wings, at least for a time. Westmoreland was very reassuring to him about doing this, very considerate and supportive, even recalling an accident of his own "on my 11th jump at which time I ended up in the hospital for 10 days." Letter, Westmoreland to Palmer, 23 November 1949, Box 1, WPSCL.

6 Many correspondents had trouble with Katherine Van Deusen Westmoreland's nickname "Kitsy." Among variations over the years there appeared Kitsie, Kitzy (from Maxwell Taylor), Kitsi, Kittsy, Kittsie, Kittsey, Kitzi (in *Assembly* magazine's Westmoreland obituary), Kitzie, Kittzie, Kitsey (Admiral McCain), Kitsye, and even Kits.

7 "How Two Patriots Saw Their Call to Duty," *Parade* (7 July 1985), pp. 4ff.

8 As quoted in Levona Page, "Westy," *The State* (Columbia, S.C.) (21 January 1974), Box 65, WPSCL. There was a similar age difference between Westmoreland's parents, his father being nearly a decade older than his mother.

9 Kitsy later told an interviewer about the brief courtship: "I met West when I was a student at WC (UNC-G)," Women's College of the University of North Carolina at Greensboro. "He was a full colonel commanding some paratroopers at Fort Bragg, and he drove that winding road from the east to see me every weekend in Greensboro. He was as attractive then as he is now, and all the girls in my dorm would line up at the windows to admire him when he came to call. My housemother was right there at the windows looking at him too. She used to say to me, 'Kitsy, that colonel is much too old for you, but he's just the right age for me.'" As quoted in Linetta Pritchard, "General's Wife," *Raleigh Times* (11 November 1971), Box 10, WPSCL.

10 Years later Kitsy received a letter from a girlhood friend. "One of the most significant episodes of our wonderful childhood," she wrote, "is to recall your saying so often to me, especially at the Hunter Trials and other horse shows at good old Fort Sill, 'You see that lieutenant, Patty? I'm going to marry him some day.' And I used to laugh at you. Imagine! Then I received a wedding invitation, and at the bottom of such a formal invitation were your inscribed words, 'Ha! Ha! I told you so.'" Letter, Patricia Metcalf to Kitsy, 20 February 1968, Box 38, WPSCL.

11 Letter, Maj. Gen. Louis A. Craig to Westmoreland, 8 October 1951, Box 1, WPSCL.

5. JAPAN AND KOREA

1 Significantly, the assumption of command order is signed "Colonel, Artillery."

2 Westmoreland, *A Soldier Reports*, p. 30.

3 Memorandum, Westmoreland to Subordinate Commanders, Subject: Command Inspection, 10 September 1952, Box 1, WPSCL.

4 Westmoreland Oral History (Cameron/Funderburk), MHI.

5 In notes dictated for preparation of his memoirs Westmoreland refers twice in discussing this matter to the "Chowon" Valley, but General Frederick Kroesen, who then as a major commanded the displaced unit, recalled that

they were then "above Kumwha" and, in a contemporaneous letter to Westmoreland, Major General Barriger notes that Westmoreland's outfit was then "defending the approaches to Kumwha." In his history of the 187th RCT Fred Waterhouse notes that the unit had "closed into defensive positions in the Chorwon Valley." And in an e-mail to the author General Kroesen recalled that "the Chorwan Valley is one of three principal approaches from N to S, the one that leads to Seoul. Kumwha, now in the DMZ, heads Kumwha Valley which leads to Chorwan Valley."

6   Westmoreland Paper on Korean War Experience, Box 41, WPSCL. A history of the unit notes that Barriger had "a surname which Westmoreland, to the General's annoyance, insisted on mispronouncing 'Barringer.'" Waterhouse, *The Rakkasans*, p. 100. Barriger was awarded the Distinguished Service Medal for his command of the division.

7   Kroesen interview, 18 September 2009.

8   Westmoreland was now continuing a familiar pattern of harsh criticism of many of those he had worked under, going back to his first battery commander at Fort Sill, his battalion commander in Hawaii, the division artillery commander in Europe during World War II, and others. After Korea he stated: "The division commanders were adequate but not impressive. I disagreed with both of them. I wasn't one not to argue with senior officers. I was no 'yes man,' and no politician." Oral History (Cameron/Funderburk), MHI.

9   Westmoreland, speech to Rakkasan reunion, c. 1980s, Box 50, WPSCL.

10 Costa, *Diamond in the Rough*, p. 204.

11 Posted by Lt. Col. Robert Frank on a West Point eulogy page for Westmoreland.

12 Letter, Westmoreland to Col. R. H. Tucker, 9 April 1955, Box 1, WPSCL.

13 Letter, Westmoreland to Col. R. L. Ashworth, 18 February 1954, Box 1, WPSCL.

14 Letter, Westmoreland to SFC Paul E. Blenis, 22 December 1953, Box 1, WPSCL.

15 Flanagan, "A New Perspective," *Army* (July 1993), p. 49.

6. PENTAGON

1   Furgurson, *Westmoreland*, p. 228.

2   Westmoreland, Taylor Profile, Box 41, WPSCL.

3   Westmoreland Oral History (Ganderson), MHI.

4   DePuy Oral History, MHI.

5   Westmoreland, "Our Twentieth Century Army," Lecture, Airborne Conference, 7 May 1957, as quoted in Linn, *Echo of Battle*, p. 170.

6   In his memoirs Westmoreland wrote that the Vice Chief of Staff, General Williston Palmer, "directed that Christmas cards not be exchanged among officers and civilian officials on the staff who associated with each other daily," but that is not the way others remembered it, and the order Westmoreland issued had read "by direction of the Chief of Staff."

7   Maj. Gen. Winant Sidle interview, 6 June 1999. Sidle thought that West-

moreland was effective as SGS, "but maybe not as effective as some others might have been." Of the decree that there would be no local Christmas cards: "Westy would think on that plane."

7. DIVISION COMMAND

1   Hon. Martin R. Hoffmann interview, 18 March 2010. Hoffmann was Secretary of the Army during 1975–1977.
2   Brig. Gen. Weldon F. Honeycutt telephone interview, 12 December 2009.
3   Maj. Gen. John K. Singlaub interview, 23 November 2005.
4   Westmoreland, *A Soldier Reports*, p. 30. Under a photograph of Westmoreland in the *Nashville Banner* on 24 April 1958 the caption read: "He jumped last."
5   Westmoreland, *A Soldier Reports*, p. 30.
6   Westmoreland's jump log, on file with his papers at the University of South Carolina, shows him making his first jump with the division on 8 April 1958, then his next jump on 23 April, the date of the accident. The next jump entered is on 28 April. In his memoirs Westmoreland says that he jumped first the day after the accident, apparently misremembering the chronology. Many years later, on a radio program during a book tour, Westmoreland still maintained that "the next day I jumped by myself." Mike Miller Radio Show, 16 December 1976, WPSCL. Actually it was five days later, and his aide accompanied him on that jump. Westmoreland Parachute Log, WPSCL.
7   E-mail, Hon. Martin R. Hoffmann to Sorley, 19 March 2010.
8   Palmer interview, 1 December 1997.
9   Hon. Martin R. Hoffmann interview, 18 March 2010.
10   Westmoreland Oral History (Cameron/Funderburk), MHI.
11   "How Two Patriots Saw Their Call to Duty," *Parade* (7 July 1985), pp. 4ff.
12   Millett was a good choice for the assignment. Said another officer who served in the division at that time, "He was a little bit crazy, but had a lot of charisma and was a guy you would want with you in a fight." Col. Franklin A. Hart interview, 14 December 2009.
13   E-mail, McKenna to Sorley, 1 May 2005.
14   Westmoreland Oral History (Cameron/Funderburk), MHI.
15   Subsequently Westmoreland sought to get a report on Operation Overdrive published in the *Harvard Business Review*. "Consensus of the editors and the Editorial Board is that it is too limited in its interpretation for our broad audience of more than 65,000 subscribers," responded that publication.
16   Linn, *Echo of Battle*, p. 178.
17   Col. John H. VonDerBruegge telephone interview, 22 April 1995. VonDerBruegge was the briefer. "The critique of the exercise was in the Bragg theater," he said. "I came in and found Westmoreland rehearsing his speech to a blank wall." General Howze told an interviewer, in considerable contrast to the rather uncritical view of airborne soldiers held and frequently stated by Westmoreland, that when he took command of the 82nd Airborne Division he was "rather shocked . . . at what then was the attitude of the airborne. And that is that their interest in military operations stopped—not completely,

but to a considerable extent—as of the time they hit the ground and got out of their parachutes." Howze Oral History, MHI.

18 Text of the talk as printed in Westmoreland, "The How of STRAC," *Army* (December 1958), pp. 61–62.

19 Westmoreland Oral History (Ganderson), MHI. Westmoreland's change of heart may not have had much influence on termination of the Pentomic division organization, which occurred two years after he submitted his critical letter. When Garrison Davidson was serving in the Weapons System Evaluation Group on the Army Staff, he sent for the background studies underpinning the decision to create the Pentomic division. "We were quite disappointed," said Davidson. "The foundation was nothing more than a series of unsupported opinions." Having gone through the material, he concluded that "the 'widely heralded but short lived Pentomic division' was a directed verdict with no substantive study by the powers that were in the Army hierarchy." See Davidson, *Grandpa Gar*, p. 128. Once Maxwell Taylor was no longer Chief of Staff, the rapid demise of the Pentomic division was virtually assured. Promulgated in 1958, the concept was supplanted just four years later.

20 As quoted in John G. Hubbell and David Reed, "The Man for the Job in Vietnam," *Reader's Digest* (January 1966), pp. 55–60.

21 Westmoreland Oral History (Ganderson), MHI.

22 Lt. Gen. Dave Palmer interview, 1 August 2006. Former Secretary of Defense James Schlesinger recalled a conversation he once had with Lieutenant General Vernon Walters. "I asked him whether he thought Westmoreland had peaked at major general," said Schlesinger. "No," replied Walters, "colonel." Schlesinger telephone interview, 8 February 2004.

8. SUPERINTENDENT

1 Lt. Gen. Robert M. Cannon, Commanding General of Sixth Army, wrote to Westmoreland: "For many reasons the Kaydets might not think too highly of Westy—but they will certainly give him a 3.0 for having Kitsy as Mrs. Supt." The 3.0 referred to a maximum grade in the Military Academy marking system of that day.

2 Gen. John R. Galvin interview, 2 May 2001. Galvin also recalled what a fine memory for names and people Westmoreland had. "It was not until later," he said, "in Vietnam and then in the Pentagon, that I began to see him in a different light."

3 Remarks, 24 August 1961, Box 44, WPSCL. The portion relating to rain is lined through on the text, so may not have been delivered.

4 Creighton, *A Different Path*, p. 112. "Sam" Wetzel was Robert Lewis Wetzel, USMA 1952, who earned a Silver Star as a battalion commander in Vietnam and reached the rank of lieutenant general, commanding a corps in Germany, before his retirement.

5 Anecdote posted by Sheridan as a personal eulogy to Westmoreland on a West Point website, 31 July 2005.

6 Hon. Stephen Ailes interview, 25 October 1989.

7   Lt. Gen. John Norton interview, 13 June 2001.
8   Westmoreland Oral History (Ganderson), MHI.
9   Crackel, *West Point*, p. 380.
10   Lt. Gen. Charles Simmons telephone interview, 21 September 2007.
11   Col. Roger Nye, a long-serving Professor of History at West Point, com-
     piled an important study, *The Inadvertent Demise of the Traditional Academy*.
     Addressing the expansion of the Corps, he wrote that it "created great pres-
     sure to change the traditional process of admitting cadets, which amounted
     to 'let those who want to follow a military career find us and qualify for en-
     try.' The burden was shifted to the Academy to find enough qualified candi-
     dates, motivate them towards a military career, and retain them through
     graduation into long-term career service. When the Academy appeared to be
     failing in this process, Washington authorities intervened, usually accompa-
     nied by media criticism."
12   Westmoreland Oral History (Ganderson), MHI.
13   The exclamation point is included in that document.
14   In a radio interview some years later Westmoreland said that being Superin-
     tendent at West Point was "my favorite assignment, but my favorite *place* was
     Hawaii."

9. VIETNAM

1   Ailes interview, 25 October 1989.
2   Westmoreland interview with Dorothy McSweeney, 8 February 1969, Box
     52, WPSCL.
3   These appointments were evidence of an aggressive youth movement in the
     McNamara Pentagon. Passing over twelve four-star generals, Harold K.
     Johnson, chosen as the new Army Chief of Staff, was 32 on the current list of
     lieutenant generals, Westmoreland was 33, and Creighton Abrams, 34.
4   When Charles MacDonald was ghostwriting Westmoreland's memoirs, he
     put in a mention of how important Taylor's mentoring and patronage had
     been to Westmoreland's career. Westmoreland took it out. "He didn't want
     to admit that he owed his success to anybody else," said MacDonald. Dr.
     Walter S. Poole, long a member of the Joint Chiefs of Staff History Office,
     looked at the other side of the relationship and wrote that "despite his dis-
     claimers, Taylor was responsible for choosing Harkins and Westmoreland,
     both of whom proved quite unequal to their tasks." Letter, Poole to Maj. H.
     R. McMaster, 30 July 1996, copy provided to Sorley by Brig. Gen. David
     Armstrong. Later, said Gen. Bruce Palmer Jr., after Westmoreland and Tay-
     lor differed over the introduction of U.S. ground forces into Vietnam, "Max-
     well Taylor disowned him. 'He's *not* my protégé!'" Palmer interview, 26 Sep-
     tember 1994.
5   Jordan telephone interview, 6 November 2009.
6   Letter, Maj. Gen. William Yarborough to Westmoreland, 26 February 1964,
     Westmoreland History Backup File #3, CMH.
7   Ibid.
8   Westmoreland, Taylor Profile, Box 41, WPSCL.

9 Gen. Harold K. Johnson notes, JCS meeting, 171400 January 1964, Johnson Papers, MHI.

10 McMaster, "Dereliction of Duty," *Air Force Magazine* (January 1998), p. 71.

11 Gibbons, *U.S. Government and the Vietnam War*, Part III, pp. 2–3, recounting a conversation presidential aide Bill Moyers had with LBJ following a meeting of high-level advisors on 24 November 1963.

12 Notes of 021400 March 1964 JCS meeting, Harold K. Johnson Papers, MHI.

13 As quoted in Charlton and Moncrieff, *Many Reasons Why*, p. 135.

14 Westmoreland, *A Soldier Reports*, p. 49.

15 McMaster, *Dereliction of Duty*, p. 86. Clifton and Westmoreland were West Point classmates.

16 Westmoreland interview by Maj. Paul Miles, 10 October 1970, WPSCL.

17 Recalled Westmoreland's aide Capt. Dave Palmer of this trip: "While we were there we played tennis on some grass courts at an old British club in Kualà Lumpur. The 'ballboys' were girls, bare-breasted. It was a memorable game." Palmer interview, 1 February 2006.

18 Montgomery interview, 5 February 2009. The press of business did, however, sometimes keep Westmoreland from getting to the dictation until quite some time after the events described, offering him considerable advantage of hindsight. A notable example occurred after the onset of the 1968 Tet Offensive when, describing a matter that took place on 23 January 1968, he added that "this arrangement continued until early March."

19 MacDonald, *Outline History*, p. 44.

20 Palmer interview, 26 September 1994.

21 Letter, Wheeler to Westmoreland, 17 September 1964, Westmoreland History Backup File #8, Westmoreland Papers, CMH.

22 Ibid.

23 Message, Westmoreland to Brig. Gen. E. C. Dunn, MAC 6468, 151200Z December 1964, quoting his own message "sent to all senior advisors on 8 December." Box 23, WPCMH.

24 Westmoreland received a letter of condolence from Senator Strom Thurmond. "Your father was one of South Carolina's finest citizens," he wrote. "When I was Governor, he was Chairman of the Board of Visitors at the Citadel, and we always had a warm friendship and a good working relationship."

25 Westmoreland, *A Soldier Reports*, p. 89.

26 Later it was speculated that the Brink attack had been triggered prematurely, that it had really been intended to take place the following day when the Bob Hope troupe would arrive at the hotel. Hope, always on the alert for grist for his mill, told GIs that "a funny thing happened on the way in from the airport—I passed my hotel going the other way." And, he claimed, "When I landed at Tan Son Nhut [the airport serving Saigon] I got a nineteen-gun salute. One of them was ours." Hope and his troupe returned at Christmas 1965 for another round of shows, in the process demonstrating that there were limits to Westmoreland's power and authority. When they flew over from Bangkok to do a show at Tan Son Nhut, Hope was very impatient with the delay in getting started. His advance man explained that the technicians

had arrived on the same plane with the cast, not before, and that it took an hour and a half to set up. Hope and Westmoreland were standing there, and Hope protested that Westmoreland had promised they could have any support they needed, so they could have had two planes instead of only one. Hope's staffer explained the realities: "Your friend the General says we can have them, but my friend the Sergeant says we can't." Hope, *Five Women I Love*, p. 95.

27 Westmoreland, News Conference, Saigon, 10 June 1968.
28 Westmoreland, *A Soldier Reports*, p. 9.

10. FORCES BUILDUP

1   Bui Diem, "Reflections on the Vietnam War," in Head and Grinter, *Looking Back on the Vietnam War*, p. 246. Emphasis in the original.

2   Ambassador Ellsworth Bunker Oral History (unpublished transcript), interviewed by Stephen Young. Bunker felt so strongly about this that he even commented on it when accepting the Thayer Award at West Point on 8 May 1970: "We did not in the beginning, I think, fully understand the complexities of this kind of warfare. Prior to Tet 1968 we underestimated the capabilities of the enemy. And we were slow in equipping our Vietnamese allies while the enemy was being equipped by the Soviets and Chinese with a wide range of the most sophisticated weapons." Thayer Award Acceptance Speech, West Point, N.Y., 8 May 1970.

3   *Pentagon Papers*, III:396.

4   Remarks, 10th Mountain Division Symposium, National Defense University, Fort McNair, Washington, D.C., 1 February 2006.

5   Gen. Richard Stilwell Oral History, MHI.

6   Lt. Gen. Julian Ewell interview, 24 April 1997.

7   Message, Taylor to President, Embtel #2057, 6 January 1965, Westmoreland History Backup File #12, WPCMH.

8   Collins, *Development and Training of the South Vietnamese Army*, p. 128.

9   Tolson, *Airmobility*, p. 83.

10 As quoted in Maurer, *Strange Ground*, p. 449.

11 Westmoreland, *A Soldier Reports*, p. 118.

12 Westmoreland, Memorandum for John McNaughton, 6 February 1965, History File Backup #13, WPCMH.

13 Westmoreland, *A Soldier Reports*, p. 45.

14 Message, Westmoreland to Wheeler, MAC 3275, 261000Z June 1965, Box 24, WPCMH.

15 Gibbons, *The U.S. Government and the Vietnam War*, III:235.

16 McNamara, *In Retrospect*, p. 192.

17 Ibid., p. 188.

18 As quoted in Gibbons, *The U.S. Government and the Vietnam War*, IV:16.

19 It would almost be possible to charge General Douglas MacArthur with losing the Vietnam War. A major inhibiting concern throughout the conflict was the possibility of Chinese intervention, and this was in some (probably large) measure a legacy of MacArthur's disastrously wrong prediction that

the Chinese would not enter the Korean War when United Nations forces drove north to the Yalu. Not only a sense of the unpredictability of Chinese actions, but loss of confidence in the advice of American military leadership, were products of the MacArthur fiasco.

20 Rosson, "Four Periods," p. 208.

21 Gibbons, 3:369.

22 Pimlott, *Vietnam*, p. 48.

23 Lt. Gen. Harry W. O. Kinnard interview, 25 July 2001. See also Coleman, *Pleiku*, p. 45, and Kinnard's Army Aviation Senior Officer Oral History Program interview, MHI.

24 Kinnard interview, 25 July 2001.

25 Westmoreland History File #2, 29 August 1965, WPCMH.

26 Ibid.

27 Westmoreland History File, Box 8, 6 November 1965, WPCMH.

28 Lt. Gen. Harry W. O. Kinnard Oral History, MHI.

29 Lt. Gen. Harold Moore, Remarks, "Rendezvous with War Symposium," College of William and Mary, Williamsburg, Va., April 2000, as reported in *The VVA Veteran* (October/November 2001).

30 McNamara, *In Retrospect*, p. 213.

31 Ibid.

32 Ibid., p. 221.

33 Ibid., pp. 222, 224–225.

34 As quoted in Berman, *Lyndon Johnson's War*, p. 40.

35 Westmoreland, *Face the Nation*, CBS Television Network, 19 December 1971.

36 Glenn, *Reading Athena's Dance Card* (draft), pp. 291–292.

37 Westmoreland, *A Soldier Reports*, p. 417.

38 Debriefing report, CG 9th Infantry Division, "Impressions of a Division Commander in Vietnam," 17 September 1969, MHI. Soon after the war ended an academician visited the Air Force Academy, where during dinner he listened to three Air Force officers and an Army major discuss the recent conflict. The rotation policy that kept people in their jobs for only six months came up, prompting one of the Air Force officers to observe that "if you tried to run a business like that, it would go under." After a pause the Army officer responded: "Ours did." Edward M. Coffman in "Commentary," *Second Indochina War Symposium*, ed. John Schlight, p. 187.

39 Memorandum, DCSPER-DRD to Vice Chief of Staff, Subject: Study of the 12-Month Vietnam Tour, 29 June 1970, Box 41, WPCMH.

40 "By July 1967," Army Chief of Staff General Harold K. Johnson told his generals in the 31 May 1966 issue of the *Weekly Summary*, "more than 40 percent of our officers and more than 70 percent of our enlisted men will have less than two years of service." Everyone should recognize, he counseled, "that this is not just a local problem but is true Army-wide."

41 Gen. Harold K. Johnson, Recording: "The Military Professional," U.S. Army War College, 4 November 1980, MHI. Emphasis in the original.

42 Tom Johnson's Notes of Meetings, Box 1, 13 August 1966, LBJ Library. Quoted with permission.

43 Transcript 66-117, News Conference, Johnson City, Tex., 14 August 1966, USIS Tokyo.

44 Westmoreland, "Commentary & Reply," *Parameters* (Winter 1991–1992), pp. 106–108.

45 Westmoreland, "The Long Haul, with an Escape Route," *Washington Times* (8 October 1990).

46 In Charlton and Moncrieff, *Many Reasons Why*, p. 143.

11. SEARCH AND DESTROY

1   Westmoreland interview, British Broadcasting Corporation, 23 October 1977, Box 52, WPSCL.

2   Westmoreland interview, 27 April 1981, WGBH Interview Collection, Healey Library, University of Massachusetts at Boston.

3   Westmoreland interview with Dorothy Pierce McSweeney, 8 February 1989, Box 52, WPSCL.

4   Remarks, Andrew Krepinevich, Fort Leavenworth Conference, 10 August 2006.

5   Haig, *Inner Circles*, p. 161.

6   Williamson, unpublished memoir, n.d.

7   *Commander's Combat Notes*, Number 73, Headquarters, 173rd Airborne Brigade (Separate), 5 August 1965. Army Chief of Staff Johnson found these commentaries so valuable that at one point he sent Williamson this cable: "If you've stopped writing the combat notes, start again. If you've taken me off distribution, put me back on."

8   As quoted in Gibbons, *The U.S. Government and the Vietnam War*, III:153.

9   Secretary of the Navy Paul H. Nitze as quoted in Message, Krulak to Greene, 170615Z July 1966, Gen. Wallace M. Greene Jr. Papers, Marine Corps Historical Center.

10 McNamara, *In Retrospect*, p. 207.

11 Message, Sharp to Wheeler, CINCPAC 220725Z September 1965, as quoted in Gibbons, IV:75.

12 Later Westmoreland, coaching his ghostwriter in the preparation of his memoirs, described Admiral Sharp as "a man of runt stature." Interview, Charles B. MacDonald with Westmoreland, 17 June 1973, Box 31, Westmoreland Papers, LBJ Library.

13 As quoted in Kahin, *Intervention*, p. 378.

14 Lt. Gen. Fred C. Weyand, Senior Officer Debriefing Report, CG II Field Force, Vietnam, 4 October 1968, MHI.

15 Murphy, *Semper Fi*, pp. 41–42.

16 As quoted in Pettit, *Experts*, p. 210.

17 Message, Westmoreland to Wheeler and Sharp, MAC 3240, 241220Z June 1965, Box 24, WPCMH.

18 Moore and Galloway, *We Were Soldiers Once*, p. 185.

19 Ibid., p. 339. The figures depend on the time span included. For the more

restricted intense encounter at Landing Zones X-Ray and Albany, the JCS History states that "the enemy lost 1,286 men in the Ia Drang Valley, the US had 217 killed and 232 wounded."

20 Moore and Galloway, *We Were Soldiers Once*, p. 345.

21 Related by Sen. Hollings to Dr. Roger Cirillo at Normandy in 1994.

22 Stanton, *The 1st Cav in Vietnam*, p. 65.

23 Message, Army Chief of Staff Gen. Harold K. Johnson to Westmoreland, WDC 10453, 010105Z December 1965, Box 25, WPCMH. These operational losses were compounded by another major problem, one that Gen. Wheeler remonstrated with Westmoreland about. A recent meeting of the Vietnam Coordinating Committee, a Washington entity, had learned from Agency for International Development officers of the "effects of graft, corruption, and VC economic penetration of US/GVN stockpiles of materials in transit or storage. While some loss to theft has always been known," said Wheeler, "the cumulative effect now coming to light is staggering and implies large scale, organized operations to falsify documents, divert shipments, and siphon off large amounts of materials for private profit and enemy use." Message, Wheeler to Westmoreland and Sharp, JCS 4161-65, 011142Z November 1965, WPCMH.

24 Bergerud, *Red Thunder*, p. 64.

25 Halberstam as quoted in Dorland, *Legacy of Discord*, p. 63.

26 Moore and Galloway, *We Were Soldiers Once*, p. 319.

27 Class of 1936 Notes, *Assembly* (Winter 1965), p. 71.

28 Maj. Gen. Ellis Williamson interview, 1 July 1999.

29 Ibid.

30 Brig. Gen. Douglas Kinnard, "Westmoreland's and McNamara's War," p. 42.

31 Sidle interview, 6 June 1999.

32 History File 11, 19 November 1966, WPCMH.

33 McNamara testimony, Westmoreland vs. CBS libel trial, in *Vietnam: A Documentary Collection*, p. 4911.

34 *Newsday* (11 November 1966).

35 Westmoreland, Press Conference, Honolulu, 5 February 1966, Box 1, WPCMH.

36 As quoted in Maurer, *Strange Ground*, p. 453.

37 As quoted in Mark Perry, "The Resurrection of John Paul Vann," *Veteran* (July 1988), p. 31.

38 Gen. Bruce Palmer Jr. Oral History Interview: General Creighton Abrams Story, 29 May 1975, MHI.

39 Millett, "Why the Army and the Marine Corps Should Be Friends," *Parameters* (Winter 1994–1995), p. 33.

40 Message, Krulak to CG III MAF, 270218Z October 1965, marked "Marine Corps Eyes Only," General Wallace M. Greene Jr. Papers, Marine Corps Historical Center.

41 Westmoreland, *A Soldier Reports*, p. 165.

42 Quoted in Shulimson and Wells, "First In, First Out," *Marine Corps Gazette* (January 1984), p. 39.

43 Johnson Oral History Interview, MHI.

44 Message, Johnson to Westmoreland, WDC 13029, 021205Z October 1967, MHI.

45 Message, Westmoreland to Senior Advisor, IV Corps, and DCG, USARV, MAC 34269, 180905Z October 1967, Box 14, WPCMH.

46 Debriefing Report, Maj. Gen. Charles P. Stone, CG 4th Infantry Division (4 January 1968–30 November 1968), CMH.

47 Kinnard, *The War Managers*, p. 45. Westmoreland was one of those receiving the survey instrument. "I am happy to fill out your questionnaire," he wrote to Kinnard. "It is returned herewith." Later, when he learned of some of the results, Westmoreland changed his position to being harshly critical of the research effort. After making strenuous (but unavailing) efforts to pry out of Kinnard the names of some of the respondents who had denounced his approach, Westmoreland told Kinnard he should have showed the manuscript to him before publishing it. Kinnard, by then an academician, explained to Westmoreland that as a professor he didn't even clear what he published with his department chairman.

48 Gen. Arthur E. Brown Jr. interview, 3 September 1994.

49 Lt. Gen. Ngo Quang Truong, *Territorial Forces*, p. 134.

50 JCS History, II:38–41.

51 Davidson, LBJ Library Oral History, as quoted in Moyar, *Phoenix and the Birds of Prey*, p. 49.

52 Westmoreland Oral History (Ganderson), MHI.

53 Ibid.

54 Davidson, *Vietnam at War*, p. 407.

55 As quoted in Stewart Harris, "U.S. Forces Prepared," *London Times*, c. 1 April 1965, Box 66, WPSCL. The article is datelined 31 March 1965.

56 Dennis Chamberland, "Interview: Westmoreland," *Naval Institute Proceedings* (July 1986), p. 48.

57 OJCS, *Intensification*, p. C-12.

58 For an extended description of the PROVN Study, see Sorley, *Honorable Warrior*, pp. 227–241.

59 As quoted in Nagl, *Learning to Eat Soup with a Knife*, p. 127, and in many other sources.

60 Dr. Herbert Y. Schandler interviews, 23 and 26 February 1996.

61 Address, Colby Military Writers' Symposium, Norwich University, Northfield, Vt., 5 April 2001.

62 William E. Colby Interview, 16 July 1981, WBGH Interview Collection, Healey Library, University of Massachusetts at Boston.

63 Chester Cooper wrote perceptively of the Tuesday Lunch that it "had much the character of a cabal: no agenda, no minutes, no regular subsequent communication or follow-up with staff officers and subordinate officials," all at least in part the result of LBJ's "almost pathological fear of leaks." *The Lost Crusade*, p. 414. Gen. Wheeler was not included regularly until after the August 1967 Stennis subcommittee hearings made an issue of his exclusion. Then, according to his associate Lt. Gen. Harry Lemley, Wheeler "used to complain that they always had liver, and it was tough liver." Lemley Oral History, MHI.

64 Enthoven and Smith, *How Much Is Enough?*, p. 294.

65 Ibid., p. 295. Lt. Gen. Dave Palmer, Westmoreland's former aide, likewise raised the question of "why [Westmoreland's] search-and-destroy tactics seemed in the end to have required so much to do so little." *Summons of the Trumpet*, p. 135.

66 Paper, "Lessons in Strategy," identified as "Based on discussion by Historian with General Westmoreland, May 1968," in Westmoreland History File #32, WPCMH.

12. ATMOSPHERICS

1 *Time* (7 January 1966), p. 13.

2 "Man of the Year," *Time* (7 January 1966), pp. 15–21. Hopper was a gossip columnist known for her collection of large and flamboyant hats.

3 Ibid., p. 20.

4 Palmer, *Summons of the Trumpet*, p. 103.

5 Daniel Cragg interview, 10 May 1993.

6 Interview with Charles B. MacDonald, 18 June 1973, Box 31, WPLBJ.

7 Quoted in Marc Phillip Yablonka, "Personality," *Vietnam* (August 2000), p. 18.

8 Westmoreland History File 13, WPCMH.

9 Letter to the Editor, *Vietnam* (October 2004), p. 56.

10 Westmoreland, *A Soldier Reports*, p. 277.

11 Ibid.

12 Arnett, *Live from the Battlefield*, p. 212. Arnett's account may not be accurate. To another reporter, Keyes Beech, Westmoreland explained that he had resigned from the club because his headquarters had moved to Tan Son Nhut Air Base and it was no longer convenient to play downtown. And, near the end of his tour in Vietnam, Westmoreland stated the same rationale to his former deputy, Lt. Gen. John Heintges. "Yes, I did resign from the Circle Sportif," he wrote. "It had nothing to do with criticism and, in fact, was initiated before any criticism arose. I simply found that there were better and more convenient facilities available here on Tan Son Nhut." He clearly had not, in any case, given up the game.

13 Westmoreland Oral History (Ganderson), MHI.

14 Westmoreland testimony, p. 3580, Westmoreland vs. CBS Microfiche File, USAWC Library.

15 Col. Robert M. Cook Oral History, MHI.

16 In his memoirs Westmoreland still maintained that "from the first I emphasized a large and active inspector-general system." *A Soldier Reports*, p. 284.

17 Remarks, Lt. Gen. Herron N. Maples, USAWC, 2 May 1974.

18 Cook Oral History draft, MHI.

19 Ibid.

20 Cook interview, 24 March 1988.

21 Letter, Westmoreland to Dickinson, 9 July 1969, Box 8, WPSCL.

22 Message to John H. Hesterly, 15 September 2005, copy forwarded to Sorley by Hesterly.

23 Lt. Gen. Walter F. Ulmer Jr. telephone interview, 12 August 1997.

24 E-mail, Buckingham to Sorley, 1 June 2010.

25 Westmoreland, *A Soldier Reports*, pp. 59–60. Westmoreland's aide at the time, Dave Palmer, found this incident highly amusing and submitted it to *Reader's Digest*, using the pseudonym Owen Clemmer, and it was duly published.

26 Remarks, U.S. Air Force Academy History Symposium, 15 October 1992.

27 Col. Carl C. Ulsaker telephone interview, 12 July 1999.

28 David Maraniss, *They Marched into Sunlight*, p. 468.

13. BODY COUNT

1 As quoted in Braestrup, *Big Story*, II:163.

2 Message, Westmoreland to Wheeler, MAC 4114, 121245Z August 1965, History Backup File #17, WPCMH. In his memoirs Westmoreland states regarding body count that he "directed several detailed studies which determined as well as anybody could that the count probably erred on the side of caution." *A Soldier Reports*, p. 273. John Mueller later recalled, in a powerful article about the search for a breaking point in Vietnam, a MACV study that had analyzed some seventy captured enemy documents and then concluded that the reported body count for 1966 had been accurate to within 1.8 percent. Then, he noted, OSD Systems Analysis reviewed the same documents and concluded that MACV's "enemy body count was overstated by at least 30 percent." "Search for a Breaking Point," p. 504.

3 Kinnard, *The War Managers*, p. 75. After Kinnard's book was published Westmoreland tried to get him to reveal the names of officers who made comments such as these. Lyndon Johnson was of course also deeply interested in body count. In a sad postscript to the war related by Doris Kearns, when the LBJ Library opened in Austin, Johnson wanted more people to visit it than any other presidential library, and he asked the staff to give him daily attendance figures. And then, wrote Kearns, "knowing that Johnson would be angry at them if the figures were low, the staff—in a painful similarity to another staff in another place—tended, gradually at first and then more and more regularly, to escalate the body count." Kearns, *Lyndon Johnson*, pp. 364–365.

4 Westmoreland Interview, *Face the Nation*, CBS Television Network (19 December 1971), CMH.

5 McNamara Deposition, p. 112, *Vietnam: A Documentary Collection*, USAWC Library.

6 Bunker Oral History, LBJ Library, II:22.

7 As quoted in Charlton and Moncrieff, *Many Reasons Why*, p. 144.

8 The other objectives were: "1. Increase the population in secure areas to 60% from 50%. 2. Increase the critical roads and RR open for use to 50 from 20%. 3. Increase the destruction of VC/PAVN base areas to 40–50% from 10–20%. 4. Ensure the defense of all military bases, political and population centers and food-producing areas now under govt. control. 5. Provide the military security needed for pacification of the four selected high-priority areas—increasing the pacified population in those areas by 235,000." *Pentagon Papers*, IV:625.

9   Holbrooke interview, 7 July 1983, WBGH Interview Collection, Healey Library, University of Massachusetts at Boston.

10   Westmoreland, *A Soldier Reports*, p. 207.

11   Ibid., p. 159.

12   *Pentagon Papers*, II:548.

13   Ibid., II:554.

14   Gibbons, IV:358.

15   Memorandum, Westmoreland to Commander, 2nd Air Division. No date on copy viewed, but the document is in the 1966 folder, Box 4, DePuy Papers, MHI.

16   Letter, Westmoreland to Maj. Gen. Lewis Walt, 14 August 1965, History Backup File #17, WPCMH.

17   DePuy Oral History, MHI.

18   Zorthian interview, 15 June 2007.

19   Wilson, *Washington Post* (5 September 1974).

20   General Walter T. Kerwin Jr., Oral History Interview, MHI.

21   Don Moser, "Starched, Courtly Man Gambles to Win the War," *Life* (11 November 1966).

22   As quoted in ibid.

23   Westmoreland interview (Felter/Gritz), c. 1974, Box 52, WPSCL.

24   Lecture, "The Press in Vietnam," University of New Brunswick, Canada, 8 December 1972, Box 45, WPSCL.

25   Parker, *Last Man Out*, pp. 167–168.

26   Westmoreland History File #22, 5 December 1967, WPCMH.

27   Bunker, "Lost Victory," p. 87. Westmoreland wrote in his memoirs that "by depicting evacuation of the fortified village of Ben Suc in the Iron Triangle early in 1967 as an act of inhumanity rather than the essential and—under the circumstances—beneficent act that it was," Schell "added to the misunderstanding." *A Soldier Reports*, p. 153.

28   Letter, Westmoreland to Gen. James H. Polk, 3 February 1967, Box 5, WPSCL.

29   As quoted in Dorland, *Legacy of Discord*, p. 4.

30   Westmoreland, *A Soldier Reports*, pp. 250, 251.

31   Letter, Jordan to Westmoreland, 16 June 1980, Box 41, WPSCL.

32   PBS Telecast, "Ethics in America: Under Orders, Under Arms," 31 October 1987.

14. M-16 RIFLES

1   Westmoreland's marginal note in the copy of Douglas Kinnard's book *The War Managers* sent to him by Kinnard, then extensively marked up by Westmoreland and returned to Kinnard, p. 23.

2   Nguyen Cao Ky, *Buddha's Child*, p. 336.

3   Truong, *RVNAF and US Operational Cooperation and Coordination*, p. 166.

4   Lung, *Strategy and Tactics*, p. 73.

5   Khuyen, *RVNAF Logistics*, p. 57.

6   Westmoreland dictated notes, Box 41, WPSCL.

7   Westmoreland, "General John Throckmorton," material dictated for mem-

oirs preparation, Box 41, WPSCL. Wrote former Army Chief of Military History Brigadier General James Lawton Collins Jr.: "After 1965 the increasing U.S. buildup slowly pushed Vietnamese armed forces materiel needs into the background. In December 1966 the Secretary of Defense directed that the issue of M16's to South Vietnam Army and Republic of Korea (ROK) forces be deferred and that the allocations previously planned for these forces be redirected to U.S. units." *Development and Training of the South Vietnamese Army*, p. 101.

8   *Time* (19 April 1968).
9   Gen. Harold Keith Johnson CMH interview, 20 November 1970.
10 Ibid.
11 Gen. Frank S. Besson Jr. Oral History, MHI.
12 Collins, *Development and Training of the South Vietnamese Army*, p. 101.
13 Weyand, Debriefing Report, Headquarters, II Field Force, Vietnam, 4 October 1968, Box 15, WPCMH.
14 Message, Wheeler to Sharp and Westmoreland, JCS 6767-66, 4 November 1966, Box 4, WPCMH.
15 Record of Chief of Staff Fonecon with Walt Rostow, 20 July 1971, History File #40, WPCMH.
16 Westmoreland Paper, "M-16 Rifle," Box 41, WPSCL.
17 Ibid.
18 Ibid.
19 Letter, Westmoreland to Tom Johnson (Executive Assistant to LBJ), 10 April 1970, History File #37, WPCMH.
20 Westmoreland, *A Soldier Reports*, pp. 202–203. In describing Clarke as "a retired World War II general," Westmoreland was apparently trying to portray him as old and irrelevant, since he had to know that, whereas Clarke did in World War II earn a battlefield promotion to brigadier, along with the Distinguished Service Cross and three Silver Stars, as a brilliant commander of tank forces in Europe, he served for many years thereafter, rising to the rank of four-star general and command of a corps in the Korean War (when Westmoreland was there as a brigadier), command of Continental Army Command (when Westmoreland was commanding a subordinate division and had briefed General Clarke on Operation Overdrive), and command of U.S. Army, Europe.
21 Letter, Gen. Bruce C. Clarke to Brig. Gen. Hal C. Pattison, 29 December 1969, Clarke Papers, MHI.
22 Letter, Clarke to Westmoreland, 19 February 1976, Box 15, WPSCL. Mayborn was editor and publisher of the Temple (Tex.) *Daily Telegram*.
23 "General Bruce C. Clarke's Report on Visit to Vietnam (Draft)," Box 6, WPSCL. Clarke was in Vietnam 7–13 February 1968.
24 Letter, Clarke to Westmoreland, 15 March 1968, Box 6, WPSCL.
25 Fonecon, Westmoreland with Bunker, 2005 hours, 16 February 1968, Box 37, WPCMH.
26 Letter, Adm. John S. McCain Jr. to Clarke, 22 February 1969, Box 7, WPSCL.
27 Written responses to questions posed by Townsend Hoopes, 28 June 1969, Box 52, WPSCL.

28 Schandler in McNamara et al., *Argument Without End*, p. 351.

29 Westmoreland, *A Soldier Reports*, p. 243.

30 Military History Institute of Vietnam, *Official History*, p. 193.

31 Telecon, Gen. John R. Galvin, 2 January 2005.

32 Weyand, Debriefing Report, Headquarters, II Field Force, Vietnam, 15 July 1968.

33 Schwarzkopf, *It Doesn't Take a Hero*, p. 126.

34 Kinnard, *The War Managers*, p. 176. More than one response could be marked, and an equal number—91 percent—also chose "defining the objectives."

35 As quoted in Andrew F. Krepinevich, "Vietnam: Evaluating the Ground War, 1965–1968," in Showalter and Albert, eds., *An American Dilemma*, p. 99. Krepinevich cited a 17 June 1982 interview with Taylor.

36 Message, Abrams to Johnson, MAC 5307, 040950Z June 1967, CMH.

37 Westmoreland Marine Corps Oral History, 4 April 1983.

38 Westmoreland, *A Soldier Reports*, p. 328.

39 Charlton and Moncrieff, *Many Reasons Why*, p. 145.

40 Army Chief of Staff's *Weekly Summary* (21 May 1968).

41 Westmoreland, *A Soldier Reports*, p. 309. Emphasis supplied.

42 Stilwell interview, 26 January 1989.

43 Kinnard, *The War Managers*, p. 144.

44 Vien and Khuyen, *Reflections*, p. 80.

45 Clarke, *Advice and Support*, p. 278.

46 *JCS History 1969–1970*, p. 177.

15. PROGRESS OFFENSIVE

1 Message, Wheeler to Westmoreland, CJCS 1810-67, 092252Z March 1967, Joint Exhibit 231, Westmoreland vs. CBS Microfiche File, USAWC Library.

2 Message, Wheeler to Westmoreland, JCS 1843-67, 110036Z March 1967, Joint Exhibit 233, Westmoreland vs. CBS Microfiche File, USAWC Library.

3 Ibid.

4 Ibid.

5 Message, Westmoreland to Sharp, MAC 2715, 220302Z March 1967, Box 27, WPCMH.

6 Paper, Third Working Group, Report by the J-3 to the Joint Chiefs of Staff on Courses of Action for Southeast Asia, JCS 2343/646-7, 13 January 1967, p. 4, WPCMH.

7 Gibbons, IV:578.

8 Address, Associated Press Managing Editors Luncheon, New York City, 24 April 1967, Box 1, WPCMH.

9 Gen. Donald V. Bennett Oral History, MHI.

10 McNaughton Notes on Discussions with the President, 27 April 1967, Westmoreland vs. CBS Exhibit 1400, Microfiche File, USAWC Library.

11 Zaffiri, *Westmoreland*, p. 5.

12 Speech transcript filed with Westmoreland History File #15, WPCMH.

13 Letter, Westmoreland to William T. Kyle, date not recorded, WPSCL.

14 Message, Wheeler to Westmoreland, JCS 2218-67, 241543Z March 1967, Box 27, WPCMH.

15 Kinnard, "Adventures in Two Worlds," p. 54.

16 Later Westmoreland would tell a college audience that in April 1967 he had been "called to Washington by the President to report to Congress and meet with the press to explain the war to increase public support." In explaining an optimistic press briefing he gave the following August, he told Charles Mac-Donald that "commanders must show optimism, else gloom and doom pervades the command. [I was] trying to give the US public a picture of confidence to offset negativism of the press."

17 Memorandum, Westmoreland to Dan M. Burt, Re: June, 1967, dated 14 February 1985, Box 19, WPSCL.

18 Lecture, "The Press in Vietnam," University of New Brunswick, Canada, 8 December 1972, Box 45, WPSCL.

19 Westmoreland v. CBS Microfiche, p. 29955, USAWC Library.

20 Transcript, Westmoreland Background Briefing, 29 June 1967, History File #18, Box 13, WPCMH. In Washington, on 26 May 1967, Army Chief of Staff General Harold K. Johnson had written to a friend in Vietnam: "One of our major difficulties continues to be failure to get blunt, outspoken views on what is right and what is wrong with our performance in Vietnam. . . ." HKJ Papers, MHI.

21 *Pentagon Papers*, IV:518.

22 McNamara, *In Retrospect*, p. 237. Years later, when Westmoreland was deposed in connection with his libel suit against CBS, he said: "When I came back in April [1967], if you'll read all of my statements, I was not very optimistic. I was confident, uh—there was no suggestion in April about any crossover point." Another trial document, a McNamara Chronology labeled as Exhibit 488, states regarding the 27 April 1967 White House meeting: "Westy tells the President that the Crossover Point has been virtually reached, and that the enemy force has leveled off at 285,000."

23 McNamara, Remarks to the Press Upon Departing Vietnam, Saigon, 11 July 1967, Exhibit 1807, Westmoreland vs. CBS Microfiche File, USAWC Library.

24 Gen. Walter T. Kerwin Jr., Oral History Interview, 22 March 1980, MHI. See also Kerwin Oral History, 13 September 1985, LBJ Library.

25 George McArthur interview, 18 September 1989.

26 Chaisson, Letter of 20 May 1967, Hoover Institution Archives, Stanford University.

27 Cyrus R. Vance interview, 11 May 1989.

28 Keyes Beech interview, 28 September 1989.

29 Davidson in 5 March 1969 Brief, as quoted in Sorley, *Vietnam Chronicles*, p. 138.

30 "The McNamara-Westmoreland Clash," *Washington Post* (20 July 1967).

31 Actually there is some confirmatory evidence that LBJ might have been shrewdly correct in keeping Westmoreland where he was. Said Gen. Bruce Palmer Jr., "It began to dawn on them [the politicians in Washington] that Westy was a political threat. It was obvious to me that Westy was bitten by the presidential bug as early as the spring of 1967. They wouldn't want

Westy back in the US under those circumstances." Palmer interview, 25 July 2000.

32 As quoted in Senate Committee on Foreign Relations, *Vietnam: Policy and Prospects*, 1970, p. 441.

33 Quoted in Chester Cooper, *Lost Crusade*, p. 506.

34 MFR, Brig. Gen. J. R. Chaisson, MACV Commanders' Conference, 13 May 1967, dated 21 May 1967, Box 13, WPCMH.

35 Message, Wheeler to Westmoreland and Sharp, JCS 7126, 301429Z August 1967, Box 28, WPCMH.

36 Message, Palmer to Johnson, ARV 1522, 190930Z August 1967, MHI.

37 As quoted in Berman, *Lyndon Johnson's War*, pp. 57–58.

38 Murray Fromson, "Name That Source," *New York Times* (11 December 2006).

39 Ibid. Apple, said Fromson, considered the stalemate piece "the most important story he'd ever done."

40 *Pentagon Papers*, II:403.

41 Transcript, *Meet the Press*, NBC Television Network, 19 November 1967, Box 1, WPCMH. When, two years later, the United States did indeed begin to withdraw its forces from Vietnam, Westmoreland tried to claim credit for an accurate forecast, glossing over the fact that his prediction had clearly been based on an expectation of successful prosecution of the war such that withdrawals could begin, whereas when they did in fact take place it was for an exactly opposite reason, that we had given up on the war.

42 Westmoreland, *A Soldier Reports*, p. 234.

43 Westmoreland Address, National Press Club, Washington, D.C., 21 November 1967, Box 1, WPCMH.

44 Ibid.

45 Braestrup in Willenson, *The Bad War*, p. 190.

46 Message, Johnson to Abrams, WDC 15663, 221857Z November 1967.

47 Palmer, *The 25-Year War*, p. 75.

48 Gen. Walter T. Kerwin Jr. interview, 2 April 1999.

49 *MACV: The Joint Command in the Years of Withdrawal*, p. 20.

50 Gen. Donn A. Starry interview, 1 September 1989. Starry also described this episode in "Remarks on Joint Operations" at the Army War College on 14 October 1986. See Sorley, *Press On!*, I:424–425. There were confirmatory accounts from others. Herbert Schandler said in *Argument Without End*, a collaborative work written with Robert McNamara, that Westmoreland had asserted, asking for forty-four battalions, and then another twenty-four, that this force would enable him "to take the offensive that year and, with 'appropriate' (but unspecified) additional reinforcements, . . . defeat the enemy by the end of 1967." Certainly that was McNamara's understanding of what had been predicted, as he wrote in his earlier book *In Retrospect*, referring to Westmoreland's 1 September 1965 paper, "Concept of Operations in the Republic of Vietnam," a description of a multiphased prosecution of the war. "Phase 3 would kick in 'to destroy or render militarily ineffective the remaining organized VC units and their base areas,'" McNamara quoted from the document. "It would begin July 1, 1966, and run through December 31, 1967." Westmoreland later countered that "this common reading of his 1965

timetable is inaccurate," but if so it appears he badly misled the Secretary of Defense at the time. Another who heard that optimistic prediction was Air Force Major General Edward Mechenbier, in June 1967 a young fighter pilot flying F-4Cs out of Danang. Westmoreland came to visit. "The war will be over in two months," Mechenbier remembers hearing Westmoreland tell them. Two days later he was shot down over North Vietnam. When he arrived at the "Hanoi Hilton," he told the other prisoners what Westmoreland had said, "and there was jubilation," at least for a time.

51 Starry interview, 31 May 1995.

52 *The Pentagon Papers as Published by The New York Times*, pp. 474–475.

53 As quoted in Bilton and Sim, *Four Hours in My Lai*, p. 46.

54 Clarke, *Advice and Support*, p. 521. Dr. Clarke later became the Army's Director of Military History.

55 Halberstam, "Return to Vietnam," *Harper's Magazine* (December 1967), as reprinted in *Vietnam Roundup* (24 November 1967), U.S. Information Agency, pp. 1–12. Quotation from p. 4.

56 CBS Reports, "The Uncounted Enemy," 23 January 1982, Transcript, p. 21.

57 As quoted in Tom Johnson, "Notes of the President's Meeting with General Earle Wheeler, JCS and General Creighton Abrams, March 26, 1968, Family Dining Room," LBJ Library. Quoted with permission.

16. ORDER OF BATTLE

1 When Gen. McChristian arrived in Vietnam to become the MACV J-2 in mid-1965, he asked his predecessor, Marine Maj. Gen. Carl Youngdale, to brief him on his order of battle files. "He asked me, 'What do you mean by order of battle?'" said McChristian. "They didn't have any." Affidavit, Westmoreland vs. CBS libel suit, 20 April 1984. Also McChristian interview, 26 April 1999. In 1968 Maj. Gen. Youngdale returned to Vietnam, this time to command the 1st Marine Division.

2 Memorandum, 26 August 1966, Joint Exhibit 217, Westmoreland vs. CBS Microfiche File, USAWC Library.

3 As quoted in Berman, *Lyndon Johnson's War*, p. 22.

4 Westmoreland, "Memorandum of Law," 20 July 1984, p. 15.

5 McChristian, Trial Transcript, pp. 9018 and 9024–9025, Westmoreland vs. CBS Microfiche File, USAWC Library.

6 Allen, *None So Blind*, p. 245.

7 Remarks, Press Conference, Honolulu, 5 February 1966, Box 1, WPCMH.

8 As quoted in Cubbage, "Westmoreland vs. CBS," p. 79.

9 BDM Corporation, *Study of Strategic Lessons Learned*, VI: page not recorded.

10 Ibid., VI:9–31.

11 Maj. Gen. George Godding interview, 26 February 1996.

12 As quoted in Palmer, "US Intelligence and Vietnam," p. 58.

13 Hawkins testimony, Westmoreland vs. CBS Microfiche File, p. 9510, USAWC Library.

14 McChristian Affidavit, p. 5, Trial Exhibit 1837, Westmoreland vs. CBS Mi-

crofiche File, USAWC Library. There is nothing in the Westmoreland history file on this meeting, an omission Westmoreland was quizzed about during the later libel trial. His response was that this was nothing very significant and therefore did not warrant a mention. For those familiar with the history files, in which such matters as wearing of the white uniform, tennis court injuries, water skiing arrangements, and other such ephemera get frequent mention, the omission of this meeting is puzzling.

15 McChristian interview, 26 April 1999.

16 McChristian Affidavit, pp. 4–5, Trial Exhibit 1837, Westmoreland vs. CBS Microfiche File, USAWC Library.

17 Westmoreland vs. CBS Microfiche File, p. 29967, USAWC Library.

18 Allen, *None So Blind*, p. 247.

19 Harold P. Ford, "Why Were CIA Analysts So Doubtful About Vietnam?" Society for Military History Annual Meeting, McLean, Va., 18 April 1996.

20 As quoted in Allen, *None So Blind*, p. 248.

21 Maj. Gen. Joseph A. McChristian interview, 26 April 1999.

22 As quoted in Helms, *A Look Over My Shoulder*, p. 324.

23 As quoted in Ford, *CIA and the Vietnam Policymakers*, p. 99.

24 Helms, *A Look Over My Shoulder*, pp. 325–326.

25 Ibid., p. 326.

26 Ford, *CIA and the Vietnam Policymakers*, p. 87.

27 Message, Davidson to Godding, Joint Exhibit 251, Godding document collection.

28 As quoted in Jones, *War Without Windows*, pp. 101–102.

29 Message, Davidson to Godding, Joint Exhibit 251, Godding document collection. Davidson, while totally committed to carrying out Westmoreland's wishes, was not all that admiring of him. "Westmoreland used to have a theory—he had a lot of theories which had no basis in fact," he later said. See Sorley, *Vietnam Chronicles*, p. 138.

30 Hawkins Testimony, Westmoreland vs. CBS Microfiche File, pp. 9525–9526, USAWC Library.

31 Hawkins deposition, 21 September 1983.

32 As quoted in Berman, *Lyndon Johnson's War*, p. 120.

17. KHE SANH

1 Message, Westmoreland to Wheeler, MAC 01049, 22 January 1968, CMH.

2 Westmoreland, "Westmoreland in Vietnam: Pulverizing the Boulder," *Army* (February 1976), p. 41.

3 As quoted in Cosmas, *MACV*, p. 39.

4 Col. Reamer W. Argo interview, 14 October 2003.

5 Westmoreland, *A Soldier Reports*, p. 338.

6 Krulak, *First to Fight*, p. 215.

7 Lehrack, *No Shining Armor*, p. 141.

8 Message, Westmoreland to Cushman, MAC 02128, 150148Z February 1968, CMH.

9   DePuy, "Our Experience in Vietnam," *Army* (June 1987), p. 40.

10   An additional estimated 3,600 tons of munitions were delivered by artillery positioned at Khe Sanh and elsewhere within range, but that was not a particularly happy part of the story. "We were outgunned, totally outgunned," said Major Jim Stanton of the 26th Marines Fire Support Coordination Center. "Their 130mm guns could sit out beyond our range and shoot us up something awful. We had nothing that could reach them, not even our two towed 155s. Their 130mm guns had a 27,000-meter maximum range and our 155mm guns had a 14,000-meter maximum range. Even their 100mm guns and 122mm rockets outshot us. They parked everything 1,000 meters outside of our artillery fan. We knew where they were, but we couldn't reach them from inside the combat base with our guns and howitzers." As quoted in Hammel, *Assault on Khe Sanh*, p. 177.

11   Westmoreland, *A Soldier Reports*, p. 336.

12   Ibid., p. 338.

13   Ibid.

14   *Time* (16 February 1968), p. 19.

15   Message, Westmoreland to Sharp Info Wheeler, MAC 01060, 230138Z January 1968, Box 31, WPCMH.

16   Westmoreland, *A Soldier Reports*, p. 335.

17   Ibid.

18   As quoted in Laura Palmer, "The General," *MHQ* (Autumn 1988), p. 33.

19   Sharp in his portion of *Report on the War in Vietnam*, p. 8.

20   Westmoreland, *A Soldier Reports*, p. 456.

21   Maj. Gen. Rathvon McC. Tompkins, Marine Corps Oral History, 13 April 1973, MCHC.

22   Letter, Brig. Gen. John R. Chaisson to his wife, 17 April 1968, Chaisson Papers, Hoover Institution Archives, Stanford University.

23   Gen. Walter T. Kerwin Jr. interview, 1 March 1989.

24   Gen. Walter T. Kerwin Jr. Oral History, 13 September 1985, LBJ Library.

25   Message, Sharp to Wheeler, 181231Z June 1968, CMH.

26   Lt. Gen. Charles A. Corcoran interview, 14 September 1989.

27   Message, Abrams to Wheeler and Sharp, MAC 8515, 261202Z June 1968, CMH.

18. TET 1968

1   Message, Westmoreland to Sharp and Wheeler, MAC 01108, 231328Z January 1968, Box 31, WPCMH.

2   Davidson, *Vietnam at War*, pp. 555–556.

3   Maj. Gen. Rathvon McC. Tompkins Marine Corps Oral History, 13 April 1973.

4   Davidson, *Vietnam at War*, p. 557.

5   Message, Westmoreland to Cushman, MAC 02707, 260936Z February 1968, Box 29, WPCMH.

6   Message, Westmoreland to Wheeler, MAC 01011, 220052Z January 1968.

7   Gen. Kerwin read this letter to Sorley in a telecon of 15 January 2007.

8   Westmoreland, *A Soldier Reports*, p. 342.

9   Westmoreland Marine Corps Oral History, 4 April 1983, MCHC.

10   Per Merle Pribbenow, "In some areas, the last-minute General Staff cable ordering that the offensive be launched on the night of the last day of the Lunar New Year was misunderstood because of a one-day difference between North and South Vietnam in the date of the beginning of the Lunar New Year. As a result, a number of provinces in Central Vietnam launched their attacks one day early." In "General Vo Nguyen Giap," p. 19. Westmoreland was apparently not overly concerned by these first attacks since, the official Army history notes, the next day (30 January) he "played a mid-day game of tennis." Cosmas, *MACV*, p. 56.

11   As quoted in Mann, *Grand Delusion*, p. 569.

12   As dictated for his history notes for 23 January 1968 and following, Box 8, WPCMH.

13   North, "VC Assault," pp. 46–47. Thus began the Year of the Monkey. Most accounts of the Tet Offensive state that within a matter of days enemy attacks had been turned back everywhere but in Saigon and Hue. Thus it is worth noting that over a month after the offensive began Westmoreland cabled Wheeler to advise that "major cities which still have Main Force units in the central area include: Hue, Danang, Hoi An, Pleiku, Kontum, Saigon, My Tho and Con Tho." MAC 02960, 021148Z March 1968, HKJ Papers, MHI.

14   Bradford, "Perfume River," p. 15.

15   Col. Ted Kanamine, Oral History interview: Abrams Story, MHI.

16   Brig. Gen. Zeb B. Bradford Jr. interview, 12 October 1989.

17   Ibid.

18   Westmoreland History Notes, Box 8, WPCMH. This is one of a number of instances in which, referring on a given date to events well in the future, Westmoreland clearly has the benefit of hindsight in shaping his dictated entries.

19   Weyand interview, 27 September 1999.

20   Mangold and Penycate, *The Tunnels of Cu Chi*, p. 175.

21   *MACV Command History 1967*, p. 108, CMH.

22   Westmoreland, *A Soldier Reports*, p. 318.

23   Lt. Gen. Fred C. Weyand, Debriefing Report, 4 October 1968, copy in Westmoreland History File #34, Box 15, WPCMH.

24   AP datelined Saigon 26 December 1967, Westmoreland History File #27, WPCMH.

25   Richard Dudman, "Westmoreland Shares the Blame," *St. Louis Post-Dispatch* (16 June 1968).

26   As quoted in Palmer, *Summons of the Trumpet*, p. 203.

27   While the MACV report called 1967 a "Year of Progress," Conrad Gibbons's brilliant Congressional Research Service compilation and commentary titled the relevant chapter "A Year of Reckoning."

28   As quoted in Peter Osnos, "Westmoreland," *Washington Post* (1 April 1972).

29   Ky, *Twenty Years and Twenty Days*, p. 158.

30   Maj. Gen. Nguyen Duy Hinh, *Vietnamization*, p. 5.

31   Lung, *The General Offensives*, p. 37.

32 Meeting of 10 February 1968, Tom Johnson's Notes, Box 2, LBJ Library. Cited with permission.

33 In *Vietnam: A Television History*, Segment 7: Tet 1968, MacArthur Foundation Video Classics Project.

34 *Newsweek* (19 February 1968).

35 Col. Fred B. Schoomaker interview, 3 March 1997.

36 Letter, Westmoreland to Tim Young, 10 September 1991, Box 17, WPSCL.

37 Westmoreland, *Report of the Chief of Staff*, p. 4.

38 Westmoreland History File #34, Remarks to Army Personnel Serving Outside DA, 9 November 1968, Box 15, WPCMH.

39 Memorandum, "Enemy Strategy," Argo to Westmoreland, 11 May 1968, Westmoreland History File #32, WPCMH.

40 George Crile interview with McNamara, 16 June 1981, Joint Exhibit 33, Westmoreland vs. CBS Microfiche File, USAWC Library.

41 As quoted in Mann, *Grand Delusion*, p. 568.

42 As quoted in Hammel, *Assault on Khe Sanh*, p. 10.

43 Simpson, *Inside the Green Berets*, p. 117.

44 Combat After Action Report, Battle of Lang Vei, 22 March 1968, VNIT 138, CMH.

45 Prados, *Hidden History*, p. 172.

46 Pisor, *End of the Line*, p. 195.

47 Westmoreland, *A Soldier Reports*, p. 341.

48 Ibid.

49 Col. Jonathan F. Ladd Oral History: Abrams Story Collection, MHI. Westmoreland's account and that given by Col. Ladd are mutually exclusive.

50 Westmoreland Marine Corps Oral History, 4 April 1983, MCHC.

51 Pisor, *End of the Line*, p. 198.

52 Gen. Robert E. Cushman Jr. Marine Corps Oral History, December 1982, MCHC.

53 Message, Abrams to Cushman, MAC 7462, 061125Z June 1968, Rosson Papers, CMH.

54 Brig. Gen. Charles A. Corcoran Oral History, MHI.

55 *Washington Post* (27 March 1968).

56 Ibid.

57 Thompson, "Viet Reds' Drive," *Washington Post*, 11 February 1968, as reprinted from *London Sunday Times*, Box 4, WPCMH.

58 Col. John Barrie Williams, Joint Exhibit 45A, Westmoreland vs. CBS Microfiche File, USAWC Library.

59 As quoted in CBS Memorandum to Dismiss, 23 May 1984, citing Joint Exhibit 305, Westmoreland vs. CBS Microfiche File, USAWC Library.

60 Gen. William E. DePuy Oral History, MHI.

61 Charles W. Corddry, "Westmoreland," *Baltimore Sun* (23 March 1968).

19. TROOP REQUEST

1 Dates mentioned in press accounts can be confusing. Saigon (in the H or Hotel time zone in the international system) was thirteen hours ahead of

Washington (on R or Romeo time). But the relative timing and sequencing of military messages can always be easily determined since, regardless of the local time where they originate, all are marked with the DTG (date-time group) in Z or Zulu time, meaning Greenwich Mean Time. Thus, for example, a message marked 090021Z February 1968 would have been dispatched on 9 February 1968 at 21 minutes past midnight GMT.

2   Gen. William C. Westmoreland, "The Origins of the Post-Tet 1968 Plans for Additional American Forces in RVN." Unpublished paper, April 1970. Major Paul Miles prepared this White Paper for Westmoreland in an effort to show that Westmoreland had not actually "requested" the 206,000 troops in the wake of the Tet Offensive.

3   Message, Wheeler to Westmoreland Info Sharp, JCS 01529, 080448Z February 1968, Box 29, WPCMH. Westmoreland later told historian Herbert Schandler: "It seemed to me that for political reasons or otherwise, the president and the Joint Chiefs of Staff were anxious to send me reinforcements. . . . My first thought was not to ask for any, but the signals from Washington got stronger." Schandler, *Unmaking of a President*, p. 97, citing a 16 September 1973 interview with Westmoreland.

4   Message, Wheeler to Westmoreland Info Sharp, JCS 01590, 090021Z February 1968, Box 29, WPCMH.

5   Message, Westmoreland to Wheeler, MAC 01858, 091633Z February 1968, HKJ Papers, MHI.

6   Message, Westmoreland to Wheeler Info Sharp, MAC 02018, 121823Z February 1968, Box 29, WPCMH.

7   Message, Wheeler to Westmoreland Info Sharp, JCS 01695, 120108Z February 1968, Box 29, WPCMH.

8   Message, Westmoreland to Wheeler, MAC 01975, 120612Z February 1968, Box 29, WPCMH. Recounted Westmoreland in his White Paper on the matter: "[M]y formal request for reinforcements was not made until 12 February. Only after extensive consultations among General Wheeler, Admiral Sharp, and myself concerning the utilization of reinforcements and their logistical support did I send the following message to General Wheeler: 'I need reinforcements in terms of combat elements.'"

9   As quoted in Perry, *Four Stars*, p. 187, citing a Perry interview with Westmoreland.

10 Kinnard, *The War Managers*, p. 79.

11 McNamara, *In Retrospect*, p. 315.

12 As quoted in Henry, "February, 1968," p. 23.

13 Westmoreland, *A Soldier Reports*, p. 357.

14 As quoted in Willenson, *The Bad War*, p. 97.

15 Clifford, *Counsel to the President*, p. 481.

16 Ibid.

17 As quoted in Westmoreland, *A Soldier Reports*, p. 358.

18 Chronology, Exhibit 488, Westmoreland vs. CBS libel trial. The following day Wheeler cabled Westmoreland that "there is strong resistance from all quarters to putting more ground force units in South Vietnam." Regarding prospects for a reserve call-up, said Wheeler, "you should not count upon an

affirmative decision for such additional forces. With this cheerless counsel I
will sign off." Message, Wheeler to Westmoreland Info Sharp, JCS 2767,
090130Z March 1968, WPCMH.

19   Johnson, *The Vantage Point*, p. 407.

20   As reported in Brandon, *Anatomy of Error*, p. 241.

21   Westmoreland *White Paper*.

22   Clifford in Segment 7: Tet 1968, *Vietnam: A Television History*, MacArthur
     Foundation Library Video Classics Project.

23   Tom Johnson, "Notes of the President's Meeting with General Earle
     Wheeler, JCS and General Creighton Abrams, March 26, 1968, Family Din-
     ing Room," LBJ Library. As revealed in the body of the notes, Secretary of
     State Rusk was also present. Quoted with permission.

24   Hanson W. Baldwin, "Westy's Side: His Memoirs of a Bitter War," *Army*
     (January 1976), p. 58.

25   Karnow, *Vietnam: A History*, p. 551.

26   In Gittinger, *The Johnson Years*, p. 88.

27   Gelb and Betts, *The Irony of Vietnam*, pp. 173–174.

28   Letter, Chaisson to wife, 26 February 1968, Chaisson Papers, Hoover Insti-
     tution Archives, Stanford University.

29   Amb. Ellsworth Bunker Oral History, II:15, LBJ Library.

30   Clifford, *Counsel to the President*, pp. 479–482. Emphasis in the original.

31   Ibid., p. 496. Emphasis in the original.

32   Taylor, *Swords and Plowshares*, p. 389.

33   Lt. Gen. Elmer H. Almquist Jr. interview, 13 June 1989. The acronym indi-
     cates where Almquist was assigned on the Army Staff, the Office of the Dep-
     uty Chief of Staff for Operations.

34   Lt. Gen. Robert E. Pursley (USAF) served as Military Assistant to a succes-
     sion of Secretaries of Defense. Regarding the 206,000 troop request, Pursley
     said that "the characterization and the interpretation of that was so clearly a
     request that, if not one, Westmoreland would have stepped in to correct it.
     Certainly that's what Clark Clifford thought it was. Clifford was just horri-
     fied by the 206,000 troop request. All the people who were going to make
     determinations about it thought it was a request." Pursley interview, 15 No-
     vember 2007.

35   Westmoreland interview with Sam Donaldson on *This Week with David
     Brinkley*, ABC-TV (Channel 7 in Washington), 28 April 1985. Westmore-
     land prefaced the admission by saying it was "on the assumption that we
     would call up the reserves and that there would be a change of strategy."

36   MacDonald, *Outline History*, p. 72.

37   Cosmas, *MACV*, p. 104. Daniel Davidson was a member of the Harriman
     delegation to the Paris peace talks. He remembered that there the North
     Vietnamese referred to Westmoreland as "Limoges." Asking around to find
     out what that meant, Davidson learned that Limoges was a city in France
     "where they assigned military commanders who had failed. In France, if you
     wanted to put a commander out of play you made him commander of the
     garrison at Limoges—a city equidistant from the French frontiers and
     therefore unlikely to involve him in further fighting. So it was a very sophis-
     ticated analysis of what had happened to Westmoreland." As quoted in Appy,

Patriots, pp. 463–464. While Limoges is not literally in the center of France, it is close enough for the image to work.

20. HEADING HOME

1   Nearly a decade later Westmoreland was still trying to combat the notion that he had been relieved as a result of Tet. He wrote a hot letter to Charles Van Doren, editor of *Webster's American Biographies*, saying "the Tet offensive was not the reason for my transfer as implied. Six months earlier, after four years in Vietnam and three years away from my family, I was told that I would be transferred to another assignment, the details of which were discussed with me at that earlier date." Van Doren's reply was not conciliatory. "On carefully reading your biography," he said, "I cannot see what you are complaining about. What did we say that you object to?" And: "If we made any errors, I am sorry. But errors are easy to make. You yourself misspelled my name twice in your letter, even though you must have had it before you as you wrote." Both letters may be found in Box 15, WPSCL.

2   Lt. Gen. Phillip B. Davidson Jr. telephone interview, 25 October 1995.

3   Letter, Dr. Edwin A. Deagle to his sons, 23 April 1995, copy provided Sorley by Deagle. Some years later Deagle again encountered Bundy, by then heading up the Ford Foundation. "Do you remember that little talk we had at Harvard?" Bundy asked. "Yes, sir, I do," responded Deagle. "You were absolutely right," Bundy told him. "We were wrong, and I've been regretting it ever since." Deagle interview, 12 November 2002. In the event, of course, Westmoreland had been named Chief of Staff, Abrams took over in Vietnam, and Goodpaster was sent out as deputy to Abrams. General Lyman Lemnitzer continued as NATO commander.

4   Thayer Award Address, West Point, N.Y., as printed in the *Congressional Record* (28 May 1970), pp. E4731–4733. Quoted passage found on p. E4732.

5   Gen. Bruce Palmer Jr. in Willenson, *The Bad War*, p. 158.

6   Gene Roberts, "Victory Doubted," *New York Times* (11 June 1968).

7   "Top General Flying Home," *Chicago Daily News* (12 June 1968). Royce Brier, writing in the *San Francisco Chronicle* (14 June 1968), commented on Westmoreland's contention that the shelling of downtown Saigon had no military significance. "The validity of this pronouncement," Brier observed, "depends on your immediate view of the event. It is true enough there is no 'military significance' in the death of several hundred civilians, with up to a thousand wounded, and the wrecking of scores of city streets. These people aren't doing anything to put down the Viet Cong rebellion, as most of them are women and children and skinny cyclists, anyway." But: "If enemy forces ringed Washington with rocket artillery which was not dislodged in three weeks, and were dropping rockets daily a few blocks from the White House and the big hotels, it would hardly be considered of no 'military significance,' even in the Pentagon."

8   That unconcern represented a change of outlook on Westmoreland's part now that he was being called to account for the attacks. Earlier, during a 1966 National Day parade, the enemy had fired fourteen recoilless rifle

9 Bunker Reporting Cables #52 (23 May 1968) and #53 (29 May 1968), Indochina Archive, University of California, Berkeley.

10 Brig. Gen. S.L.A. Marshall, *Philadelphia Inquirer* (5 May 1968).

11 As quoted in Col. Harry G. Summers Jr., "Troops to Equal Any," *Vietnam* (February 1988), p. 24.

12 Westmoreland, *A Soldier Reports*, p. 261.

13 Ibid. The quotation is rendered rather differently by Count Yorck von Wartenburg in his *Napoleon as a General:* "Every commander-in-chief who takes upon himself to execute a plan which he considers bad or ruinous is culpable; he ought to remonstrate, to insist upon alterations, and, if necessary, rather resign, than become the means of the defeat of the force entrusted to him." I:51.

14 Westmoreland, *A Soldier Reports*, p. 262.

15 MFR, D. Lambertson, "MACV Briefing for National Assembly Members," 13 May 1968, Westmoreland History File #32, WPCMH.

16 As per the *New York Times* article of 10 June 1968.

17 McNamara, *Argument Without End*, pp. 385–386.

18 As quoted in Beech, *Not Without the Americans*, p. 307.

19 As quoted in Gibbons, IV:50, citing CRS interview of 1 August 1988.

20 Weigley, "Review of Eliot A. Cohen's *Supreme Command*," *Journal of Military History* (October 2002), pp. 1275–1276, quotation from p. 1276.

21 Williams Oral History, LBJ Library, 16 March 1981.

22 Westmoreland, *A Soldier Reports*, p. 46.

21. CHIEF OF STAFF

1 Message, Abrams to Westmoreland, MAC 8886, 031158ZJUL1968, WPCMH.

2 Message, Abrams to Westmoreland, MAC 8958, 050021ZJUL1968, WPCMH. In a later oral history interview Westmoreland was, surprisingly, unsure of his late brother-in-law Fred's class at West Point, saying "he was in the Class of 1952 or 1953." In a later letter to Lieutenant General "Hank" Emerson, commander of the brigade in which Fred had been serving as a battalion commander when he was killed, Westmoreland remembered that Fred "was devoted to Kitsy and to his family," adding churlishly of Fred's widow that "unfortunately, Caroline was not an asset to him and we, and you, knew that."

3 Westmoreland's confirmation hearing before the Senate Armed Services Committee had taken place on 4 June, during a brief visit to the United States (when he also attended the graduation of his daughter Stevie from Bradford Junior College). The SASC, including Senator Stephen Young of Ohio, unanimously endorsed Westmoreland's nomination, which two days later was approved by the full Senate. Before the hearings General Wheeler

had warned Westmoreland that Young "has been outspokenly critical of you, and he may attempt to embarrass you on such subjects as being surprised at Tet, unnecessary use of force resulting in inordinate damage to civilian property, loss of civilian life, and creation of refugee problems."

4   Maj. Gen. William B. Steele telephone interview, 1 March 2010.

5   Col. Paul Miles interview, 26 July 2000.

6   Westmoreland interview (Felter and Gritz), c. 1974, Box 52, WPSCL.

7   Remarks, Westmoreland to Army Personnel Serving Outside DA, 9 November 1968, Westmoreland History File #34, Box 15, WPCMH.

8   Johnson, *White House Diary*, pp. 718–719.

9   Interview, Dr. James E. Hewes Jr. with Col. S. V. Edgar, 8 December 1981, CMH. During the critical period when the plan was briefed to the Secretary of the Army and the Secretary of Defense for their approval Westmoreland was on an extended trip to the Pacific area.

10   Robert Froehlke interview, 3 October 1988.

11   Gen. Bruce Palmer Jr. telephone interview, 20 January 2000.

12   Gen. William McCaffrey interview, 16 January 1989.

13   Col. William Greynolds interview, 22 March 2000.

14   Letter, Ailes to Johnson, 20 August 1968, Johnson Papers, MHI.

15   Gen. Ferdinand J. Chesarek Oral History, MHI.

16   Col. Reamer W. Argo interview, 14 October 2003.

17   E-mail, Buckingham to Sorley, 1 June 2010.

18   Peers, *My Lai Inquiry*, p. 254.

19   Secretary of the Army Stanley Resor, reported the *Washington Post*, "conceded that 'a great deal of information suggesting that a possible tragedy of serious proportions had occurred at My Lai was either known to General Koster or was readily available in the operational logs and other records of the division.'" Koster obituary, 10 February 2006.

20   Peers, *My Lai Inquiry*, p. 223.

21   Ambrose in Anderson, *Facing My Lai*, p. 190.

22   Lewy, "Vietnam," *Commentary* (February 1978), p. 46.

23   Memorandum, Lt. Gen. W. R. Peers to Westmoreland, Subject: The Son My Incident, 18 March 1970, Exhibit 1600, Westmoreland vs. CBS Microfiche File, USAWC Library.

24   Goldstein et al., *My Lai Massacre*, p. 7.

25   Memorandum, Army General Counsel Robert E. Jordan III to SecArmy, 2 October 1970, Westmoreland History File #38, Box 16, WPCMH.

26   Memorandum for Record, Secretary of the Army Stanley R. Resor, "Charge Against General William C. Westmoreland," 14 October 1970, Westmoreland History File #38, Box 16, WPCMH.

27   Buckley, "Observations on Calley Reaction," *Washington Star* (7 April 1971).

28   Letter, Maj. Gen. Russel B. Reynolds to Col. John H. Tucker Jr., 6 December 1970, Box 10, WPSCL. Reynolds died two days after sending this letter.

29   Hoopes, *Limits of Intervention*, pp. 62–63.

30   Record of Telephone Conversation, Westmoreland with Brig. Gen. Pattison, 1350 hours on 5 January 1970, Box 40, WPCMH.

31   DePuy Oral History, MHI.

32 Gen. William E. DePuy, "Our Experience in Vietnam," *Army* (June 1987), p. 32.

33 As quoted in Maurer, *Strange Ground*, p. 453.

34 As quoted in Royal United Services Institution, *Lessons from the Vietnam War*, pp. 2–3.

35 Despite his full schedule, Westmoreland handled for himself a number of routine matters that aides or executive officers could easily have dealt with, continuing a practice from his years in Vietnam. Heading for Okinawa, for example, Westmoreland cabled ahead to say that, "if possible, I would like you to incorporate a round of golf into my itinerary."

36 Sell, "Younger Military Critics," *Washington Post* (25 March 1968).

37 Col. James Barbara interview, 17 November 2006.

38 Buckingham, "Ethics and the Senior Officer," p. 31.

39 Ibid., p. 32.

40 As Sewall, by then a retired major general, recalled this incident many years later, he added with feeling that it had made him profoundly thankful for the American system of government in which civilians control the military.

41 As quoted in Peers, *My Lai Inquiry*, p. 249.

42 Lt. Gen. Walter F. Ulmer Jr. telephone interview, 12 August 1997.

43 Lt. Gen. Walter F. Ulmer Jr. Oral History, MHI. In 1993, asked by a correspondent about the study, Westmoreland responded, "I vaguely recall requesting such a study."

22. SHAPING THE RECORD

1 Dr. Robert E. Morris, Vietnam War Conference, University of Virginia School of Law, Charlottesville, Va., 28–29 April 2000.

2 See Shulimson, *U.S. Marines in Vietnam: An Expanding War, 1966*, p. 14.

3 Describing the holdings of the U.S. Army Center of Military History, the Army Director of Military History, Dr. Jeffrey Clarke, identified the William C. Westmoreland Papers as "the largest and most important" collection of a single officer, "an extensive group of approximately 18 linear feet of papers assembled and maintained by General Westmoreland and his personal staff during his tenure as MACV commander. Most significant," he added, "is his diary, or history, contained in about thirty loosely bound volumes detailing his daily activities and decisions, and often the thoughts behind the decisions. Appended to the entries are copies of pertinent incoming and outgoing messages, memorandums, reports, and other documents that Westmoreland considered important at the time the entries were made." *Advice and Support*, pp. 529–530.

4 The JCS historians got it right, noting that "the United States had included the strengthening of the Republic of Vietnam Armed Forces (RVNAF) among its objectives since the beginning of its involvement in South Vietnam, but in the period 1965 through early 1968, major US attention was devoted primarily to the conduct of combat operations. It was only after the 1968 Tet offensive, when President Johnson ruled out a further US troop increase in South Vietnam, that the United States undertook serious prepara-

tions for eventual South Vietnamese assumption of the combat effort." U.S. Joint Chiefs of Staff, *The History of the Joint Chiefs of Staff,* p. 177.

5 Letter, Westmoreland to Hobart Lewis, 25 October 1969, Box 39, WPSCL.

6 This put Keiser in a difficult position, as his wife was with child and expected to deliver any day. Fortunately, after spending a couple of days in Honolulu massaging the document, Keiser made it back before the birth occurred. Things didn't go as well subsequently, however, since two days later Westmoreland sent him back to Hawaii for more of the same and, before he could again get back home, his wife delivered their third child. She was not, and *remained* not, happy about what had transpired.

7 A year and a half after the *Report* was published, the Government Printing Office reported having sold 6,588 copies of the 7,208 printed.

8 Braestrup, "Vietnam as History," *Wilson Quarterly* (Spring 1978), p. 180.

9 Message, Maj. Gen. Willard Pearson to Westmoreland, Subject: Post Mortem on Vietnam Strategy, 6 September 1968, Westmoreland History File 34, WPCMH.

10 Rosson, *Assessment of Influence,* pp. ii–iii.

11 Letter, Col. Rod Paschall to Westmoreland, 13 February 1987, Box 13, WPSCL. As of June 2010 the Military History Institute had no record of any other senior officers having requested destruction of their interview tapes.

23. VOLUNTEER ARMY

1 Col. Jack R. Butler, "The All-Volunteer Armed Force," *Parameters* (Vol. II, No. 1, 1972), page not recorded.

2 Brian Doherty, "The Life and Times of Milton Friedman," *Reason* (March 2007), as quoted in *Washington Times* (2 February 2007).

3 Gen. Bruce Palmer Jr. interview, 15 April 1988.

4 Col. Jack R. Butler, "The All-Volunteer Armed Force," *Parameters* (Vol. II, No. 1, 1972), pp. 24, 26.

5 As quoted in Griffith, *U.S. Army's Transition,* p. 271n1.

6 Van Atta, *With Honor,* p. 151.

7 In reporting this speech the *Baltimore Sun* (14 October 1970) noted that "only last spring, General Westmoreland testified before the House and Senate Armed Services Committees that the change to an all-volunteer force 'may be impractical for some time to come.'"

8 Butler, "All-Volunteer," p. 27.

9 As described by Lt. Gen. George I. Forsythe in Oral History, MHI.

10 Charles B. MacDonald interview of Lt. Gen. George Forsythe, 16 June 1973, Box 31, WPLBJ.

11 Gen. Walter T. Kerwin Jr. Oral History, MHI.

12 Lt. Gen. George I. Forsythe Oral History, MHI.

13 Telecon, Brig. Gen. James L. Anderson, 14 January 2010.

14 L. James Binder, "Well Done, Westy!" in *Army* (August 1972), pp. 9–10.

15 As quoted in John L. Moore, "National Security," *National Journal* (12 November 1971), p. 2452.

16 Gen. Donald V. Bennett interview, 23 August 1994.

17 As quoted in Griffith, *U.S. Army's Transition,* p. 177.

24. VIETNAM DRAWDOWN

1   Westmoreland added a relevant comment to the paperback edition of his memoirs, a sentence that did not appear in the original book, referring to the period when Melvin Laird was Secretary of Defense: "Despite the fact that I had four and one-half years of experience in Vietnam, my advice was seldom sought." *A Soldier Reports*, p. 511 of the May 1980 Dell paperback edition.

2   Palmer, *The 25-Year War*, p. 91.

3   Gen. Donn A. Starry, "Review of *Thunderbolt*," *Armor* (September-October 1992), pp. 50–51.

4   Telecon, Palmer with Westmoreland, 1720 hours on 28 May 1969, Box 37, WPCMH.

5   Starry, "Review of *Thunderbolt*." Years later Starry explained the effects of Westmoreland's insistence on bringing out individuals rather than units in a talk, "Recruiting and the Soldier." "What happened to the Army in the last months of Vietnam was not that the ethical value system of the officer and NCO corps collapsed, as some have alleged. Rather it was that, in redeployment from Vietnam, the centralized individual replacement system demanded redeployment of individuals, not units. Those who remained were reassigned to remaining units. As the pace of redeployment quickened this constant shuffling insured lack of cohesion in the residual force—in the leadership and amongst the soldiers. Careerism there may have been, and may still be, but the root problem was that the sense of community was destroyed. There simply was no cohesion. In that hostile environment the leadership was overloaded, and it behaved accordingly." As quoted in Sorley, *Press On!*, p. 706.

6   Gen. Maxwell R. Thurman interview, 2 August 1995.

7   Halberstam, *The Best and the Brightest*, p. 549.

8   Lt. Gen. Sidney B. Berry interview, 3 May 1989.

9   Lock-Pullan, "Inward Looking Time," p. 493. See also Spiller, *In the School of War*, pp. 226–227. In 2010 Kelvin Crow, an Army historian at the Combined Arms Center at Fort Leavenworth, conducted extensive inquiries into the validity of reports of the alleged booing incidents at Fort Benning and Fort Leavenworth. His tentative conclusion was that "the 'Revolt of the Officers' is an urban legend, perhaps arising from the discontent of the era. While it catches 'the tenor of the times,' it does not represent an actual event, or has become so exaggerated over time as to be best regarded as a 'war story.'" Crow also contacted Dr. Roger Spiller, who had alluded to such episodes in his work cited above. Now, Crow reported, Spiller acknowledged that the account "could be an urban legend" but observed that "the fact that it became an urban legend is telling." Crow prefaced his summation of the research with what he identified as an "Italian saying": "If it's not true, it ought to be." Attachment to e-mail, Crow to Sorley, 2 September 2010.

10   In discussion with Sorley at the Society for Military History Annual Meeting in Charleston, South Carolina, on 25 February 2005.

11   Dr. Alan Gropman telephone interview, 29 September 2009.

12   Surut e-mail to Sorley, 2 October 2009.

13   Letter, Grum to Sorley, 5 July 1999.

14 Westmoreland Oral History (Ganderson), MHI.
15 As quoted in Charles R. Smith, *U.S. Marines in Vietnam: High Mobility and Standdown 1969*, p. 10n.
16 Westmoreland Marine Corps Oral History.
17 Westmoreland Oral History (Ganderson), MHI.
18 William McG. Morrison telephone interview, 30 September 2009.
19 Notes, James Westmoreland (identified as "Freshman, Son of Gen. W. C. Westmoreland, Chief of Staff, USA"), Box 38, WPSCL.

25. DEPARTURE

1   Department of the Army, *Historical Summary: Fiscal Year 1970*, p. 152.
2   Moskos, "The Enlisted Ranks," in Keeley, *The All-Volunteer Force*, p. 40.
3   Message, Woolnough to Westmoreland, MRO 1269, 101929Z September 1970, Box 36, WPCMH.
4   Message from the Chief of Staff, Cumulative Mandatory Training Requirements, 10 May 1972, Lt. Gen. F. J. Brown Papers.
5   Remarks, CONARC Commanders Conference, Fort Monroe, Va., 17 May 1972.
6   Memorandum of Conversation, Subject: General Westmoreland's Conversation with Colonel Warner, evening, 2 April 1972, Volney F. Warner Papers, MHI. Handwritten at the bottom of this document is an indication that Brig. Gen. Dunn had recently mentioned an ambassadorial position to Westmoreland. At that point Westmoreland had no discernible "civilian plans."
7   In anticipation of having the portrait painted, Westmoreland took Herbert Abrams with him on an extended January 1972 trip to the Pacific, including Vietnam, so Abrams could "study" his intended subject. Abrams made a reported 4,000 slides during the trip.
8   Davidson, *Vietnam at War*, p. 378.
9   Gen. Walter T. Kerwin Jr. Oral History, 13 September 1985, LBJ Library.
10 Palmer, *The 25-Year War*, p. 124.

26. IN RETIREMENT

1   "Like many others before them, the Westmorelands have found that the old Charleston families do not easily welcome newcomers to the social fold," wrote Levona Page. "Westy," *The State* (Columbia, S.C.) (21 January 1974).
2   Orr Kelly, "Westmoreland's Place of Honor," Washington Close-Up column, *Washington Star*, c. 3 July 1972, Box 64, WPSCL.
3   Bunker responded that he would put Westmoreland's name forward, whereupon Westmoreland wrote again: "Your interest in my interest is deeply appreciated."
4   Seymour M. Hersh, "The Decline and Near Fall of the U.S. Army," *Saturday Review* (December 1972), p. 58.
5   Ambrose, *Nixon*, p. 632.
6   As quoted in Randolph, *Powerful and Brutal Weapons*, p. 381n1.
7   Kissinger, *Ending the Vietnam War*, p. 356.

8   Westmoreland, MFR, "President Nixon's Address to the Nation, 7 October," dated 8 October 1970, History File 38, WPCMH.

9   Over a decade later Westmoreland wrote to the Chief of Protocol in the Office of the Army Chief of Staff to request fifty more copies of his official photo. "Tell the photolab this will probably be my final request," he said.

10   McCaffrey interview, 16 January 1989.

11   As quoted in Valerie Wieland, "The Inevitable General," *Vietnam* (December 2003), pp. 34–41, 64. Quotation on p. 41.

12   Letter, Westmoreland to Haig, 3 February 1973, Box 14, WPSCL. Westmoreland was extremely unhappy when Nixon chose Haig for the four-star position. "I think a serious mistake was made by the President in nominating such a young and inexperienced officer as Vice Chief of Staff who came into the White House a little more than three years ago as a colonel," he wrote to another officer. "Who advised the President in this regard, I do not know, but I suspect his advice came from sources unfamiliar with the Army and not aware of the implications of his decision. In fact, the advice he received was utterly naïve."

13   Westmoreland, "Vietnam in Perspective," *Military Review* (January 1979), pp. 34, 35.

14   Westmoreland, "Vietnam Blunders," *Honolulu Advertiser* (26 March 1978).

15   Remarks, Hampden-Sydney College Symposium, September 1993, as reported in *Richmond Times-Dispatch* (18 September 1993).

27. MEMOIRS

1   Enclosure to Letter, Westmoreland to Charles MacDonald, 5 June 1974, Box 41, WPSCL.

2   Memorandum, Maj. Paul L. Miles Jr., Research Assistant, to Westmoreland, 19 June 1972, Box 1, WPCMH. Note also in Box 1.

3   MacDonald comments in the discussion following his "Contrasts in Command" lecture, U.S. Army War College, 17 May 1976, MHI.

4   Westmoreland apparently was referring to the Cao Dai sect and meant to indicate that its organization was similar to that of the Catholic hierarchy.

5   Moore and Galloway, *We Are Soldiers Still*, pp. 123–124. The extent of the carnage at LZ Albany is illustrated by what happened there to Company C, 2nd Battalion, 7th Cavalry. That unit began the day with 110 officers and men. A day later only eight were present for duty, the rest dead or wounded.

6   Dr. John Carland, Ia Drang 40th Reunion Symposium, Washington, D.C., 12 November 2005.

7   Interview, MacDonald with Westmoreland, 7 June 1973, Box 31, WPLBJ.

8   MacDonald, "Contrasts in Command," 17 May 1976.

9   Ibid. MacDonald also said, regarding sales of the book (then at about 30,000), "If it doesn't sell 50,000, I don't make any money."

10   Some years earlier, when Westmoreland was about to leave Vietnam, he had written of his service there to his brother-in-law, N. Heyward Clarkson Jr.: "It has been a period of constant crisis, and I believe if I ever write a book I will name it such." That became a chapter title in the memoirs.

11   Westmoreland responded to an expatriate living in Canada, who had written asking for permission to translate portions of the book into Vietnamese, that he would have to contact the publisher since they held the copyright, apparently not realizing that he himself owned the copyright.

12   The correspondence between Westmoreland and MacDonald includes nearly a hundred potential titles, including "Mission Frustration," "Confused Conflict," "War in a Goldfish Bowl," and "We Could Have Won." Some prepublication review copies were apparently sent out using the title "War in Vain?" They then settled on "The War Nobody Won" until that choice was overtaken by events. The B/O reference is obscure. Colonel Miles suggests it may have stood for Baltimore & Ohio Railroad.

13   "Westmoreland Won't Fade Away," *Washington Star* (18 January 1976).

14   S.L.A. Marshall, "Westmoreland: His Career Rose Like a Phoenix—Until Vietnam," *Chicago Sun-Times* (25 January 1976).

28. CAMPAIGNER

1   Lee Bandy interview, 16 August 2005.

2   Analysis as reported by William E. Rome Jr., "Memo on General's Loss," *Atlanta Journal-Constitution* (18 August 1974).

3   Letter, Joseph O. Rogers Jr. to Westmoreland and others, 20 December 1976, Box 28, WPSCL. Emphasis in the original.

4   Letter, Joseph O. Rogers Jr. to Westmoreland and others, 4 January 1977, Box 15, WPSCL.

5   Ernest B. Furgurson, "Westmoreland: To Him, the War Was Unkind," *The State* (20 March 1974).

6   Letter, Dorn to Westmoreland, 24 July 1974 (enclosing a copy of the telegram), Box 14, WPSCL. It seems rather surprising that, only days after the end of the failed primary campaign, and with his house in Charleston newly finished, Westmoreland should have been actively seeking a federal job in Washington.

29. PLAINTIFF

1   Letter, Crile to Wallace, 11 May 1981, as quoted in Kowet, *Matter of Honor*, p. 55.

2   Note, Crile to Wallace, 11 May 1981, Exhibit 504, Westmoreland vs. CBS Microfiche File, USAWC Library. Apparently no honorarium was paid, but the Westmorelands were lodged at a New York hotel at the network's expense.

3   As quoted in Kowet, *Matter of Honor*, p. 57.

4   Buckley, "Vietnam: The Short Count," *New York Daily News* (28 January 1982). Buckley marveled at how Wallace was able to get people to submit to interviews that were clearly not in their own interest, speculating that in an earlier day he "would have succeeded in getting Jack the Ripper to talk to him on the subject of how London's streets were crowded with unnecessary young ladies."

5   Braestrup, "'The Uncounted Enemy,'" pp. 46ff.

6   CBS Reports, "The Uncounted Enemy," 23 January 1982, Transcript, Joint Exhibit 1, Westmoreland vs. CBS Microfiche File, USAWC Library.

7   Ibid.

8   CBS Reports, "The Uncounted Enemy," 23 January 1982, Transcript, p. 13.

9   Ibid.

10 Ibid.

11 Ibid.

12 Ibid., p. 14.

13 Westmoreland, as shown in the documentary: "I can't remember figures like that. You've . . . done some research. I haven't done any research. I'm just . . . reflecting on my memory." In a 10 February 1982 letter to the editor of *Broadcasting Magazine*, Westmoreland said that, "Having been taken by surprise and having come unprepared for the grilling on statistics that were 14 years old, I made some unfortunate slips." Letter in Box 19, WP-SCL.

14 Routing Slip with Attachment, Hannah M. Zeidlik, OCMH, 29 October 1984.

15 Vince Demma, telephone interview, 29 August 2005.

16 As it was characterized in the Benjamin Report, about which more below.

17 CBS Reports, "The Uncounted Enemy," 23 January 1982, Transcript, p. 3.

18 McChristian interview, 26 April 1999.

19 As quoted in Cubbage, "Westmoreland vs. CBS," p. 46.

20 As quoted in Kowet, *Matter of Honor,* p. 167.

21 Westmoreland vs. CBS Microfiche File, p. 35233, USAWC Library.

22 Kaiser, "Westmoreland Denounces TV Program," *Washington Post* (27 January 1982).

23 David Zucchino, "Vietnam Cables Reveal Extent of Numbers Game," *Orange County Register* (8 December 1985). In February 1967 MACV J-2 published a *MACV Order of Battle Reference Manual,* a publication subsequently endorsed by members of the Intelligence Community attending the Honolulu Order of Battle Conference that same month. The manual defined Self-Defense Forces as "a VC para-military structure responsible for the defense of hamlet and village areas controlled by the VC. . . . Duties consist of conducting propaganda, constructing fortifications, and defending home areas." For Secret Self-Defense Forces, the definition was similar, "a clandestine VC organization which performs the same general functions in GVN-controlled areas. Their operations involve intelligence collection as well as sabotage and propaganda activities."

24 Westmoreland testimony, pp. 439–440, Westmoreland vs. CBS Microfiche File, USAWC Library.

25 Deposition of General William C. Westmoreland, 28 June 1983, p. 456, Westmoreland vs. CBS Microfiche File, USAWC Library.

26 Ibid.

27 Kowet, *Matter of Honor,* p. 184. Westmoreland's sister said that he had Alzheimer's for the last ten years of his life (thus from about 1995). This account suggests the possibility of a far earlier onset of that cruel disease.

28 Kowet, *Matter of Honor,* p. 239.

29 As quoted in Benjamin, *Fair Play*, p. 177.

30 Ibid., p. 114.

31 Ibid., p. 160.

32 Ibid., p. 12.

33 Affidavit of Joseph A. McChristian, General William C. Westmoreland v. CBS Inc. et al., 21 December 1983.

34 Westmoreland vs. CBS Microfiche File, p. 4507, USAWC Library.

35 As quoted in Eleanor Randolph, "Agonizing Self-Criticism May Embarrass CBS," *Washington Post* (25 November 1984).

36 Benjamin, *Fair Play*, pp. 122–123.

37 Palmer interview, 29 April 1995.

38 Halberstam, "The Call to Duty," *Parade* (c. 1985), Box 19, WPSCL. Halberstam, who could be devastating in his profiles, was in this case sympathetic. "I believe Westy is a sort of decent man, not smart, and he is politicized. He didn't understand the war," Halberstam told Burton Benjamin. *Fair Play*, p. 159.

39 Lt. Gen. Phillip B. Davidson Jr. interview, 27 March 1995.

40 Brewin and Shaw, *Vietnam on Trial*, p. 213.

41 Ibid.

42 A telling episode occurred when General McChristian was deposed at his home in Florida. "On the eighth day," he recalled, "Westmoreland's lawyers subpoenaed me and all of my telephone records. . . . Westmoreland's lawyers then went to lunch and didn't even look at the papers. But the CBS lawyers did not go to lunch and read all the papers." McChristian interview, 26 April 1999.

43 Benjamin, *Fair Play*, p. 186.

44 Deposition of General George A. Godding, General William C. Westmoreland vs. CBS Inc., et al., 19–20 April 1983.

45 Typed notes, George Crile, Joint Exhibit 20D, pp. 20522 and 20524. Westmoreland vs. CBS Microfiche File, USAWC Library.

46 Roth, *The Juror and the General*, p. 157.

47 Letter, Westmoreland to Secretary of the Army John O. Marsh Jr., 21 March 1982, Box 19, WPSCL.

48 Westmoreland, "Comment by General W. C. Westmoreland." Copy provided Sorley by Lt. Gen. Dave R. Palmer. Later (according to this same document), in what sounds very much like a threat, Westmoreland says he called McChristian again and "cautioned him that, in his own interest, he should realize that in the hands of clever lawyers we were both expendable."

49 Westmoreland vs. CBS Microfiche File, p. 29979, USAWC Library. During the CBS broadcast Westmoreland referred to his having given McChristian command of a corps at Fort Hood. McChristian commanded a division there, not a corps, and had been given that post by Army Chief of Staff General Harold K. Johnson, not Westmoreland, who as MACV commander had no authority to assign officers anywhere outside his own organization.

50 As quoted in Benjamin, *Fair Play*, p. 69.

51 McChristian testimony on direct examination, pp. 9028–9029, Westmoreland vs. CBS Microfiche File, USAWC Library.

52 Ibid.

53 Westmoreland, *A Soldier Reports*, p. 245.

54 McChristian testimony on direct examination, p. 9027, Westmoreland vs. CBS Microfiche File, USAWC Library.

55 CBS Reports, "The Uncounted Enemy," 23 January 1982, Transcript, p. 6.

56 Westmoreland Deposition, 4 April 1983, Westmoreland vs. CBS Microfiche File, p. 103, USAWC Library. During Westmoreland's testimony in the libel trial, Boies ascertained that there was no mention in Westmoreland's dictated history notes of the meeting at which McChristian brought in the increased strength estimates. He asked Westmoreland about the omission, to which Westmoreland responded that the meeting was "inconsequential" and "insignificant," adding, "Well, I didn't have everything in my history notes." Boies noted tellingly that Westmoreland had put in those notes such things as referring a visiting congressional staff person to his surgeon for treatment of a head cold. He might well have added the numerous times Westmoreland wrote in the notes about his putting on or wearing a white uniform, playing tennis, wiring ahead to arrange for water skiing, LBJ's attending a sheep auction, and other such pedestrian matters. Omission of the McChristian meeting appeared in contrast to be deliberate and inappropriate.

57 McChristian, Westmoreland vs. CBS Microfiche File, p. 9030, USAWC Library.

58 McChristian interview, 26 April 1999.

59 Roth, *The Juror and the General*, pp. 127, 128. "Years later," wrote Harold Ford, "Graham admitted to George Allen that 'of course' he had not believed MACV's 300,000 figure but had defended it because it was 'the command position.'" Ford, *CIA and the Vietnam Policymakers*, p. 94. Discussing these matters in a 29 April 1995 interview, General Bruce Palmer Jr. commented that "Danny Graham was a fraud."

60 Stipulation of Dismissal, U.S. District Court, 17 February 1985, Westmoreland vs. CBS Microfiche File, USAWC Library.

61 Letter, Westmoreland to Editor, *American Spectator,* 19 February 1987, Box 19, WPSCL.

62 Speech, "David vs. Goliath," 1985, Box 49, WPSCL.

63 Editorial, "A General Surrenders," *New York Times* (19 February 1982).

64 Roth, *The Juror and the General*, p. 299.

65 As quoted in Brewin and Shaw, *Vietnam on Trial*, p. 355.

66 As shown in A&E "Biography: General Westmoreland," 28 February 1995. Transcript in Box 58, WPSCL.

67 George Crile III, "The Westmoreland Episode: Who Won?" International Platform Association Convention, c. 1987, Box 19, WPSCL.

68 Westmoreland, "A Case for Press Responsibility," *Mercer Law Review* (Spring 1987), p. 775.

69 Wallace, *Between You and Me*, p. 203.

30. DUSK

1 Beschloss, "Not the President's Men," *New York Times Book Review* (5 August 2007), p. 10.

2 Westmoreland, *A Soldier Reports*, p. 565 of the 1980 Dell paperback edition.
3 Atkinson, *Long Gray Line*, p. 408.
4 Ward Just interviewed James Ford, the Cadet Chaplain at West Point, who had been chosen for that post when Westmoreland was Superintendent. "There are two boats which are sinking today," said Ford in 1970, "the military and the church, and I have got a foot in both." Just asked him whether there would be a statue of Westmoreland at the Military Academy. "'No, probably not,' Ford said sadly." *Military Men*, pp. 43, 45.

EPILOGUE

1 Quotation is from Alan Levy, "Two Stars on the Fly," *Courier-Journal Magazine* (28 February 1960), USMA Archives.
2 As quoted in ibid.

SELECTED BIBLIOGRAPHY

BOOKS

Adams, Sam. *War of Numbers: An Intelligence Memoir.* South Royalton, Vt.: Steer-forth Press, 1994.

Adler, Renata. *Reckless Disregard: Westmoreland v. CBS et al.* New York: Knopf, 1986.

Allen, George W. *None So Blind: A Personal Account of the Intelligence Failure in Vietnam.* Chicago: Ivan R. Dee, 2001.

Ambrose, Stephen E. *Nixon: The Triumph of a Politician, 1962–1972.* New York: Simon & Schuster, 1989.

Anderson, David L., ed. *Facing My Lai: Moving Beyond the Massacre.* Lawrence: University Press of Kansas, 1998.

———. *Shadow on the White House: Presidents and the Vietnam War, 1945–1975.* Lawrence: University Press of Kansas, 1993.

Anderson, Jack, and Dale Van Atta. *Stormin' Norman: An American Hero.* New York: Zebra Books, 1991.

Appy, Christian G., ed. *Patriots: The Vietnam War Remembered from All Sides.* New York: Viking, 2003.

Atkinson, Rick. *The Long Gray Line.* Boston: Houghton Mifflin, 1989.

Baskir, Lawrence M., and William A. Strauss. *Chance and Circumstance: The Draft, the War and the Vietnam Generation.* New York: Vintage Books, 1978.

Bass, Jack, and Marilyn W. Thompson. *Ol' Strom: An Unauthorized Biography of Strom Thurmond.* Atlanta: Longstreet Press, 1998.

Beech, Keyes. *Not Without the Americans: A Personal History.* Garden City, N.Y.: Doubleday, 1971.

Benjamin, Burton. *Fair Play: CBS, General Westmoreland, and How a Television Documentary Went Wrong.* New York: Harper & Row, 1988.

Bergerud, Eric M. *Red Thunder, Tropic Lightning: The World of a Combat Division in Vietnam.* Boulder, Colo.: Westview, 1993.

Berman, Larry. *Lyndon Johnson's War: The Road to Stalemate in Vietnam.* New York: W. W. Norton, 1989.

Beschloss, Michael R., ed. *Taking Charge: The Johnson White House Tapes, 1963–1964.* New York: Simon & Schuster, 1997.

Betros, Lance. *West Point: Two Centuries and Beyond.* Abilene, Tex.: McWhiney Foundation Press, 2004.

Bilton, Michael, and Kevin Sim. *Four Hours in My Lai.* New York: Viking, 1992.

Blair, Anne. *There to the Bitter End: Ted Serong in Vietnam.* Crows Nest, Australia: Allen & Unwin, 2001.

Boettcher, Thomas D. *Vietnam: The Valor and the Sorrow.* Boston: Little, Brown, 1985.

Bonds, Ray, ed. *The Vietnam War: The Illustrated History of the Conflict in Southeast Asia.* New York: Crown, 1979.

Braestrup, Peter. *Big Story: How the American Press and Television Reported and Interpreted the Crisis of Tet 1968 in Vietnam and Washington,* 2 vols. Boulder, Colo.: Westview, 1977.

Brandon, Henry. *Anatomy of Error: The Inside Story of the Asian War on the Potomac, 1954–1969.* Boston: Gambit, 1969.

Brewin, Bob, and Sydney Shaw. *Vietnam on Trial: Westmoreland vs. CBS.* New York: Atheneum, 1987.

Bui Diem with David Chanoff. *In the Jaws of History.* Boston: Houghton Mifflin, 1987.

Bui Tin. *Following Ho Chi Minh: The Memoirs of a North Vietnamese Colonel.* Honolulu: University of Hawaii Press, 1995.

Carland, John M. *Stemming the Tide: May 1965 to October 1966.* Washington, D.C.: U.S. Army Center of Military History, 2000.

Charlton, Michael, and Anthony Moncrieff. *Many Reasons Why: The American Involvement in Vietnam.* London: Scolar Press, 1978.

Clarke, Jeffrey J. *Advice and Support: The Final Years, 1965–1973.* Washington, D.C.: U.S. Army Center of Military History, 1988.

Clausewitz, Carl von. *On War,* ed. Michael Howard and Peter Paret. Princeton: Princeton University Press, 1976.

Clifford, Clark, with Richard Holbrooke. *Counsel to the President: A Memoir.* New York: Random House, 1991.

Coleman, J. D. *Pleiku: The Dawn of Helicopter Warfare in Vietnam.* New York: St. Martin's Press, 1988.

Cooper, Chester L. *The Lost Crusade: America in Vietnam.* New York: Dodd, Mead & Company, 1970.

Cosmas, Graham A. *MACV: The Joint Command in the Years of Withdrawal, 1968–1973.* Washington, D.C.: U.S. Army Center of Military History, 2007.

Costa, James M. *Diamond in the Rough.* Victoria, B.C.: Trafford Publishing, 2003.

Crackel, Theodore J. *West Point: A Bicentennial History.* Lawrence: University Press of Kansas, 2002.

Creighton, Neal. *A Different Path: The Story of an Army Family.* Privately printed: Xlibris, 2008.

Davidson, Garrison H. *Grandpa Gar: The Saga of One Soldier as Told to His Grandchildren.* Privately printed, 1974.

Davidson, Lieutenant General Phillip B. *Vietnam at War: The History, 1946–1975.* Novato, Calif.: Presidio Press, 1988.

DeForest, Orrin. *Slow Burn: The Rise and Bitter Fall of American Intelligence in Vietnam.* New York: Simon & Schuster, 1990.

DePuy, William E. *Changing an Army: An Oral History of General William E. DePuy, USA Retired.* Carlisle Barracks, Pa.: U.S. Army Military History Institute, [1979].

Dixon, Norman F. *On the Psychology of Military Incompetence.* New York: Basic Books, 1976.

Dorland, Gil. *Legacy of Discord: Voices of the Vietnam War Era.* Washington, D.C.: Brassey's, 2001.

Elliott, Mai. *RAND in Southeast Asia: A History of the Vietnam War Era.* Santa Monica, Calif.: RAND Corporation, 2010.

Ellsberg, Daniel. *Secrets: A Memoir of Vietnam and the Pentagon Papers.* New York: Viking, 2002.

Enthoven, Alain C., and K. Wayne Smith. *How Much Is Enough? Shaping the Defense Program, 1961–1969.* New York: Harper & Row, 1971.

Fallaci, Oriana. *Interview with History.* New York: Liveright, 1976.

Fallows, James. *National Defense.* New York: Vintage, 1981.

Flanagan, E. M., Jr. *The Rakkasans: The Combat History of the 187th Airborne Infantry.* Novato, Calif.: Presidio Press, 1997.

Ford, Harold P. *CIA and the Vietnam Policymakers: Three Episodes 1962–1968.* Washington, D.C.: History Staff, Center for the Study of Intelligence, Central Intelligence Agency, 1998.

Furgurson, Ernest B. *Westmoreland: The Inevitable General.* Boston: Little, Brown, 1968.

Garland, Albert N., ed. *A Distant Challenge: The U.S. Infantryman in Vietnam, 1967–1972.* Nashville: Battery Press, 1983.

Gelb, Leslie H., with Richard K. Betts. *The Irony of Vietnam: The System Worked.* Washington, D.C.: Brookings Institution, 1979.

Giap, Vo Nguyen. *Big Victory, Great Task.* New York: Praeger, 1968.

Glenn, Russell. *Reading Athena's Dance Card: Men Against Fire in Vietnam.* Annapolis: Naval Institute Press, 2000.

Goldstein, Gordon M. *Lessons in Disaster: McGeorge Bundy and the Path to War in Vietnam.* New York: Times Books, 2008.

Goldstein, Joseph, Burke Marshall, and Jack Schwartz. *The My Lai Massacre and Its Cover-Up: Beyond the Reach of Law?* New York: Free Press, 1976.

Goodwin, Richard N. *Remembering America: A Voice from the Sixties.* Boston: Little, Brown, 1988.

Griffith, Robert K., Jr. *The U.S. Army's Transition to the All-Volunteer Force 1968–1974.* Washington, D.C.: U.S. Army Center of Military History, 1997.

Haig, Alexander M., Jr., with Charles McCarry. *Inner Circles: How America Changed the World: A Memoir.* New York: Warner Books, 1992.

Halberstam, David. *The Best and the Brightest.* New York: Random House, 1972.

Hammel, Eric. *The Assault on Khe Sanh: An Oral History.* New York: Warner Books, 1989.

———. *Khe Sanh: Siege in the Clouds, an Oral History.* New York: Crown, 1989.

Head, William, and Lawrence E. Grinter, eds. *Looking Back on the Vietnam War.* Westport, Conn.: Praeger, 1993.

Helms, Richard, with William Hood. *A Look Over My Shoulder: A Life in the Central Intelligence Agency.* New York: Random House, 2003.

Hendrickson, Paul. *The Living and the Dead: Robert McNamara and Five Lives of a Lost War.* New York: Knopf, 1996.

Herring, George C. *LBJ and Vietnam: A Different Kind of War.* Austin: University of Texas Press, 1994.

Hiam, C. Michael. *Who the Hell Are We Fighting? The Story of Sam Adams and the Vietnam Intelligence Wars.* Hanover, N.H.: Steerforth Press, 2006.

Hoopes, Townsend. *The Limits of Intervention.* New York: David McKay, 1969.

Hope, Bob. *Five Women I Love: Bob Hope's Vietnam Story.* Garden City, N.Y.: Doubleday, 1966.

Hunt, Richard A. *Pacification: The American Struggle for Vietnam's Hearts and Minds.* Boulder, Colo.: Westview Press, 1995.

Isaacson, Walter, and Evan Thomas. *The Wise Men: Six Friends and the World They Made: Acheson, Bohlen, Harriman, Kennan, Lovett, McCloy.* New York: Simon & Schuster, 1986.

Joes, Anthony James. *America and Guerrilla Warfare.* Lexington: University Press of Kentucky, 2000.

Johnson, Lady Bird [Claudia Alta Taylor Johnson]. *A White House Diary.* New York: Holt, Rinehart and Winston, 1970.

Johnson, Lyndon Baines. *The Vantage Point: Perspectives of the Presidency, 1963–1969.* New York: Holt, Rinehart and Winston, 1971.

Jones, Bruce E. *War Without Windows: A True Account of a Young Army Officer Trapped in an Intelligence Cover-Up in Vietnam.* New York: Vanguard Press, 1987.

Just, Ward. *Military Men.* New York: Alfred A. Knopf, 1970.

———. *To What End: Report from Vietnam.* New York: Public Affairs, 2000. Reprint of a 1968 work with new foreword by the author.

Kagan, Donald. *Thucydides: The Reinvention of History.* New York: Viking, 2009.

Kahin, George McT. *Intervention: How America Became Involved in Vietnam.* New York: Knopf, 1986.

Karnow, Stanley. *Vietnam: A History.* New York: Viking, 1983.

Kearns, Doris. *Lyndon Johnson and the American Dream.* New York: Harper & Row, 1976.

Keeley, John B., ed. *The All-Volunteer Force and American Society.* Charlottesville: University Press of Virginia, 1978.

Kennedy, Robert F. *To Seek a Newer World.* Garden City, N.Y.: Doubleday, 1967.

Kinnard, Douglas. *The Certain Trumpet: Maxwell Taylor & the American Experience in Vietnam.* Washington, D.C.: Brassey's, 1991.

———. *The War Managers.* Hanover, N.H.: University Press of New England, 1977.

Kissinger, Henry. *Ending the Vietnam War: A History of America's Involvement in and Extrication from the Vietnam War.* New York: Simon & Schuster, 2003.

———. *White House Years.* Boston: Little, Brown, 1979.

Korb, Lawrence J. *The Fall and Rise of the Pentagon: American Defense Policies in the 1970s.* Westport, Conn.: Greenwood, 1979.

———. *The Joint Chiefs of Staff: The First Twenty-Five Years.* Bloomington: Indiana University Press, 1976.

Kowet, Don. *A Matter of Honor.* New York: Macmillan, 1984.

Krepinevich, Andrew F., Jr. *The Army and Vietnam.* Baltimore: Johns Hopkins University Press, 1986.

Krulak, Victor H. *First to Fight: An Inside View of the U.S. Marine Corps.* Annapolis: Naval Institute Press, 1984.

Ky, Nguyen Cao. *Buddha's Child: My Fight to Save Vietnam.* New York: St. Martin's Press, 2002.

———. *Twenty Years and Twenty Days.* New York: Stein and Day, 1976.

Lehrack, Otto J., ed. *No Shining Armor: The Marines at War in Vietnam: An Oral History.* Lawrence: University Press of Kansas, 1992.

Lewy, Guenter. *America in Vietnam.* New York: Oxford University Press, 1978.

Linn, Brian McAllister. *The Echo of Battle: The Army's Way of War.* Cambridge, Mass.: Harvard University Press, 2007.

Mangold, Tom, and John Penycate. *The Tunnels of Cu Chi.* New York: Berkley Books, 1986.

Mann, Robert. *A Grand Delusion: America's Descent into Vietnam.* New York: Basic Books, 2001.

Maraniss, David. *They Marched into Sunlight.* New York: Simon & Schuster, 2003.

Maurer, Harry. *Strange Ground: Americans in Vietnam 1945–1975: An Oral History.* New York: Henry Holt, 1989.

McMaster, H. R. *Dereliction of Duty: Lyndon Johnson, Robert McNamara, the Joint Chiefs of Staff, and the Lies That Led to Vietnam.* New York: HarperCollins, 1997.

McNamara, Robert S., et al. *Argument Without End: In Search of Answers to the Vietnam Tragedy.* New York: Public Affairs, 1999.

McNamara, Robert S., with Brian VanDeMark. *In Retrospect: The Tragedy and Lessons of Vietnam.* New York: Times Books, 1995.

Military History Institute of Vietnam. *The Official History of the People's Army of Vietnam, 1954–1975.* Hanoi: People's Army Publishing House, 1994. Published in English as *Victory in Vietnam,* trans. Merle Pribbenow. Lawrence: University Press of Kansas, 2002.

Mittelman, Joseph B. *Eight Stars to Victory: A History of the Veteran Ninth U.S. Infantry Division.* [Washington, D.C.:] Ninth Infantry Division Association, 1948.

———. *Hold Fast.* Munich: Ninth Infantry Division, 1945.

Moore, Harold G., and Joseph L. Galloway. *We Are Soldiers Still: A Journey Back to the Battlefields of Vietnam.* New York: HarperCollins, 2008.

———. *We Were Soldiers Once . . . and Young.* New York: Random House, 1992.

Moyar, Mark. *Phoenix and the Birds of Prey.* Annapolis: Naval Institute Press, 1997.

Murphy, Edward J. *Semper Fi Vietnam: From Da Nang to the DMZ, Marine Corps Campaigns, 1965–1975.* New York: Ballantine Books, 1997.

Nagl, John A. *Learning to Eat Soup with a Knife: Counterinsurgency Lessons from Malaya and Vietnam.* Chicago: University of Chicago Press, 2002.

Neese, Harvey, and John O'Donnell, eds. *Prelude to Tragedy: Vietnam, 1960–1965.* Annapolis: Naval Institute Press, 2001.

9th Infantry Division. *Hitler's Nemesis: The 9th Infantry Division.* Paris: *Stars & Stripes*, 1944–1945.

O'Ballance, Edgar. *The Wars in Vietnam, 1954–1973.* New York: Hippocrene Books, 1975.

Oberdorfer, Don. *Tet! The Story of a Battle and Its Historic Aftermath.* Garden City, N.Y.: Doubleday, 1971.

Palmer, General Bruce, Jr. *The 25-Year War: America's Military Role in Vietnam.* Lexington: University Press of Kentucky, 1984.

Palmer, Dave Richard. *Summons of the Trumpet: A History of the Vietnam War from a Military Man's Viewpoint.* San Rafael, Calif.: Presidio Press, 1978.

Parker, James E., Jr. *Last Man Out: A Personal Account of the Vietnam War.* New York: Ballantine Books, 1996.

Peers, W. R. *The My Lai Inquiry.* New York: Norton, 1979.

Perry, Mark. *Four Stars.* Boston: Houghton Mifflin, 1989.

Pettit, Clyde Edwin. *The Experts.* Secaucus, N.J.: Lyle Stuart, 1975.

Pham Kim Vinh. *In Their Defense: U.S. Soldiers in the Vietnam War.* Phoenix: Sphinx Publishing, 1985.

Phillips, William R. *Night of the Silver Stars: The Battle of Lang Vei.* Annapolis: Naval Institute Press, 1997.

Pimlott, John. *Vietnam: The Decisive Battles.* New York: Macmillan, 1990.

Pisor, Robert. *The End of the Line: The Siege of Khe Sanh.* New York: W. W. Norton, 1982.

Prados, John. *The Hidden History of the Vietnam War.* Chicago: Ivan R. Dee, 1995.

Randolph, Stephen P. *Powerful and Brutal Weapons: Nixon, Kissinger, and the Easter Offensive.* Cambridge, Mass.: Harvard University Press, 2007.

Roth, M. Patricia. *The Juror and the General: An Eyewitness Account of the Libel Trial of the Century: Westmoreland vs. CBS.* New York: William Morrow, 1986.

Santoli, Al. *Leading the Way: How Vietnam Veterans Rebuilt the U.S. Military: An Oral History.* New York: Ballantine Books, 1993.

Schandler, Herbert Y. *The Unmaking of a President: Lyndon Johnson and Vietnam.* Princeton: Princeton University Press, 1977.

Schell, Jonathan. *The Real War: The Classic Reporting on the Vietnam War with a New Essay.* New York: Pantheon Books, 1987.

Schwarzkopf, H. Norman. *It Doesn't Take a Hero: The Autobiography.* New York: Bantam Books, 1992.

Shapley, Deborah. *Promise and Power: The Life and Times of Robert McNamara.* Boston: Little, Brown, 1993.

Sharp, U. S. Grant. *Strategy for Defeat: Vietnam in Retrospect.* San Rafael, Calif.: Presidio Press, 1978.

Showalter, Dennis E., and John G. Albert, eds. *An American Dilemma: Vietnam, 1964–1973.* Chicago: Imprint Publications, 1993.

Shulimson, Jack. *U.S. Marines in Vietnam: An Expanding War, 1966.* Washington, D.C.: History and Museums Division, Headquarters, U.S. Marine Corps, 1982.

——— et al. *U.S. Marines in Vietnam: The Defining Year, 1968.* Washington, D.C.: History and Museums Division, Headquarters, U.S. Marine Corps, 1997.

———— and Major Charles M. Johnson. *U.S. Marines in Vietnam: The Landing and the Buildup 1965.* Washington, D.C.: History and Museums Division, Headquarters, U.S. Marine Corps, 1978.

Simpson, Charles M., III. *Inside the Green Berets: The First Thirty Years: A History of the U.S. Army Special Forces.* New York: Berkley Books, 1984.

Smith, Charles R. *U.S. Marines in Vietnam: High Mobility and Standdown 1969.* Washington, D.C.: U.S. Marine Corps, 1988.

Sorley, Lewis. *A Better War: The Unexamined Victories and Final Tragedy of America's Last Years in Vietnam.* New York: Harcourt Brace, 1999.

————. *Honorable Warrior: General Harold K. Johnson and the Ethics of Command.* Lawrence: University Press of Kansas, 1998.

————, ed. *Press On! Selected Works of General Donn A. Starry,* 2 vols. Fort Leavenworth, Kans.: Combat Studies Institute Press, 2009.

————, ed. *Vietnam Chronicles: The Abrams Tapes, 1968–1972.* Lubbock: Texas Tech University Press, 2004.

Spiller, Roger. *An Instinct for War: Scenes from the Battlefields of History.* Cambridge, Mass.: Harvard University Press, 2005.

————. *In the School of War.* Lincoln: University of Nebraska Press, 2010.

Stanton, Shelby L. *The 1st Cav in Vietnam: Anatomy of a Division.* Novato, Calif.: Presidio Press, 1999. Paperback. First published 1987.

Starry, Donn A. *Mounted Combat in Vietnam.* Washington, D.C.: Department of the Army, 1978. Reprinted commercially as *Armored Combat in Vietnam* (New York: Arno Press, 1980).

Taylor, John M. *General Maxwell Taylor: The Sword and the Pen.* New York: Doubleday, 1989.

Taylor, Maxwell D. *Swords and Plowshares: A Memoir.* New York: W. W. Norton, 1972.

Thi, Lam Quang. *The Twenty-Five Year Century: A South Vietnamese General Remembers.* Denton: University of North Texas Press, 2001.

Thompson, Annis G. *The Greatest Airlift: The Story of Combat Cargo.* Tokyo: Dai-Nippon Printing, 1954.

Tucker, Spencer C., ed. *Encyclopedia of the Vietnam War: A Political, Social, and Military History.* Santa Barbara, Calif.: ABC-CLIO, 1998.

Van Atta, Dale. *With Honor: Melvin Laird in War, Peace, and Politics.* Madison: University of Wisconsin Press, 2008.

Wallace, Mike. *Between You and Me: A Memoir.* New York: Hyperion, 2005.

Waterhouse, Fred J., with Colonel Bill Weber. *The Rakkasans,* 2nd ed. Paducah, Ky.: Turner Publishing Company, 1997.

Wells, Tom. *The War Within: America's Battle over Vietnam.* Berkeley: University of California Press, 1994.

Westmoreland, William C. *A Soldier Reports.* Garden City, N.Y.: Doubleday, 1976.

Whitehouse, Charles S. *Then and Now: Memoirs of C. S. Whitehouse.* Marshall, Va.: privately printed, 2001.

Willenson, Kim, et al. *The Bad War: An Oral History of the Vietnam War.* New York: New American Library, 1987.

Zaffiri, Samuel. *Westmoreland: A Biography of General William C. Westmoreland.* New York: William Morrow, 1994.

BOOK CHAPTERS

MacDonald, Charles B. "The Military Build-Up in North and South," in Ray Bonds, ed., *The Vietnam War: The Illustrated History of the Conflict in Southeast Asia.* New York: Crown, 1979.

Moskos, Charles C., Jr. "The Enlisted Ranks," in John B. Keeley, ed., *The All-Volunteer Force and American Society.* Charlottesville: University Press of Virginia, 1978.

ARTICLES

Atkeson, Edward B. "Vietnam Examination Is Not a Closed Book," *Army* (August 1995), pp. 11–16.

Boroff, David. "West Point: Ancient Incubator for a New Breed," *Harper's* (December 1962), pp. 51–59.

Braestrup, Peter. "Covering the Vietnam War," *Nieman Reports* (December 1969), pp. 8–13.

———. "'The Uncounted Enemy: A Vietnam Deception': A Dissenting View," *Washington Journalism Review* (April 1982), pp. 46–48.

———. "Vietnam as History," *Wilson Quarterly* (Spring 1978), pp. 178–187.

Brower, Charles F., IV. "Strategic Reassessment in Vietnam: The Westmoreland 'Alternate Strategy' of 1967–1968," *Naval War College Review* (Spring 1991), pp. 20–51.

Buckingham, Clay T. "Ethics and the Senior Officer: Institutional Tensions," *Parameters* (Autumn 1985), pp. 23–32.

Bui Tin. "How North Vietnam Won the War," *Wall Street Journal* (3 August 1995). Interview by Stephen Young.

Butler, Jack R. "The All-Volunteer Armed Force—Its Feasibility and Implications," *Parameters* (Vol. II, No. 1, 1972), pp. 17–29.

Cole, Bernard D. "A Noglow in Vietnam, 1968: Air Power at the Battle of Khe Sanh," *Journal of Military History* (January 2000), pp. 141–158.

DePuy, General William E. "Our Experience in Vietnam," *Army* (June 1987), pp. 28–41.

———. "Vietnam: What We Might Have Done and Why We Didn't Do It," *Army* (February 1986), pp. 22–40.

Flanagan, E. M., Jr., "Before the Battle: Commanding: A New Perspective," *Army* (July 1993), pp. 49–50.

Henry, John B., II. "February, 1968," *Foreign Policy* (Fall 1971), pp. 2–33.

Lewy, Guenter. "Vietnam: New Light on the Question of American Guilt," *Commentary* (February 1978), pp. 29–47.

Lock-Pullan, Richard. "'An Inward Looking Time': The United States Army, 1973–1976," *Journal of Military History* (April 2003), pp. 483–511.

McGinniss, Joe. "Winning Hearts and Minds in South Carolina," *Harper's* (April 1974), pp. 65–72.

McMaster, H. R. "Dereliction of Duty," *Air Force Magazine* (January 1998), pp. 70–73.

Millett, Allan R. "Why the Army and the Marine Corps Should Be Friends," *Parameters* (Winter 1994–1995), pp. 30–40.

Mueller, John E. "The Search for the 'Breaking Point' in Vietnam," *International Studies Quarterly* (December 1980), pp. 497–519.

North, Don. "VC Assault on the U.S. Embassy," *Vietnam* (February 2000), pp. 38–47, 72.

Palmer, Laura. "The General, at Ease: An Interview with Westmoreland," *MHQ: The Quarterly Journal of Military History* (Autumn 1988), pp. 30–35.

Perry, Mark. "The Resurrection of John Paul Vann," *Veteran* (July 1988), pp. 31–32.

Pribbenow, Merle L., II. "General Vo Nguyen Giap and the Mysterious Evolution of the Plan for the 1968 Tet Offensive," *Journal of Vietnamese Studies* (Vol. 3, Issue 2), pp. 1–33.

Raughter, John. "The Inevitable General," *American Legion Magazine* (September 2003), pp. 16–18.

Shaw, Donald P., and Zane E. Finkelstein. "Westmoreland vs. CBS," *Commentary* (August 1984), pp. 31–37.

Shulimson, Jack, and Edward F. Wells. "First In, First Out," *Marine Corps Gazette* (January 1984), pp. 36–46.

Summers, Harry G., Jr. "Troops to Equal Any: Interview of General Fred C. Weyand," *Vietnam* (February 1988), pp. 21–25.

Westmoreland, William C. "A Case for Press Responsibility," *Mercer Law Review* (Spring 1987), pp. 771–777.

———. "Riding to Battle," *Army* (April 1993), pp. 43–44.

———. "Vietnam in Perspective," *Military Review* (January 1979), pp. 34–43.

Wieland, Valerie. "The Inevitable General," *Vietnam* (December 2003), pp. 34–41, 64. Interview of General William C. Westmoreland.

PAMPHLETS AND STUDIES

Collins, James Lawton, Jr. *The Development and Training of the South Vietnamese Army, 1950–1972.* Vietnam Studies. Washington, D.C.: Department of the Army, 1975.

Doughty, Robert A. *The Evolution of US Army Tactical Doctrine, 1946–76.* Leavenworth Papers #1. Fort Leavenworth, Kans.: Combat Studies Institute, U.S. Army Command and General Staff College, August 1979.

Hinh, Nguyen Duy. *Vietnamization and the Cease-Fire.* Indochina Monograph. Washington, D.C.: U.S. Army Center of Military History, 1980.

Joint Chiefs of Staff, Organization of. *Intensification of the Military Operations in Vietnam: Concept and Appraisal.* Report of Ad Hoc Study Group. Washington, D.C.: OJCS, 14 July 1965.

Khuyen, Dong Van. *RVNAF Logistics.* Indochina Monograph. Washington, D.C.: U.S. Army Center of Military History, 1980.

Lung, Hoang Ngoc. *The General Offensives of 1968–69.* Indochina Monograph. Washington, D.C.: U.S. Army Center of Military History, 1981.

———. *Strategy and Tactics.* Indochina Monograph. Washington, D.C.: U.S. Army Center of Military History, 1980.

McChristian, Joseph A. *The Role of Intelligence, 1965–1967.* Vietnam Studies. Washington, D.C.: Department of the Army, 1974.

Palmer, Bruce, Jr. "US Intelligence and Vietnam," *Studies in Intelligence* (Special Issue). Washington, D.C.: Central Intelligence Agency, 1984.

Palmer, Dave Richard. *Readings in Current Military History*. West Point, N.Y.: USMA Department of Military Art and Engineering, 1969.

Pike, Douglas. "A Look Back at the Vietnam War: The View from Hanoi." Paper written for the Vietnam War Symposium, Wilson Center, Washington, D.C., 7–8 January 1983.

———. *Vietnam War: View from the Other Side*. Saigon: U.S. Information Service, December 1967.

Rosson, W. B. *Assessment of Influence Exerted on Military Operations by Other Than Military Considerations*. Headquarters, U.S. Army Pacific, 1970. Published Washington, D.C.: U.S. Army Center of Military History, 1993.

Schlight, John, ed. *Second Indochina War Symposium: Papers and Commentary*. Washington, D.C.: U.S. Army Center of Military History, 1986.

Tolson, John J. *Airmobility 1961–1971*. Vietnam Studies. Washington, D.C.: Department of the Army, 1973.

Truong, Ngo Quang. *RVNAF and US Operational Cooperation and Coordination*. Indochina Monograph. Washington, D.C.: U.S. Army Center of Military History, 1980.

———. *Territorial Forces*. Indochina Monograph. Washington, D.C.: U.S. Army Center of Military History, 1978.

Vien, Cao Van, and Dong Van Khuyen. *Reflections on the Vietnam War*. Indochina Monograph. Washington, D.C.: U.S. Army Center of Military History, 1980.

DOCUMENTS

BDM Corporation. *A Study of Strategic Lessons Learned in Vietnam: Omnibus Executive Summary*. McLean, Va.: BDM Corporation, 1980.

Department of the Army. *Historical Summary: Fiscal Year 1970*. Washington, D.C.: U.S. Army Center of Military History, 1973.

Gibbons, William Conrad, ed. *The U.S. Government and the Vietnam War: Executive and Legislative Roles and Relationships*, Parts I–IV. Washington, D.C.: Congressional Research Service, Library of Congress, 1984–1994.

Gittinger, Ted, ed. *The Johnson Years: A Vietnam Roundtable*. Austin, Tex.: Lyndon Baines Johnson Library, 1993.

Hitler's Nemesis: The 9th Infantry Division. Paris: *Stars & Stripes*, 1944–1945.

Joint Chiefs of Staff. *The History of the Joint Chiefs of Staff: The Joint Chiefs of Staff and the War in Vietnam, 1960–1968*, Parts I–III. Washington, D.C.: Historical Division, Joint Secretariat, Joint Chiefs of Staff, various dates.

———. *The History of the Joint Chiefs of Staff: The Joint Chiefs of Staff and the War in Vietnam, 1969–1970*. Washington, D.C.: Historical Division, Joint Secretariat, Joint Chiefs of Staff, 26 April 1976.

MacDonald, Charles B. *An Outline History of U.S. Policy toward Vietnam*. Washington, D.C.: U.S. Army Center of Military History, April 1978.

Military Assistance Command, Vietnam. *MACV Command History 1967*. Headquarters, USMACV, 16 September 1968.

New York Times. *The Pentagon Papers as Published by The New York Times*. New York: Quadrangle Books, 1971.

Official Register of the Officers and Cadets, United States Military Academy for the Academic Year Ending June 30, 1933. West Point, N.Y.: USMA Printing Office, 1933. Also for other years.

The Pentagon Papers: The Defense Department History of United States Decisionmaking on Vietnam. Senator Gravel Edition. 5 vols. Boston: Beacon Press, 1971.

Pike, Douglas, ed. *The Bunker Papers: Reports to the President from Vietnam, 1967–1973.* 3 vols. Berkeley: Institute of East Asian Studies, University of California, 1990.

PROVN Study (Program for the Pacification and Long-Term Development of Vietnam). Washington, D.C.: Department of the Army, 1 March 1966.

Royal United Service Institution. *Lessons from the Vietnam War.* Whitehall: RUSI, 12 February 1969.

Sharp, U.S.G., and W. C. Westmoreland. *Report on the War in Vietnam.* Washington, D.C.: U.S. Government Printing Office, 1969.

U.S. Army Center of Military History. *Department of the Army Historical Summary, Fiscal Year 1972.* Washington, D.C.: CMH, 1974. And other years.

U.S. Congress. House. Committee on Armed Services. *Report of the Special Subcommittee on the M-16 Rifle Program,* 19 October 1967. Washington, D.C.: U.S. Government Printing Office, 1967.

———. *Report by the Special Subcommittee on the M-16 Rifle Program,* 26 September 1968. Washington, D.C.: U.S. Government Printing Office, 1968.

U.S. Congress. Senate. Committee on Foreign Relations. *Vietnam: Policy and Prospects, 1970.* Washington, D.C.: Government Printing Office, 1970.

U.S. Military Academy. *Official Register of the Officers and Cadets, United States Military Academy for 1936.* West Point: USMA, 1936. Also other years.

U.S. Military Assistance Command, Vietnam. *MACV Order of Battle Reference Manual.* Headquarters, USMACV OACSI, 12 February 1967.

Vietnam, A Documentary Collection: Westmoreland v. CBS. New York: Clearwater Publishing, 1985. Microform.

Westmoreland, William C. "Plaintiff General William C. Westmoreland's Memorandum of Law in Opposition to Defendant CBS's Motion to Dismiss and for Summary Judgment." United States District Court, Southern District of New York, 20 July 1984.

———. *Report of the Chief of Staff of the United States Army, 1 July 1968 to 30 June 1972.* Washington, D.C.: Department of the Army, 1977.

West Point Association of Graduates. *2000 Register of Graduates and Former Cadets.* West Point, N.Y.: Association of Graduates, 2000. Also other years.

West Point Class of 1936. *The Howitzer.* West Point, N.Y.: Class of 1936, 1936. Also other years.

UNPUBLISHED MATERIALS

Benjamin, Burton. "An Examination," 8 July 1982. Internal CBS evaluation of the CBS Reports program "The Uncounted Enemy: A Vietnam Deception."

Bernard, Carl. "The War in Vietnam: Observations and Reflections of a Province Senior Advisor." Mimeograph, October 1969.

Bradford, Zeb B., Jr. "A Memoir of the Tet Offensive." Draft manuscript.

————. "Perfume River: A Vietnam Memoir." Unpublished manuscript, 13 May 1997.

Bunker, Ellsworth. "Lost Victory." Unpublished draft book manuscript.

————. Oral history transcript. Interviewed by Stephen Young.

Chaisson, John R. Letters to his wife. Hoover Institution Archives, Stanford University.

Cubbage, T. L., II. "Westmoreland vs. CBS: Was Intelligence Corrupted by Policy Demands?" Paper presented at the Second USAWC Conference on Intelligence and Military Operations, Carlisle Barracks, Pa., May 1987.

Hamilton, Andrew. "The 'Uncounted Enemy' in Vietnam: A Reconsideration."

Kinnard, Douglas. "Adventures in Two Worlds." Unpublished draft memoir.

————. "The War Managers Revisited." Unpublished book chapter.

MacDonald, Charles. "Contrasts in Command: Vietnam—Westmoreland and Abrams," U.S. Army War College Remarks, 17 May 1976. Recording, Audio-Visual Collection, U.S. Army Military History Institute, Carlisle Barracks, Pa.

Nye, Colonel Roger H. "The Inadvertent Demise of the Traditional Academy, 1945–1995."

Rosson, William B. "Four Periods of American Involvement in Vietnam: Development and Implementation of Policy, Strategy and Programs, Described and Analyzed on the Basis of Service Experience at Progressively Senior Levels." Unpublished doctoral dissertation, New College, Oxford Trinity, 1979.

Salet, Eugene. Unpublished memoir, MHI.

Scoville, Thomas W., and Charles B. MacDonald. "Interview with Ambassador Robert W. Komer and Colonel Robert M. Montague," 6–7 November 1969. Typescript ms. Office of the Chief of Military History, Department of the Army.

Todd, W. Russell "Vietnam."

Westmoreland, William C. "The Origins of the Post-Tet 1968 Plans for Additional American Forces in RVN." Unpublished paper, April 1970. White Paper prepared for Westmoreland by Major Paul L. Miles Jr.

Weyand, Frederick C. "Notes on Vietnam Experience." Typescript.

Williamson, Ellis W. Unpublished memoir.

OTHER SOURCES

Interviews Conducted

Honorable Stephen Ailes; Lieutenant General Elmer H. Almquist Jr.; Brigadier General James L. Anderson; Colonel Reamer W. Argo; Major General Gordon H. Austin; Lieutenant General Charles W. Bagnal; Patsy Bagnal; Lee Bandy; Colonel James Barbara; Major General Edward Bautz; Lieutenant General Robert Beckel (USAF); Keyes Beech; General Donald V. Bennett; Lieutenant General Sidney B. Berry; Betty Wheeler Besson; Tony Bliss Jr.; Claire Boatwright; Brigadier General Zeb B. Bradford Jr.; General Arthur E. Brown Jr.; Lieutenant General Frederic J. Brown III; Major General Clay T. Buckingham; Kevin Buckley; Major General Larry D. Budge; Colonel Victor T. Bullock; Lieutenant General William S. Carpenter Jr.; Brigadier General Lawrence H.

Caruthers Jr.; Colonel Edward Gustave Aaron Chalgren Jr.; General Ferdinand J. Chesarek; Lieutenant General Richard A. Chilcoat; Margaret Westmoreland Clarkson; Lieutenant General Charles G. Cleveland (USAF); Colonel William F. Cody; Colonel Robert M. Cook; George Crile; Lieutenant General John Crosby; Lieutenant General John H. Cushman; Lieutenant General Phillip B. Davidson Jr.; General Michael S. Davison; Dr. Edwin A. Deagle Jr.; Vincent H. Demma; General William E. DePuy; Colonel Roger Donlon; Lieutenant General David K. Doyle; Brigadier General Karl W. Eikenberry; Major General James N. Ellis; Daniel Ellsberg; Colonel Charles Endress; Lieutenant General Julian J. Ewell; Lieutenant Colonel Robert Finkenaur; Colonel Zane E. Finkelstein; Lieutenant General Edward M. Flanagan Jr.; Lieutenant General Eugene P. Forrester; General John W. Foss; Honorable Robert Froehlke; Ernest B. Furgurson; Brigadier General Gerald E. Galloway Jr.; General John R. Galvin; Major General George A. Godding; General Andrew J. Goodpaster; General Paul F. Gorman; Brigadier General Michael J. L. Greene; General Wallace M. Greene Jr. (USMC); Colonel William Greynolds; Lieutenant General Thomas N. Griffin Jr.; Dr. Alan Gropman; Colonel David Hackworth; Arthur T. Hadley; General Ralph E. Haines Jr.; David Halberstam; Brigadier General Ed Y. Hall (S.C. State Guard); Richard Halloran; Andrew Hamilton; Colonel Franklin A. Hart; Herbert Hartsook; Brigadier General James A. Herbert; Colonel Morris J. Herbert; Seymour Hersh; Honorable Martin R. Hoffmann; Brigadier General Elizabeth P. Hoisington; Ambassador Richard C. Holbrooke; Ambassador John H. Holdridge; Christopher Holmes; Brigadier General Weldon F. Honeycutt; Colonel George D. Jacobson; Brigadier General Amos A. Jordan Jr.; Lieutenant General James G. Kalergis; General Donald R. Keith; General Walter T. Kerwin Jr.; Brigadier General Douglas Kinnard; Lieutenant General Harry W. O. Kinnard; General William A. Knowlton; Ambassador Robert W. Komer; Richard D. Kovar; Dr. Fritz G. A. Kraemer; General Frederick J. Kroesen; Lieutenant Colonel Charles A. Krohn; Honorable Melvin R. Laird; Ambassador William Leonhart; General William J. Livsey; Colonel Donald S. Marshall; Colonel Lloyd J. Matthews; George McArthur; Lieutenant General William J. McCaffrey; Major General Joseph A. McChristian; Colonel Robert B. McCue; Frank McCullough; Ambassador Clay McManaway; Honorable Robert S. McNamara; Lieutenant General John B. McPherson (USAF); John Merriam; General Edward C. Meyer; Lieutenant Colonel Michael D. Mierau; General Frank T. Mildren; Colonel Paul Miles; Brigadier General Robert M. Montague Jr.; Chief Warrant Officer 2 Charlie M. Montgomery Jr.; Admiral Thomas H. Moorer (USN); Colonel George Morgan; Captain William McG. Morrison (USNR); Dr. Charles Moskos; Lieutenant General John Norton; Walter M. Oates; General Joseph T. Palastra Jr.; General Bruce Palmer Jr.; Lieutenant General Dave R. Palmer; Rufus C. Phillips III; Dr. Douglas Pike; Lieutenant General Beverley E. Powell; Lieutenant General Robert E. Pursley (USAF); Charles D. Ravenel; Honorable Stanley R. Resor; Elisabeth Robe; General William B. Rosson; Walt Rostow; Lieutenant General Edward L. Rowny; Dr. Herbert Y. Schandler; Benjamin F. Schemmer; Dr. James R. Schlesinger; Colonel Fred B. Schoomaker; Lieutenant General Robert L. Schweitzer; Major General John O. B. Sewell; Colonel

Donald P. Shaw; William Seth Shepard; Major General Winant Sidle; Lieutenant General Charles J. Simmons; Major General John K. Singlaub; Colonel Kenneth V. Smith; Dr. Don M. Snider; General Donn A. Starry; Major General William B. Steele; General Richard G. Stilwell; Major General Adrian St. John II; Lieutenant General Orwin C. Talbott; Lieutenant General Herbert R. Temple Jr.; General Maxwell R. Thurman; Lieutenant General Ngo Quang Truong (ARVN); Lieutenant General Walter F. Ulmer Jr.; Colonel Carl C. Ulsaker; Sergeant Major of the Army Leon L. Van Autreve; Honorable Cyrus R. Vance; Lieutenant General Dale A. Vesser; Colonel John H. VonDerBruegge; General Sam W. Walker; General Volney F. Warner; James R. Westmoreland; General William C. Westmoreland; General Frederick C. Weyand; Blair Weymouth; Ambassador Charles Whitehouse; Nils F. Wikner; Major General Ellis W. Williamson; George Wilson; Major General Stephen R. Woods; John Lawrence Worrall; Lieutenant General William P. Yarborough; Stephen B. Young; Barry Zorthian.

U.S. *Army Military History Institute Oral History Interviews Consulted*

General Donald V. Bennett; General Frank S. Besson Jr.; Lieutenant General Charles A. Corcoran (Abrams Story Interview); General William E. DePuy; General Ralph E. Haines Jr.; General Hamilton H. Howze; General Harold K. Johnson; Colonel Ted Kanamine (Abrams Story Interview); General Donald R. Keith; Lieutenant General Harry W. O. Kinnard; General William A. Knowlton; Colonel Jonathan F. Ladd (Abrams Story Interview); General William B. Rosson; General John L. Throckmorton; Lieutenant General Walter F. Ulmer Jr.; General Volney F. Warner; General William C. Westmoreland (Cameron/Funderburk); General William C. Westmoreland (Ganderson).

Other *Oral Histories and Interviews Consulted*

LBJ LIBRARY

Ambassador Ellsworth Bunker; Ambassador W. Averell Harriman; General Walter T. Kerwin Jr.; Ambassador Robert Komer; General John P. McConnell (USAF); Honorable Stanley R. Resor; General Earle G. Wheeler; Lieutenant General Samuel T. Williams.

MARINE CORPS ORAL HISTORY COLLECTION

Lieutenant General John A. Chaisson; Major General Rathvon McC. Tompkins; General William C. Westmoreland (USA).

U.S. ARMY CENTER OF MILITARY HISTORY

Ambassador Robert Komer and Colonel Robert M. Montague; Professor Douglas Pike; Major General Ellis W. Williamson.

WBGH COLLECTION, HEALEY LIBRARY, UNIVERSITY OF MASSACHUSETTS AT BOSTON

William E. Colby; Ambassador Richard Holbrooke; Honorable Melvin Laird; Ambassador Henry Cabot Lodge; General Edward C. Meyer; Ambassador John Negroponte; Dr. Walt Rostow; General William C. Westmoreland.

INDEX

ABOUT THE AUTHOR

Lewis Sorley is a soldier, historian, biographer, and scholar of the Vietnam War. A third-generation graduate of West Point, he also holds a doctorate from the Johns Hopkins University. His two decades of service in the United States Army included command of tank and armored cavalry units, teaching at West Point and the Army War College, and Pentagon staff duty. Later he was a senior civilian official of the Central Intelligence Agency. This is his ninth book.